# Fundamentals of Tool Design

## Fifth Edition

# Fundamentals of Tool Design

## Fifth Edition

**Chief Technical Reviewer**
David Spitler
Professor Emeritus
Industrial and Technology Engineering Department
Central Michigan University

**Contributors**
Jeff Lantrip
Engineer/Scientist, Manufacturing R&D
The Boeing Company

Dr. John G. Nee, CMfgE
Professor Emeritus
Industrial and Technology Engineering Department
Central Michigan University

David A. Smith
President
Smith & Associates

Society of
Manufacturing
Engineers

Dearborn, Michigan

Library of Congress Catalog Card Number: 2003103269
International Standard Book Number: 0-87263-650-X

Additional copies may be obtained by contacting:
Society of Manufacturing Engineers
Customer Service
One SME Drive, P.O. Box 930
Dearborn, Michigan 48121
1-800-733-4763
www.sme.org

SME staff who participated in producing this book:
Michael McLelland, Staff Editor
Rosemary Csizmadia, Production Supervisor
Kathye Quirk, Graphic Designer/Cover Design
Frances Kania, Administrative Coordinator
Jon Newberg, Production Editor

Printed in the United States of America

# Reviewers

Billy B. Boren, CMfgT
Manufacturing Engineer
Eaton Corporation
Shawnee, OK

George S. Fako
Product Engineer
CAD & CAE Applications
International Truck and Engine Corporation
Melrose Park, IL

Kevin Fenner
Applications Engineer
Edison Welding Institute
Columbus, OH

Todd Grimm
President
T.A. Grimm & Associates
Edgewood, KY

Peter Hall
Principle Engineer
Edison Welding Institute
Columbus, OH

Douglas L. Ketron
Senior Engineer
Edison Welding Institute
Columbus, OH

Tony Laker
Senior Technical Systems Specialist
International Truck and Engine Corporation
Melrose Park, IL

Scott Mayer
Manager
Engineering Systems, Engine Group
International Truck and Engine Corporation
Melrose Park, IL

G. Mukundan, Ph.D.
Research and Technology Manager
CAx Research and Development
DaimlerChrysler Corporation
Auburn Hills, MI

George Ritter, Ph.D.
Principle Engineer
Edison Welding Institute
Columbus, OH

Tom Roberts
Director of Engineering
Simkar Corporation
Philadelphia, PA

Greg A. Ross
Technical Specialist
Hagemeyer-Cameron and Barkley Company
West Columbia, SC

Alvin Sabroff
President
Metalworking Consultant Group
Mentor, OH

Jim Schlusemann
Director, Information Technology
International Truck and Engine Corporation
Melrose Park, IL

David A. Smith
President
Smith & Associates
Monroe, MI

James L. Stephens, PE
Senior Engineer
CNH Global NV
Racine, WI

Israel Stol
Senior Manufacturing Specialist
Alcoa Technical Center
Pittsburgh, PA

Pamela Waterman
Author
EngineeringInk Technical Editorial Services
Mesa, AZ

# Past Contributors

Wilford H. Abraham
James Albrecht
David Ardayfio
William B. Ardis
Ross L. Beaulieu
Seth J. Beck
Louis H. Benson
Bernard R. Better
Joseph S. Blachutshannon
Ernest G. Boyd
D. Caddell
Peter Carbone
Bernard Cardinal
Carl H. Cedarblad
Paul E. Charrette
Charles E. Clark
Walter F. Coles
Harry Conn
Paul Dalla Guardia
Charles DeRooler, Ph.D.
Edward T. Drabik
Fred R. Drake
Alan B. Draper
Duane Dunlap, Ph.D.
Francis L. Edmondson
William A. Ehlert
Keith D. Entrekin
Jack Fickers
Joseph C. Fogarty
Norman G. Foster
Howard A. Frank
Raymond E. Gariss
Roger L. Geer
Ralph D. Glick

Floyd D. Goar
David Goestsch, Ph.D.
William H. Gourlie
R. H. Grind
Earl Harp
Fred D. Hitter
Edward G. Hoffman, Ph.D.
Howard Holder
Paul Jacobs, Ph.D.
David Johnson
Donald G. Johnson
Kenneth M. Kell
Donald R. King
Joseph A. Klancink
Donald Koch
Anthony R. Konecny
Joseph C. Kopeck
Robert C. Kristofek
Martin Kuklinski
Everett Laitala
Brian Lambert, Ph.D.
Clarence E. Lane
Robert M. Larson
Charles O. Lofgren
Paul A. Longo
Robert Lown
Harry J. Lund
Lincoln Mager
Louis J. Mahlmeister
James C. Mangus
Dan J. McKeon
Raymond H. Meckley
John Mitchell
Karl H. Moltrecht

Joseph Mundbrot
Robert E. Nauth
Alvin G. Neumann
Elsayed A. Orady, Ph.D.
Harry B. Osborn, Jr.
Carl Oxford, Jr.
Clayton F. Paquette
Roy B. Perkins
Ralph L. Perlewitz
Willis J. Potthoff
Robert J. Quillici
Fred T. Richter
Edward S. Roth
Donald M. Satava
George H. Sheppard
David A. Smith
Richard A. Smith
Stanley J. Snorek
Fred L. Spaulding
John D. Sprinkel

Gilbert Stafford
Ronald F. Steward
Paul Suksi
Sundberg Ferar Product Development Staff
Frank Swaney
Howard R. Swanson
James L. Thomas
Edward A. Tobler
Thomas Ury
Edwin M. Vaughn
Clifford C. Vogt
John T. Vukelich
Emmett J. Welky
Ernie W. Wheeler
Charles W. Williams
Beverly D. Wilson
Frank W. Wilson
Gerald C. Woythal
Lester C. Youngberg
Raymond J. Zale

# Table of Contents

# Abbreviations

| | | | |
|---|---|---|---|
| 3D | three-dimensional | CAT | computer-aided tomography |
| 3DP | three-dimensional printing | CBD | chronic beryllium disease (berylliosis) |
| | | CBN | cubic boron nitride |
| **A** | | CCD | charged couple device |
| | | CE | concurrent engineering |
| AC | alternating current | CFR | Code of Federal Regulations |
| A | ampere | CIM | computer-integrated manufacturing |
| AFM | abrasive flow machining | CGA | circle grid analysis |
| AGD | American Gage Design | CHM | chemical machining |
| AISI | American Iron and Steel Institute | cm | centimeter |
| AJM | abrasive-jet machining | CMM | coordinate measuring machine |
| AMSA | American Metal Stamping Association | CNC | computer numerical control |
| AMT | Association for Manufacturing Technology | | |
| | | **D** | |
| ANSI | American National Standards Institute | dB | decibel |
| | | DC | direct current |
| ASCII | American standard code for information interchange | DCC | direct computer control |
| | | DFM | design for manufacturability |
| ASM | American Society for Metals | dia. | diameter |
| ASME | American Society of Mechanical Engineers | DOE | design of experiments |
| | | DOF | depth of field |
| ASTM | American Society for Testing and Materials | DSCP | direct-shell production casting |
| | | DXF | drawing exchange format |
| **B** | | | |
| | | **E** | |
| Bhn | Brinell hardness number | EBM | electron beam machining |
| | | ECM | electrochemical machining |
| **C** | | EDM | electrical discharge machining |
| | | EH&S | environmental, health, and safety |
| C | Celsius | ELP | electropolishing |
| CAD | computer-aided design | EPA | Environmental Protection Agency |
| CAM | computer-aided manufacturing | Eq. | equation |
| CAMI | Coated Abrasive Manufacturers' Institute | | |

## F

| | |
|---|---|
| F | Fahrenheit |
| FDM | fused deposition modeling |
| fig. | figure |
| FIM | full indicator movement |
| FLD | forming limit diagram |
| FMS | flexible manufacturing system |
| FOV | field of view |
| ft | foot/feet |
| ft³/min | cubic feet per minute |

## G

| | |
|---|---|
| g | gram |
| gal | gallon |
| GD&T | geometric dimensioning and tolerancing |
| GMAW | gas metal-arc welding |

## H

| | |
|---|---|
| hp | horsepower |
| hr | hour |
| HSS | high-speed steel |
| Hz | hertz |

## I

| | |
|---|---|
| IC | integrated circuit |
| ID | inside diameter |
| IGES | initial graphics exchange specification |
| in. | inch |
| ipm | inches per minute |
| ipr | inches per revolution |
| ISO | International Organization for Standardization |

## J

| | |
|---|---|
| J | joule |
| JIT | just-in-time |

## K

| | |
|---|---|
| kg | kilogram |
| kHz | kilohertz |
| km | kilometer |
| kN | kilo Newton |
| kPa | kilo Pascal |
| ksi | 1,000 pounds per square inch |
| kW | kilowatt |

## L

| | |
|---|---|
| lb | pound |
| lbf | pound force |
| LBM | laser beam machining |
| LED | light-emitting diode |
| LMC | least material condition |
| LVDT | linear variable displacement transformer |

## M

| | |
|---|---|
| m | meter |
| μin. | microinch |
| μm | micrometer |
| mg | milligram |
| mi | mile |
| MIL | military specification |
| min | minute |
| mm | millimeter |
| MMC | maximum material condition |
| MN | mega Newton |
| MPa | mega Pascal |
| ms | millisecond |
| MSCMM | multi-sensor coordinate measuring machine |

## N

| | |
|---|---|
| N | Newton |
| NC, N/C | numerical control |
| Nd:YAG | neodymium: yttrium aluminum garnet |
| NEMA | National Electrical Manufacturers Association |
| NFPA | National Fire Protection Association |
| NIST | National Institute for Standards and Technology |
| nm | nanometer |
| N/m | Newton/meter |
| NSMPA | National Screw Machine Products Association |
| NTIS | National Technical Information Service |

## O

| | |
|---|---|
| OBI | open-back inclinable (press) |
| OBS | open-back stationary (press) |
| OD | outside diameter |

OHD       overhead drive
ohm       unit of electrical resistance
OSHA      Occupational Safety and Health Ad-
          ministration
oz        ounce

## P

PAM       plasma-arc machining
PCBN      polycrystalline cubic boron nitride
PCD       polycrystalline diamond
PCM       photochemical machining
PH        precipitation hardening
ppm       parts per million
psi       pounds per square inch
PWB       printed wiring board

## Q

qt        quart

## R

R         radius
Ra        roughness average
RCRA      Resource Conservation and Recovery
          Act
rev       revolution
RFS       regardless of feature size
$R_{max}$ maximum roughness depth
RMS       root mean square
rpm       revolutions per minute
RP&M      rapid prototyping and manufacturing

## S

s, sec.   second
SAE       Society of Automotive Engineers
sfm, sfpm surface feet per minute
sfpm      surface feet per minute
SiC       silicon carbide
SLS       selective laser sintering
SME       Society of Manufacturing Engineers
SPC       statistical process control
SPM       strokes per minute
STEP      standard for the exchange of product
STL       stereolithography

## T

TEM       thermal energy method
TiC       titanium carbide
TiN       titanium nitride
TP        true position

## U

U.K.      United Kingdom
UL        Underwriters' Laboratories
U.S.      United States
USM       ultrasonic machining
UV        ultraviolet

## V

V         volt

## W

W         watt
WAM       waterjet abrasive machining
WBTC      Worldwide Burr Technology Commit-
          tee
WJM       waterjet machining

## Y

yd        yard
yr        year

## MATH SYMBOLS

$\sim$    about equal to
$^{\circ}$ degree
/         divided by or per
$>$       greater than
$\geq$    greater than or equal to
$\alpha$  Greek alpha
$\beta$   Greek beta
$\Delta$  Greek delta
$\epsilon$ Greek epsilon
$\gamma$  Greek gamma
$\lambda$ Greek lambda
$\mu$     Greek mu
$\Omega$  Greek omega
$\phi$    Greek phi

$\pi$  Greek pi
$\sigma$  Greek sigma
$\theta$  Greek theta
$\vartheta$  Greek theta (lower case)
$<$  less than
$\leq$  less than or equal to
$-$  minus
$\%$  percent
$+$  plus
$\pm$  plus or minus
$\times$  times

# Preface

The field of tool design is one of the most diverse areas of manufacturing engineering. From simple cutting tools and workholders to complex cutting dies, computer software applications, and rapid prototyping/manufacturing technology, the field of tool design has evolved into an individual discipline requiring imagination and innovation to solve today's complex tooling conundrums.

In larger facilities, the tool designer often has a specific area of concentration. In smaller shops, the designer may have complete design responsibility along with supervising and directing the toolroom. In either case, today's tool designers must be familiar with all aspects of tool design to perform their responsibilities effectively.

This book, the *Fundamentals of Tool Design*, is a means to that end. For the Society of Manufacturing Engineers, this book is a touchstone—it brings to ground the SME mission of gathering and sharing information. For those who reviewed chapters, it was a positive experience to share the insight acquired throughout their careers and an opportunity to smooth the road a little for those coming after them. For those who will read this book, whether in a classroom or on a shop floor, it might be considered an educational autobahn, the confluence of myriad roads and avenues, entrances, and exits that leads one to the wellspring of knowledge, and what is a better destination than that?

*David Spitler*

# Tool Design

Tool design is a specialized area of manufacturing engineering comprising the analysis, planning, design, construction, and application of tools, methods, and procedures necessary to increase manufacturing productivity. To carry out these responsibilities, today's tool designer must have a working knowledge of machine shop practices, toolmaking procedures, machine tool design, manufacturing procedures and methods, as well as the more conventional engineering disciplines of planning, designing, engineering graphics and drawing, and cost analysis.

## OBJECTIVES

The main objective of tool design is to increase production while maintaining quality and lowering costs. To this end, the tool designer must realize the following goals:

- Reduce the overall cost to manufacture a product by making acceptable parts at the lowest cost.
- Increase the production rate by designing tools to produce parts as quickly as possible.
- Maintain quality by designing tools to consistently produce parts with the required precision.
- Reduce the cost of special tooling by making every design as cost-effective and efficient as possible.
- Design tools to be safe and easy to operate.

Every design must be created with these objectives in mind. No matter how well a tool functions or produces parts, if it costs more to make the tool than it saves in production, its usefulness is questionable. Likewise, if a tool cannot maintain the desired degree of repeatability from one part to the next, it is of no value in production.

## TOOL DESIGNER RESPONSIBILITIES

Typically, tool designers are responsible for creating a wide variety of special tools. Whether these tools are an end product or merely an aid to manufacturing, the tool designer must be familiar with:

- cutting tools, toolholders, and cutting fluids;
- machine tools, including modified or special types;
- jigs and fixtures;
- gages and measuring instruments;
- dies for sheet-metal cutting and forming;
- dies for forging, upsetting, cold finishing, and extrusion, and
- fixtures and accessories for welding, riveting, and other mechanical fastening.

In addition, the tool designer must be familiar with other engineering disciplines, such as metallurgy, electronics, computers, and machine design. These engineering disciplines affect the design of special tools.

In most cases, the size of the employer or the type of product will determine the exact duties of each designer. Larger companies with several product lines may employ many tool designers. In this situation, each designer may have an area of specialization, such as die design, jig and fixture design, gage design, or any similar tool design area. In smaller companies, however, one tool designer may have to do all tool designs, as well as other tasks in manufacturing.

## THE DESIGN PROCESS

While the specifics of designing each type of tool will be discussed in the subsequent chapters of this text, a few basic principles and procedures are introduced here. The design process consists of five basic steps:

1. statement and analysis of the problem;
2. analysis of the requirements;
3. development of initial ideas;
4. development of design alternatives, and
5. finalization of design ideas.

While these five steps are separated for this discussion, in practice, each overlaps the others. For example, when stating the problem, the requirements also must be kept in mind to properly define and determine the problem or task to be performed. Likewise, when determining the initial design ideas, the alternative designs are also developed. So, like many other aspects of manufacturing, tool design is actually an ongoing process of creative problem solving.

### Statement of the Problem

The first step in the design of any tool is to define the problem as it exists without tooling. This may simply be an assessment of what the proposed tool is expected to do, such as drill four holes. Or, it may be an actual problem encountered in production where tooling may be beneficial, such as low-volume production caused by a bottleneck in assembly. Once the exact extent of the problem has been determined, the problem can be analyzed and resolved throughout the remaining steps of the design process.

### Analysis of the Requirements

After the problem has been isolated, the requirements, including function, quality, cost, due date, and other related specifics can be used to determine the parameters within which the designer must work. Every tool that is designed must:

- perform certain functions;
- meet certain minimum precision requirements;
- keep costs to a minimum;
- be available when the production schedule requires it;
- be operated safely;
- meet various other requirements such as adaptability to the machine on which it is to be used, and
- have an acceptable working life.

*Table 1-1* illustrates a method of applying these criteria to the process of choosing a tool design. Rarely, if ever, will one tool design be best in each of these five areas. The tool designer's task in this situation is to weigh all the factors and select the tool that best meets the criteria and the task to be performed.

### Development of Initial Ideas

Initial design ideas are normally conceived after an examination of the preliminary data. This data consists of the part print, process sheet, engineering notes, production schedules, and other related information. While evaluating this information, the designer should take notes to ensure nothing is forgotten during the initial evaluation. Should the designer need more information than that furnished with the design

Table 1-1. Basic pattern for tool analysis

| Create Alternatives | Analyze in terms of: | | | | |
|---|---|---|---|---|---|
| | Function | Quality | Cost | Date | Auxiliary |
| A | x | x | x | x | x |
| B | x | x | x | x | x |
| C | x | x | x | x | x |
| • | • | • | • | • | • |
| • | • | • | • | • | • |
| • | • | • | • | • | • |
| n | x | x | x | x | x |

package, the planner responsible for the tool request should be consulted to determine the required additional material. In many cases, the designer and planner will be working in a team environment and must jointly develop parameters during the initial design.

## Development of Design Alternatives

During the initial concept phase of design, many ideas will occur to the designer and/or the team. As these ideas are developed they should be written down so they are not lost or forgotten. There are always several ways to do any job. As each method is developed and analyzed, the information should be added to the list shown in *Table 1-1*.

## Finalization of Design Ideas

Once the initial design ideas and alternatives are determined, the tool designer must analyze each element to determine the best way to proceed toward the final tool design. As stated earlier, rarely is one tool alternative a clear favorite. Rather, the tool designer must evaluate the strong points of each alternative and weigh them against the weak points of the design. For instance, one tool design may have a high production rate, but the cost of the tool may be very high. On the other hand, a second tool may have a medium production rate and will cost much less to build. In this case, the value of production over cost must be evaluated to determine the best design for the job. If the job is a long-term production run, the first tool may pay for itself in increased production volume. If, however, the production run is short or is a one-time run, the second tool may work best by sacrificing production speed for a reduced tool cost. Seldom will one tool be able to meet all the expectations of the tool designer. In most cases, the best design is a compromise of the basic criteria of function, quality, cost, due date, safety, and other requirements.

## ECONOMICS OF DESIGN

The tool designer must know enough economics to determine, for example, whether tempo-

rary tooling would suffice even though funds are provided for more expensive permanent tooling. He or she should be able to check the design plan well enough to initiate or defend a planning decision to write off the tooling on a single run as opposed to a write-off distributed against probable future reruns. The tool designer should have an opinion backed by economic proof of certain changes that would make optimum use of the tools.

Many aspects of manufacturing economics are beyond the scope of this book, but excellent literature is available.

## Combined Operations

Analysis may sometimes show that operations can be cost-effectively combined. Tooling costs, production costs, or both can thus be reduced. *Table 1-2* illustrates a case where the cost of combined tools was less than the total cost of the separate tools otherwise required.

## Process-cost Comparisons

During process planning, many possible methods of manufacturing may be reduced to a few based on alternate process steps, use of available equipment, or combined operations. Under these conditions, a comparison of costs for different tools and process steps may reveal a combination resulting in the lowest total cost per part.

Let:

$N_t$ = total number of parts to be produced in a single run

$N_b$ = number of parts for which the unit costs will be equal for each of two compared methods $Y$ and $Z$ (break-even point)

$T_y$ = total tool cost for method $Y$, $

$T_z$ = total tool cost for method $Z$, $

$P_y$ = unit tool process cost for method $Y$, $

$P_z$ = unit tool process cost for method $Z$, $

$C_y, C_z$ = total unit cost for methods $Y$ and $Z$, $

Then:

$$N_b = \frac{T_y - T_z}{P_z - P_y} \qquad (1\text{-}1)$$

Table 1-2. Cost of combined versus separate operations

| Costs | Blanking Operation Alone | Forming Operation Alone | Total Blank and Form | Combined Operations |
|---|---|---|---|---|
| Tools | $40.00 | $30.00 | $70.00 | $50.00 |
| Setup | 2.00 | 2.00 | 4.00 | 3.00 |
| Maintenance | 2.00 | | 2.00 | 2.00 |
| Processing | 4.00 | 30.00 | 34.00 | 4.00 |
| Total cost | $48.00 | $62.00 | $110.00 | $59.00 |

$$C_y = \frac{P_y N_t + T_y}{N_t} \qquad (1\text{-}2a)$$

$$C_z = \frac{P_z N_t + T_z}{N_t} \qquad (1\text{-}2b)$$

For example, the aircraft-flap nose rib shown in *Figure 1-1* of 0.02 in. (0.5 mm) 2024-T Alclad was separately calculated to be formed by a hydropress, drop hammer, Marform®, steel draw die, or hand forming process. For reasons such as die life, available equipment, and handwork required, the choice narrowed down to the hydropress versus steel draw die. With the hydropress, the flanges had to be fluted, and for piece quantities over 100, a more expensive steel die costing $202 had to be used.

Actual die and processing costs for both methods are listed in *Table 1-3*. $P_y$ and $P_z$ are processing costs, and $T_y$ and $T_z$ are die costs for the steel draw die and hydropress methods, respectively.

20 in. (508 mm)

13 in. (330.2 mm)

0.020 in. (0.51 mm) 2024-T Alclad

Flutes required only when formed on hydropress

Figure 1-1. *General specifications for aircraft-flap nose rib.*

Figures in the last column of *Table 1-3* were not stated in the original report, but can properly be extrapolated on the basis of the apparent stability of $P_y$ and $P_z$ at $N_t = 500$, and assuming their stability at higher production rates.

On the basis of the listed figures at $N_t = 500$, and from *Equation 1-1*, the production at which total unit costs $C_y$ and $C_z$ will be the same for both methods is:

$$N_b = \frac{810 - 202}{1.85 - 1.05} = 760 \text{ pieces}$$

As an alternate method for calculating the break-even point between two machines (for example, a turret lathe and an automatic), the Warner & Swasey Company uses a formula based on known or estimated elements that make up production costs. The break-even-point equation is,

$$Q = \frac{pP(SL + SD - sl - sd)}{P(1 + d) - p(L + D)} \qquad (1\text{-}3)$$

where:

$Q$ = quantity of pieces at break-even point
$p$ = number of pieces produced per hour by the first machine
$P$ = number of pieces produced per hour by the second machine
$S$ = setup time required on the second machine, hours
$L$ = labor rate for the second machine, $
$D$ = hourly depreciation rate for the second machine (based on machine-hours for the base period), $
$s$ = setup time required on the first machine, hours

Table 1-3. Cost comparison of methods for producing aircraft-flap nose rib

| $N_t$ | 5 | 25 | 50 | 100 | 500 | 780[†] |
|---|---|---|---|---|---|---|
| $P_y$ | $ 3.00 | $ 1.18 | $ 1.11 | $ 1.05 | $ 1.05 | $ 1.05 |
| $P_z$ | 4.40 | 2.05 | 1.96 | 1.85 | 1.85 | 1.85 |
| $T_y$ | 810.00 | 810.00 | 810.00 | 810.00 | 810.00 | 810.00 |
| $T_z$ | 103.00 | 103.00 | 103.00 | 103.00 | 202.00 | 202.00 |
| $C_y$ | 165.00 | 33.60 | 17.30 | 9.15 | 2.67 | 2.12 |
| $C_z$ | 25.00 | 6.17 | 4.02 | 2.88 | 2.25 | 2.12 |

[†] Extrapolated

$l$ = labor rate for the first machine, $

$d$ = hourly depreciation rate for the first machine (based on machine-hours for the base period), $

For example, assume it is desired to find value $Q$ when the various factors are as follows:

$p$ = 10 pieces per hour
$P$ = 30 pieces per hour
$S$ = 6 hours
$L$ = $1.50 per hour
$D$ = $1.175 per machine-hour
$s$ = 2 hours
$l$ = $1.50 per hour
$d$ = $0.40 per machine-hour (10-year period)

Then, substituting these values in the equation,

$$Q = \frac{10 \times 30(6 \times 1.50 + 6 \times 1.175 - 2 \times 1.50 - 2 \times 0.40)}{30(1.50 + 0.40) - 10(1.50 + 1.175)} = 121 \text{ pieces}$$

or the quantity on which the cost is the same for either machine.

## Effect of Tool Material Choice on Tool Life

The tool designer can often influence the choice of cutting-tool materials. Therefore, the tool designer must know the tool-life economics involved.

Typical tool-life curves are shown in *Figure 1-2* for high-speed steel (HSS), sintered carbide, and oxide tools. The wear characteristics, as denoted by the $n$ values, show that economic life $T_c$ for carbide tools is shorter than that for HSS tools ($T_c$ equals 15 minutes for carbide versus 35 minutes for HSS). It is even more important to use

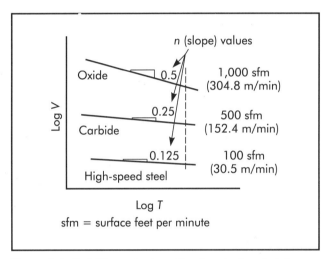

Figure 1-2. *Tool-life comparison of various tool materials.*

higher speeds and a shorter tool life for oxide tools, since $T_c$ equals five minutes.

From *Figure 1-3* it is evident that most economical metal removal demands high cutting speeds and thus higher horsepower when using carbide and oxide tooling.

## ECONOMICAL LOT SIZES

Economical lot sizes are calculated to obtain the minimum unit cost of a given part or material. This minimum is reached when the costs of planning, ordering, setting up, handling, and tooling equal the costs of storage for finished parts. These costs may be equated and the lot size determined by mathematical calculation. Depending on the number of variables to consider, the equation can range from one of relative simplicity to complex.

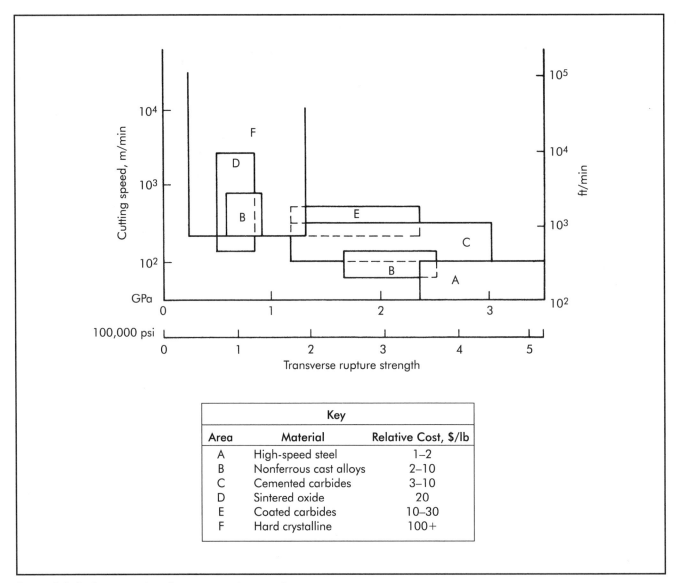

| Key | | |
|---|---|---|
| Area | Material | Relative Cost, $/lb |
| A | High-speed steel | 1–2 |
| B | Nonferrous cast alloys | 2–10 |
| C | Cemented carbides | 3–10 |
| D | Sintered oxide | 20 |
| E | Coated carbides | 10–30 |
| F | Hard crystalline | 100+ |

Figure 1-3. *Performance data for various tool materials.*

By assuming that the number of pieces required per month is a constant, the inventory increasing until the lot ordered is completely sent to stock, and decreasing uniformly with use, a relatively simple equation for calculation of economic lot size can be devised:

$$L = \sqrt{\frac{24mS}{kc(1+mv)}} \qquad (1\text{-}4)$$

where:

$L$ = lot size, pieces
$m$ = monthly consumption, pieces

$S$ = setup cost per lot, $
$k$ = annual carrying charge per dollar of inventory, $
$c$ = value of each piece, $
$v$ = ratio of machining time to lot sizes, months per piece

Labor, material, and other costs not related to lot size are omitted. In practice, reasonable values for $S$, $m$, and $c$ are difficult to obtain. These values are generally determined by the standards, sales, or methods departments.

Equations for calculating economic lot sizes can never be proven exactly because of the assumptions upon which many of the factors are based. Such an equation should be regarded as just a useful guide, to be applied with mature judgment.

## BREAK-EVEN CHARTS

Break-even charts are perhaps most widely used to determine profits based on anticipated sales. They have other uses, however, such as selecting equipment or measuring the advisability of increased automation.

To determine which of two machines is the most economical, the fixed and variable costs of each machine are plotted as shown in *Figure 1-4*. Fixed costs do not vary with respect to time or product volumes. Variable costs do vary with respect to time or product volume. The total cost is composed of the sum of the fixed and variable costs. For example, assume the initial cost for machine A is $1,500, and the unit production cost on the machine is $0.75 each. For machine B, the initial cost is $6,000 and the unit production cost is $0.15 each. The chart shows that it is more economical to purchase machine A if production never exceeds 7,500 pieces. For higher production quantities, machine B is more economical.

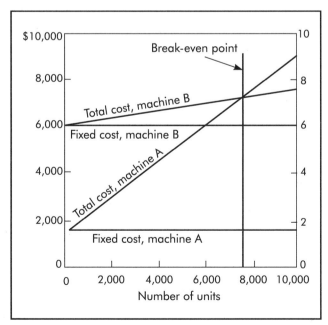

Figure 1-4. *Break-even chart for machine selection; choice is based upon volume production.*

## TOOL DRAWINGS

A tool designer must have a strong background in drafting, dimensioning, and mechanical drawing to properly present design concepts to the people who will make the proposed tool. Often, computer-aided design (CAD) technology is used for this purpose (see Chapter 12). For the most part, toolmakers and die makers do not require the same type of drawings as do the less experienced machine operators in a production department. For this reason, tool drawings can be drawn much simpler and faster to keep design costs to a minimum. The following are useful points to remember when creating tool drawings.

- Draw and dimension with due consideration for the person who will use the drawing to make the item in the tool room.
- Do not crowd views or dimensions.
- Analyze each cut, so that wherever possible the cut can be made with standard tools.
- Use only as many views as necessary to show all required detail.
- Surface roughness must be specified.
- Tolerances and fits peculiar to tools need special consideration. It is not economical to tolerance both details of a pair of mating parts as is required on production part detailing. In cases where a hole and a plug are on different details to be made and mated, the fit tolerance should be put on the male piece and the hole should carry a nominal size. This allows a toolmaker to ream the hole with a nominal-size tool and grind the plug to fit it, although nominal may vary several thousandths of an inch (mm) from exact.
- The stock list of any tool drawing should indicate all sizes required to obtain the right amount for each detail. It is necessary to allow material for finishing in almost all cases, although some finished stock is available. Whenever possible, stock sizes on hand should be used. However, in all cases, available sizes should be specified. A proper, finished detail is dependent upon starting with the right material.
- Use notes to convey ideas that cannot be communicated by conventional drawing.

Heat-treatments and finishes are usually identified as specification references rather than being spelled out on each drawing.

- Secondary operations, such as surface grinding, machining of edges, polishing, heat-treating, or similar specifications should be kept to a minimum. Only employ these operations when they are important to the overall function of the tool; otherwise, these operations will only add cost, not quality to the tool.
- Apply tolerances realistically. Overly tight tolerances can add a great deal of additional cost with little or no added value to the tool. The function of the design feature should determine the specific tolerance, not a standard title block-tolerance value.
- If tooling is to be duplicated in whole or in part for other locations or if multiple tools are required, drawings should be well-detailed.

## TOOLING LAYOUT

The actual work of creating the assembly design of equipment or tools for manufacturing processes should be done within the general framework of the following rules. This will ensure that the details of the tool will work and not interfere with any other part of the tool. These rules should be followed whether utilizing CAD or conventional techniques.

- Lay out the part in an identifying color (red is suggested).
- Lay out any cutting tools. Possible interference or other confining items should be indicated in another identifying color (blue suggested). Use of the cutting tool should not damage the machine or the fixture.
- Indicate all locating requirements for the part. There are three locating planes—use three points in one, two points in the second, and only one point in the third. This is called the 3-2-1 locate system. Do not locate on a non-machined or rough part surface such as the parting line of castings or forgings. All locators must be accessible for the simple cleaning of chips and dirt. For a more detailed explanation, see Chapters 5 and 6.

- Indicate all clamping requirements for the part. Be careful to avoid marking or deforming finished or delicate surfaces. Consider the clamping movements of the operator so that injury to the hands or unsafe situations are eliminated. Make sure it is possible to load and unload the part.
- Lay out the details with due consideration of stock sizes to minimize machining requirements.
- Use full scale in the layout if possible.
- Indicate the use of standard fixture parts (shelf items) whenever possible.
- Identify each item or design detail using balloons with leaders and arrows pointing to the detail. These should not go to a line that is common to other details.

## SAFETY

Safety laws vary greatly from one state to another. All states have some laws requiring protective guards and devices to safeguard workers.

Safety should be designed into the tooling. One of the first and least expensive requirements should be that of breaking all sharp edges and corners. More minor injuries result from these items than most others. Cutting should never be performed against a clamp because of vibration and tool chatter. Instead, parts should be nested against pins to take the cutter load. Rigidity and foolproofing should always be built into the tooling. Make drill jigs large enough to hold without the danger of spinning. Small drill jigs should always be clamped in a vise or against a bar or backstop. They should never be held with the operator's hands. Install Plexiglas® guards around all milling and flycutting operations where chips endanger workers or work areas. High-speed open cutters on production milling, drilling, turning, and jig-boring machine tools are difficult to safeguard because of the varied size and location of the workpieces to be machined.

### Punch Presses

In guarding punch presses, no single type of guard is practical for all operations. Ring guards work well on small punching setups where the ring meets an obstruction and the downward motion of the ram is stopped. Larger die sets and presses can be better protected by gate guards that must be positioned after the work is loaded and then interlocked with the clutch mechanism. Barrier guards protect workers by preventing hands and arms from being placed inside the work area. These generally are telescoping, perforated sheet-metal coverings with an opening to feed the stock. Sweep guards actually sweep clear the danger area as the press ram descends and provide protection during an accidental descent of the ram. Wire-cage guards are still another way to protect the operator's hands (the operator feeds the work through a small opening). This type of guard is useful for secondary operations. Another method of protection is to provide a space between the punch and die that is too small for the operator's hand to enter; this distance should not exceed 3/8 in. (9.5 mm). Punch presses cause many injuries every year, so employers should place a special emphasis on safety when dealing with these machines.

## Limit Switches

Limit switches can be used extensively to protect both worker and product. In punches and dies, limit switches can detect a misfeed or buckling of stock, and check the position of parts during assembly. Photoelectric equipment to protect the operator's hands or body operates when the set beam is interrupted. It is good practice for all punch presses and air or hydraulically operated tooling to be installed with a double-button interlocking protection system, requiring both buttons to be activated before the tool can be used. Other interlocking systems can include two valves in conjunction with a hand valve designed to prevent the tooling or press from operating in case one of the buttons is locked in a closed position. The device should be located in such a position or guarded in such a manner that the operator cannot operate the tooling or press while in the danger area.

## Feed Mechanisms

Feed mechanisms should be provided for all high-speed punching, machining, or assembling operations. The safety function of a feeding device is to provide a means of moving the part into the nest by gravity or mechanical action so there is no necessity for the operator to place his or her hands in the danger zone. High-speed presses equipped with automatic feeds operate at such speeds that it would be impractical as well as hazardous for an operator to attempt to feed the stock.

## Electrical Equipment

Tools and additions to machinery involving electrical equipment must be grounded. Portable electrical equipment should be checked periodically because of rough handling, and all equipment should be grounded to prevent injuries to personnel. Locks, which hold electrical switches open while tools or punches and dies are being repaired or set, help prevent accidental damage. Local regulations can be found in each state's electrical code book covering all applications.

## Other Considerations

Tooling for various industries requires different treatment to ensure safe operations. For example, special electrical controls and motors are required for industries handling explosive material. Tooling material for the chemical industry should be designed to withstand corrosive actions, and electrical equipment should be properly sealed. Careful analysis must be given to tooling for each type of application to provide maximum protection and long life.

Machining various plastics sometimes generates abrasive dust particles or poisonous fumes. In such cases, exhaust systems are required. Safety procedures should be practiced when handling certain metals. Nickel dust is a suspected carcinogen, and dust from beryllium alloys and compounds can cause chronic beryllium disease. Areas around magnesium machining should be kept clean and tools kept sharp to reduce fire hazard. Metal powders present a hazard of combustion during grinding operations. Data can be found in chemical, electrical, and industrial handbooks to help in providing tooling for these special materials.

Tooling and additions to machines must be designed so that the operator does not have to lean across a moving cutter or table. All adjustments and clamping should be easily accessible from the front or the operator's position. Consider body geometry when designing tooling—not only in terms of safety, but also production. Good ergonomic principles should always be followed.

Tooling involving welding must be guarded to prevent severe burns or eye injury from high-intensity, arc-welding rays. Safety glasses should be worn during all machining, grinding, and buffing operations. Welding curtains, proper ventilation, and efficient workstation layout should be considered as well.

All belts, chain drives, gears, sprockets, couplings, keys, and pulleys should be well guarded by sheet metal and panel guards. Guards should be strong enough to support and protect workers in the event that someone or something falls against the guard, or in case the belt or chain should break. Always provide an adequate factor of safety in the design of all tools and tooling applications.

Safety standards are available for all types of industrial applications. Personnel who are responsible for plant safety should be familiar with them. Some of the agencies handling such information are: American National Standards Institute (ANSI); National Institute of Standards and Technology (NIST); U.S. Department of Commerce; National Safety Council; National Fire Protection Association (NFPA); American Insurance Association; and the Occupational Safety and Health Administration (OSHA).

Safety is of the utmost concern when a production tool involves several complex operating mechanisms. Tooling is not restricted solely to machining operations. Automatic assembly and inserting equipment may be classified as either a tool or a machine. Today's complex tools might contain any combination of electric motors, air cylinders, hydraulic equipment, conveyors, and precision indexing tables.

## MATERIAL HANDLING IN THE WORKPLACE

It is beyond the scope of this text to detail the bulk handling of materials and parts through the factory, or the many principles that underlie the design of a workplace. Full texts are devoted to this area. It is the objective of this brief discussion to emphasize the important role that tool design plays in the total cost of an operation from the standpoint of materials handling.

To pick up a workpiece, place it into a tool, clamp the part, unclamp it, remove it, and set it aside after machining may consume more time than the actual machining. Such operations, however essential, contribute nothing of value to the product, but their performance is paid for at the same rate as the productive effort. Consequently, tooling should be designed to reduce nonproductive time and costs.

Once a tool is designed and built, the methods of handling materials into and out of it are fixed. It is essential then, that the original methods planning be carried out with care. The following two sections detail the principles that should be applied to help ensure maximum motion economies.

### Arrangement and Conditions of the Workplace

The following tips are presented to assist in the development of a satisfactory work environment (Das and Sengupta 1996).

- Obtain relevant information on the task performance, equipment, working posture, and environment through direct observation, video recording, and/or input from experienced personnel.
- Identify the appropriate user population and obtain the relevant anthropometric measurements or use the available statistical data from anthropometric surveys.
- Determine the range of work height based on the type of work to be performed. Provide an adjustable chair and a foot rest for a

seated operator, and an adjustable work surface or platform for a standing operator.

- Lay out the frequently used hand tools, controls, and bins within the normal reach space. Failing that, they may be placed within the maximum reach space. Locate the control or handle in the most advantageous position if strength is required to operate it.
- Provide adequate elbow room and clearance at waist level for free movement.
- Locate displays within the normal line of sight.
- Consider the material and information flow requirements from other functional units or employees.
- Make a scaled layout drawing of the proposed workstation to check the placement of individual components.
- Develop a mock-up of the design and conduct trials with live subjects to ascertain operator-workstation fit. Obtain feedback from these groups.
- Construct a prototype workstation based on the final design.

## Design of Hand Tools

In her article, "Hand Tool Design," Susan Nemeth provides 10 principles for the design and use of hand tools. These principles help prevent injuries and enhance the performance and quality of workmanship. The principles discuss tool design and tool use, along with workstation and job designs. This is consistent with the systems-type approach followed in this book. Principles and brief comments are discussed in the following paragraphs.

**Maintain straight wrists.** Bent wrists encourage carpal tunnel syndrome. Any wrist deviation is further aggravated by repetitive motions or large forces. The tool should be held and used in a neutral position.

**Avoid static muscle loading.** The work should be done with the arm and shoulder in a normal position to avoid excessive fatigue. This is especially true when tool weights are large or the tool is used for extended periods of time. Counterbalancing tools is a common solution.

**Avoid stress concentrations over the soft tissue of the hand.** Pressure on these tissues can obstruct blood flow and nerve function. These tools should be avoided.

**Reduce grip force requirements.** Grip forces can pressure the hands or result in tool slippage. Of special note is the distribution of force on the hand, which results in the same problems mentioned in avoiding stress concentrations.

**Maintain optimal grip span.** The optimal power grip with the fingers, palm, and thumb should span 2.5–3.5 in. (63.5–88.9 mm). For circular tool handles, such as screwdrivers, the optimum power grip is 1.25–2 in. (31.75–50.8 mm). For fingertip use, the optimum precision grip is 0.3–0.6 in. (7.62–15.24 mm).

**Avoid sharp edges, pinch points, and awkward movements.** Sharp edges cause blisters and pressure points. Pinch points can make a tool almost unusable. Awkward movements are easily found through observation or use. The movement required to open a tool is a common problem, especially when used repetitively.

**Avoid repetitive finger trigger actions.** Using a single finger to operate a trigger is harmful, especially with frequent use. Use of the thumb is preferred, since the thumb muscles are in the hand, not in the forearm (this avoids carpal tunnel syndrome). Other triggering mechanisms, such as proximity or pressure switches, are possible. Note, however, that the thumb can lock up and have similar ergonomic pain but from different sources.

**Protect hands from heat and cold.** Like soldering irons and other heat-generating tools, tools with motors can produce and transfer heat. Cold comes most often from air-powered tools with an exhaust near the hand grips.

**Avoid excessive vibration.** Vibrations cause Reynaud's syndrome, or dead fingers. The best solutions are damping or isolating the vibration, although rotating jobs is a possible way to limit exposure.

**Use gloves that fit.** Gloves that are too big or thick reduce tool control because they reduce an individual's strength and dexterity. The most economical remedies are using gloves that are the correct weight for the job and stocking gloves in several sizes.

## REFERENCES

Das, B. and Sengupta, A.K. 1996. "Computer-aided Human Modeling Programs for Workstation Design." *Ergonomics*.

Karwowsky, W., and Salvendy, G. 1998. *Ergonomics in Manufacturing*. Dearborn, MI. Society of Manufacturing Engineers.

Nemeth, Susan E. 1985. "Hand Tool Design." In *Industrial Ergonomics: A Practitioner's Guide.* David C. Alexander and B. Mustafa Pulat, eds. Norcross, GA: Industrial Engineering and Management Press.

## REVIEW QUESTIONS

1. Define the main objective of tool design.
2. Why are economic lot sizes calculated?
3. Why are break-even charts used in tool design?
4. List four points where safety should be designed into the tooling.
5. How are material handling and tool design related?
6. Define the first step in the design of any tool.
7. List three pieces of data used to identify the initial design.
8. List at least five ideas used in the development of a satisfactory work environment.
9. What are the 10 principles for the design and use of hand tools.

# 2

# Tool Materials

The material selected when creating a particular tool normally is determined by the mechanical properties necessary for that tool's proper operation. These materials should be selected only after a careful study and evaluation of the function and requirements of the proposed tool. For most applications, more than one type of material will be satisfactory, and a final choice normally will be governed by material availability and economic considerations.

The principal materials used for tools can be divided into three major categories: ferrous metals, nonferrous metals, and nonmetallic materials. Ferrous tool materials have iron as a base metal and include tool steel, alloy steel, carbon steel, and cast iron. Nonferrous materials have a base metal other than iron and include aluminum, magnesium, zinc, lead, bismuth, copper, and a variety of other metals and their alloys. Nonmetallic materials include woods, plastics, rubbers, epoxy resins, ceramics, and diamonds that do not have a metallic base. To properly select a tool's material, several physical and mechanical properties should be understood to determine how the selected materials would affect its function and operation.

## PHYSICAL PROPERTIES

The physical properties of a material control how the material will react under certain conditions. Physical properties are natural in the material and cannot be permanently altered without changing the material itself. These properties include: weight, color, thermal and electrical conductivity, rate of thermal expansion, and melting point.

## Weight

The *weight* of a material is a measure of the amount of substance contained in a specific volume, such as pound per foot or kilogram per meter. Weight is a factor to consider for ease of material handling and resisting loads and forces.

## Color

*Color* is the natural tint contained throughout the material. For example, steels are normally a silver-gray color and copper is usually a reddish brown.

## Thermal and Electrical Conductivity

*Thermal* and *electrical conductivity* measures how fast or slow a specific material conducts heat or electricity. Aluminum and copper, for example, have a high rate of thermal and electrical conductivity, while nickel and chromium have a comparatively low rate.

## Rate of Thermal Expansion

The rate of *thermal expansion* is a measure of how a material expands when exposed to heat. Materials such as aluminum, zinc, and lead have a high rate of expansion, while carbon and silicon expand very little when heated. Using materials with a low rate of thermal expansion is important when high dimensional accuracy is required. Using materials with differing rates of thermal expansion can cause problems in constructing and using tools.

## Melting Point

The *melting point* is the temperature at which a material changes from a solid to a liquid state. Materials such as tantalum and tungsten have a very high melting point, while lead and bismuth have a comparatively low melting point. The melting point is a consideration when high temperatures are involved in the use of a tool.

## MECHANICAL PROPERTIES

The mechanical properties of a material can be permanently altered by thermal or mechanical treatment. These properties include strength, hardness, wear resistance, toughness, brittleness, plasticity, ductility, malleability, and modulus of elasticity.

## Strength

*Strength* is the ability of a material to resist deformation. The most common units used to designate strength are pounds per square inch (psi) and kiloPascals (kPa). The principal categories of strength to be most concerned with when designing tools are tensile strength, compressive strength, shear strength, and yield strength.

### Tensile Strength

*Tensile strength* is the value obtained by dividing the maximum load observed during tensile testing by the specimen's cross-sectional area before testing. An example of a plot of a tensile test is shown in *Figure 2-1*. This plot shows the yield point as well as fracture points for several different materials.

Tensile strength is an important property to consider when designing large fixtures or other tooling. It is of lesser importance in tools and dies except where soft- or medium-hard ferrous or nonferrous materials are used.

The tensile tests successfully made on tool steel involve the use of drawing temperatures much

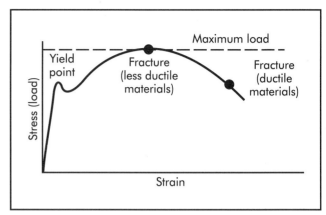

Figure 2-1. *Tensile strength.*

higher than those actually used on tools. Tool steels for hot work, fatigue, or impact applications are usually specified at lower hardness levels. Tensile properties can be obtained for tool steels from data books or vendor literature.

### Compressive Strength

*Compressive strength* plays an important part in tool design. It is the maximum stress that a metal, subjected to compression, can withstand without fracture (*Figure 2-2*).

The compressive strength test is used on hardened tool steels, especially at high hardness levels. For all ductile materials, the specimens flatten out under load, and there is no well-marked fracture. For these materials, compressive strength is usually equal to tensile strength.

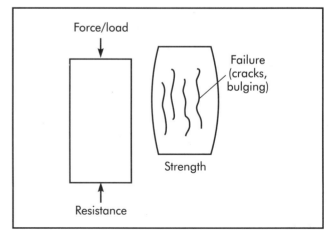

Figure 2-2. *Compressive strength.*

## Shear Strength

The shear strength of materials is significant, especially when designing machines and members subject to torsion. *Shear strength* may be defined as the value of stress necessary to cause rupture (or failure) in torsion (*Figure 2-3*).

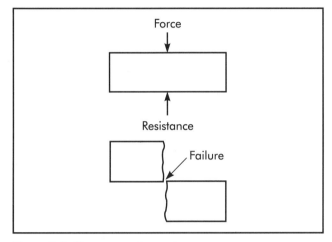

Figure 2-3. *Shear strength.*

For most steels, except tool and other highly alloyed steels, the shear strength lies between 50 and 60% of the yield strength; hence, the yield strength in tension serves fairly well as an index of shear strength.

Hardened tool steels are considered to be brittle because they deform very little before fracture in tension or bending, yet when subjected to torsion they exhibit considerable ductility. The torsion method of testing offers advantages for the toughness testing of hard tool steels. This is particularly true in those instances where very small amounts of deformation occur. Toughness is the property of absorbing considerable energy before fracture. It involves both ductility and strength.

## Yield Strength

The *yield strength* of a material generally limits its strength in application. It is the stress level at which a material will show a permanent elongation after the load is released.

To compare the elastic properties of steel, both soft and hard, a definite amount of permanent elongation is used as the criterion of yield strength. This is generally 0.2% of the 2-in. (50.8-mm) gage length used. Heat-treatment is used to alter the yield strength. If permanent deformation is not acceptable in a tool, a material (and heat-treatment) should be selected that has sufficient yield strength to resist deformation.

## Hardness

*Hardness* is the ability of the material to resist penetration or withstand abrasion. It is an important property in selecting tool materials. Hardness alone does not determine the wear or abrasion resistance of a material. In alloy steels, especially tool steels, the resistance to wear or abrasion varies with the alloy content. Different hardness scales have been developed, each of which covers a different range of hardness and is applicable for different materials.

## Rockwell Hardness

Rockwell hardness is the most widely used method for measuring the hardness of steel. The Rockwell hardness test is made using a dead weight acting through a series of levers to force a penetrator into the surface of the metal being tested. The softer the metal being tested, the deeper it will penetrate with a given load. The dial gage does not read directly into the depth of penetration, but shows arbitrary scales of Rockwell numbers. A variety of loads and penetrators can be used, each designated by a different letter, and the relative hardness or softness is measured.

Two types of penetrators are used: a diamond cone known as a brale, for hard materials such as tool steel, and a hardened steel ball for testing soft materials.

## Brinell Hardness

The Brinell hardness method of measurement is much older than the Rockwell method. It operates similarly to the Rockwell ball-test principle. In the Brinell machine, a 10 mm (0.39 in.) much larger steel ball is forced into the material being tested under a load of approximately 3,000 kg (6,600 lb). Instead of measuring the penetration, the diameter of the impression in the test piece is measured using a small hand microscope with a lens calibrated in millimeters. The measured diameter is converted into a Brinell hardness number by using a table.

The Brinell hardness measurement is most useful on soft and medium hard materials. On steels of high hardness, the impression is so small it is difficult to read; therefore, the Rockwell test is used more commonly for such materials. A comparison of the designations for each system, as well as other hardness tests, is shown in *Table 2-1*.

## Wear Resistance

*Wear resistance* is the property of a material that allows it to resist wearing by abrasive action. This property is normally linked to hardness. If a material is hard, it is usually wear-resistant as well.

## Toughness

*Toughness* is a property that permits a material to resist fracture or deformation from sudden loading or shocks. In most cases, a hard material is also tough. However, if the hardness goes beyond a certain point (normally Rockwell C 57), toughness is replaced by brittleness.

## Brittleness

*Brittleness* is the property of a material that causes it to break or fracture easily when sudden loads are applied. Most materials that are extremely hard also are very brittle. While brittleness is often thought of as a negative property in most materials, in some applications, such as shear pins, brittleness is a desirable property, especially if the failure load is consistent.

## Plasticity

*Plasticity* is the property of a material that allows it to be extensively deformed without fracture. Two general categories of plasticity are ductility and malleability.

- *Ductility* is the property of a material that allows it to be stretched or drawn with a tensional force without fracture or rupture.

- *Malleability* is the property of a material that permits it to be hammered or rolled without fracture or rupture.

## Modulus of Elasticity (Bending)

The *modulus of elasticity in bending* is a measure of the stiffness of a material. It is indicated by the slope of the line generated below the elastic limit during tensile testing. This is an important property to take into consideration when designing long tools and machine parts. Section size also must be taken into consideration.

## Modulus of Elasticity (Torsion)

The *modulus of elasticity in torsion* corresponds to the modulus of elasticity in the tensile test except that it is measured in a torsion test. The modulus is the ratio of the unit shear stress to the displacement caused by the stress per unit length in the elastic range. It is a usable value when designing shafts, taps, twist drills, or other tools working with torsion. Values should be used as a guide when available.

## FERROUS TOOL MATERIALS

Many ferrous materials can be used for tool construction. Typically, materials such as carbon, alloy, and tool steel, along with cast iron, are widely used for jigs, fixtures, dies, and similar special tools. These materials are supplied in several different forms. The most common types used for tools are hot-rolled, cold-rolled, and ground.

When steel is hot-rolled at a mill, a layer of decarburized slag, or bark, covers the entire surface of the metal. This bark should be removed when the part being made is to be hardened. If, however, the metal is to be used in an unhardened condition, the bark may be left on the part. When ordering hot-rolled materials, the designer must make allowance for the removal of the bark.

Cold-rolled steels are generally used for applications when little or no machining or welding is required. Cold-rolled bars are reasonably accurate and relatively close to size. When rolled, these steels develop internal stresses that could

Table 2-1. Approximate relations among various hardness scales for steel*

| Brinell | | | | Rockwell | | | | | |
| --- | --- | --- | --- | --- | --- | --- | --- | --- | --- |
| | Ball | | | Standard | | Superficial | | | |
| Indent Diameter, in. (mm) (6,600-lb [3,000-kg], 0.39-in. [10-mm] Ball) | Steel | Tungsten Carbide | Vickers or Firth | C 331-lb (150-kg) Diamond Brale | B 221-lb (100-kg), 1/16-in. (1.6-mm) Ball | 30-N 66-lb (30-kg) Diamond Brale | 30-T 66-lb (30-kg), 1/16-in. (1.6-mm) Ball | Knoop | Tensile Strength, ksi** (kgf/cm²) |
| 0.089 (2.25) | 745 | 840 | 1050 | 68 | | 86 | | 822 | 368 (25,873) |
| 0.091 (2.30) | 712 | 812 | 960 | 66 | | 84 | | 787 | 352 (24,748) |
| 0.093 (2.35) | 682 | 794 | 885 | 64 | | 83 | | 753 | 337 (23,693) |
| 0.094 (2.40) | 653 | 760 | 820 | 62 | | 81 | | 720 | 324 (22,780) |
| 0.097 (2.45) | 627 | 724 | 765 | 60 | | 79 | | 688 | 311 (21,866) |
| 0.098 (2.50) | 601 | 682 | 717 | 58 | | 78 | | 657 | 298 (20,952) |
| 0.100 (2.55) | 578 | 646 | 675 | 57 | | 76 | | 642 | 287 (20,178) |
| 0.102 (2.60) | 555 | 614 | 633 | 55 | 120 | 74 | | 610 | 276 (19,405) |
| 0.104 (2.65) | 534 | 578 | 598 | 53 | 119 | 72 | | 578 | 266 (18,702) |
| 0.106 (2.70) | 514 | 555 | 567 | 52 | 119 | 70 | | 563 | 256 (17,999) |
| 0.108 (2.75) | 495 | 525 | 540 | 50 | 117 | 69 | | 533 | 247 (17,366) |
| 0.110 (2.80) | 477 | 514 | 515 | 49 | 117 | 68 | | 520 | 238 (16,733) |
| 0.112 (2.85) | 461 | 477 | 494 | 47 | 116 | 67 | | 493 | 229 (16,100) |
| 0.114 (2.90) | 444 | 460 | 472 | 46 | 115 | 66 | | 480 | 220 (15,468) |
| 0.116 (2.95) | 429 | 432 | 454 | 45 | 115 | 65 | | 467 | 212 (14,905) |
| 0.118 (3.00) | 415 | 418 | 437 | 44 | 114 | 64 | | 455 | 204 (14,343) |
| 0.120 (3.05) | 401 | 401 | 420 | 42 | 113 | 63 | | 432 | 196 (13,780) |
| 0.122 (3.10) | 388 | 388 | 404 | 41 | 112 | 62 | | 421 | 189 (13,288) |
| 0.124 (3.15) | 375 | 375 | 389 | 40 | 112 | 61 | | 410 | 182 (12,796) |
| 0.126 (3.20) | 363 | 364 | 375 | 38 | 110 | 60 | | 388 | 176 (12,374) |
| 0.128 (3.25) | 352 | 352 | 363 | 37 | 110 | 59 | | 378 | 170 (11,952) |
| 0.130 (3.30) | 341 | 341 | 350 | 36 | 109 | 58 | | 368 | 165 (11,601) |
| 0.132 (3.35) | 331 | 330 | 339 | 35 | 109 | 57 | | 359 | 160 (11,249) |
| 0.134 (3.40) | 321 | 321 | 327 | 34 | 108 | 56 | | 350 | 155 (10,898) |
| 0.136 (3.45) | 311 | 311 | 316 | 33 | 108 | 55 | | 341 | 150 (10,546) |
| 0.138 (3.50) | 302 | 302 | 305 | 32 | 107 | 54 | | 332 | 146 (10,265) |
| 0.140 (3.55) | 293 | | 296 | 31 | 106 | 53 | | 323 | 142 (9,984) |
| 0.142 (3.60) | 285 | | 287 | 30 | 105 | 52 | | 315 | 138 (9,702) |
| 0.144 (3.65) | 277 | | 279 | 29 | 104 | 51 | | 307 | 134 (9,421) |
| 0.146 (3.70) | 269 | | 270 | 28 | 104 | 50 | | 299 | 131 (9,210) |
| 0.148 (3.75) | 262 | | 263 | 26 | 103 | 49 | | 285 | 128 (8,999) |
| 0.150 (3.80) | 255 | | 256 | 25 | 102 | 48 | | 278 | 125 (8,788) |
| 0.152 (3.85) | 248 | | 248 | 24 | 102 | 47 | | 272 | 122 (8,578) |
| 0.154 (3.90) | 241 | | 241 | 23 | 100 | 46 | 85 | 266 | 119 (8,367) |
| 0.156 (3.95) | 235 | | 235 | 22 | 99 | 45 | 84 | 258 | 116 (8,156) |
| 0.157 (4.00) | 229 | | 229 | 21 | 98 | 44 | 83 | 250 | 113 (7,945) |

Table 2-1. (continued)

| Brinell | | | | Rockwell | | | | | |
|---|---|---|---|---|---|---|---|---|---|
| | Ball | | | Standard | | Superficial | | | |
| Indent Diameter, in. (mm) (6,600-lb [3,000-kg], 0.39-in. [10-mm] Ball) | Steel | Tungsten Carbide | Vickers or Firth | C 331-lb (150-kg) Diamond Brale | B 221-lb (100-kg), 1/16-in. (1.6-mm) Ball | 30-N 66-lb (30-kg) Diamond Brale | 30-T 66-lb (30-kg), 1/16-in. (1.6-mm) Ball | Knoop | Tensile Strength, ksi** (kgf/cm²) |
| 0.159 (4.05) | 223 | | 223 | 20 | 97 | 43 | 82 | 243 | 110 (7,734) |
| 0.161 (4.10) | 217 | | 217 | 18 | 96 | 42 | 82 | 237 | 107 (7,523) |
| 0.163 (4.15) | 212 | | 212 | 17 | 96 | 40 | 81 | 237 | 104 (7,312) |
| 0.165 (4.20) | 207 | | 207 | 16 | 95 | 39 | 81 | 230 | 101 (7,101) |
| 0.167 (4.25) | 202 | | 202 | 15 | 94 | 38 | 80 | 224 | 99 (6,960) |
| 0.169 (4.30) | 197 | | 197 | 13 | 93 | 37 | 79 | 218 | 97 (6,820) |
| 0.171 (4.35) | 192 | | 192 | 12 | 92 | 36 | 78 | 213 | 95 (6,679) |
| 0.173 (4.40) | 187 | | 187 | 10 | 91 | 35 | 78 | 209 | 93 (6,539) |
| 0.175 (4.45) | 183 | | 183 | 9 | 90 | 34 | 77 | 205 | 91 (6,398) |
| 0.177 (4.50) | 179 | | 179 | 8 | 89 | 33 | 77 | 201 | 89 (6,257) |
| 0.179 (4.55) | 174 | | 174 | 7 | 88 | 32 | 76 | 197 | 87 (6,117) |
| 0.181 (4.60) | 170 | | 170 | 6 | 87 | 31 | 76 | 193 | 85 (5,976) |
| 0.183 (4.65) | 166 | | 166 | 4 | 86 | 30 | 75 | 189 | 83 (5,836) |
| 0.185 (4.70) | 163 | | 163 | 3 | 85 | 29 | 74 | 186 | 82 (5,765) |
| 0.187 (4.75) | 159 | | 159 | 2 | 84 | 28 | 73 | 182 | 80 5,625) |
| 0.189 (4.80) | 156 | | 156 | 1 | 83 | 27 | 73 | 179 | 78 (5,484) |
| 0.191 (4.85) | 153 | | 153 | | 82 | | 72 | 176 | 76 (5,343) |
| 0.193 (4.90) | 149 | | 149 | | 81 | | 71 | 173 | 75 (5,273) |
| 0.195 (4.95) | 146 | | 146 | | 80 | | 71 | 170 | 74 (5,203) |
| 0.197 (5.00) | 143 | | 143 | | 79 | | 70 | 167 | 72 (5,062) |
| 0.199 (5.05) | 140 | | 140 | | 78 | | 69 | 164 | 71 (4,992) |
| 0.201 (5.10) | 137 | | 137 | | 77 | | 68 | 161 | 70 (4,922) |
| 0.203 (5.15) | 134 | | 134 | | 76 | | 68 | 158 | 68 (4,781) |
| 0.205 (5.20) | 131 | | 131 | | 74 | | 67 | 154 | 66 (4,640) |
| 0.207 (5.25) | 128 | | 128 | | 73 | | 67 | 152 | 65 (4,570) |
| 0.209 (5.30) | 126 | | 126 | | 72 | | 66 | 150 | 64 (4,500) |
| 0.211 (5.35) | 124 | | 124 | | 71 | | 66 | 148 | 63 (4,429) |
| 0.213 (5.40) | 121 | | 121 | | 70 | | 65 | 146 | 62 (4,359) |
| 0.215 (5.45) | 118 | | 118 | | 69 | | 64 | 144 | 61 (4,289) |
| 0.217 (5.50) | 116 | | 116 | | 68 | | 63 | 142 | 60 (4,218) |
| 0.219 (5.55) | 114 | | 114 | | 67 | | 63 | 140 | 59 (4,148) |
| 0.221 (5.60) | 112 | | 112 | | 66 | | 62 | 139 | 58 (4,078) |
| 0.222 (5.65) | 109 | | 109 | | 65 | | 61 | 137 | 56 (3,937) |
| 0.224 (5.70) | 107 | | 107 | | 64 | | 60 | 135 | 55 (3,867) |
| 0.226 (5.75) | 105 | | 105 | | 62 | | 59 | 132 | 54 (3,797) |
| 0.228 (5.80) | 103 | | 103 | | 61 | | | 130 | 53 (3,726) |

*Figures in *italics* should be used only as a guide.
**ksi = 1,000 lb/in.² or 1,000 psi

warp or distort the part if it were extensively machined or welded. Cold-rolled bars are distinguished from hot-rolled bars by the bright, scale-free surface.

Steels are also available in a ground condition. These materials are held to very close tolerances and are available commercially in many sizes and shapes. They are normally used where a finished surface is required without additional machining. The two standard types of ground materials are "to size" and "oversize." To-size materials are ground to a specific size, such as 0.25 in. (6.4 mm), 0.50 in. (12.7 mm), or any similarly standard size. Oversize materials are normally ground 0.015 in. (0.38 mm) over the standard size.

## Carbon Steels

Carbon steels are used extensively in tool construction. They contain mostly iron and carbon, with small amounts of other alloying elements, and are the most common and least expensive types of steel used for tools. The three principal types are low-carbon, medium-carbon, and high-carbon steels. Low-carbon steel contains 0.05–0.30% carbon; medium-carbon steel contains 0.30–0.70% carbon; and high-carbon steel contains 0.70–1.50% carbon. As the carbon content is increased in carbon steel, the maximum strength, toughness, and hardness also increase when the metal is heat-treated.

Low-carbon steels are soft, tough steels that are easily machined and welded. Due to their low carbon content, these steels cannot be hardened except by case hardening. Low-carbon steels are well-suited for tool bodies, handles, die shoes, and similar situations where strength and wear resistance are not required.

Medium-carbon steels are used where greater strength and toughness are required. Since medium-carbon steels have a higher carbon content, they can be heat-treated to make studs, pins, axles, and nuts. Steels in this group are more expensive as well as more difficult to machine and weld than low-carbon steels.

High-carbon steels are the most hardenable type of carbon steel. They are used frequently for parts with which wear resistance is an important factor. Other applications where high-carbon steels are well-suited include drill bushings, locators, and wear pads. Since the carbon content of these steels is so high, parts made from high-carbon steel are normally difficult to machine and weld.

## Alloy Steels

Alloy steels are basically carbon steels with additional elements added to alter the characteristics and bring about a predictable change in the mechanical properties of the alloyed metal. Alloy steels are not normally used for most tools due to their increased cost, but some have found favor for special applications. The alloying elements used most often in steels are manganese, nickel, molybdenum, and chromium.

Another type of alloy steel frequently used for tooling applications is stainless steel. Stainless steel is a term used to describe high-chromium and nickel-chromium steels. These steels are used for tools that must resist high temperatures and corrosive atmospheres. Some high-chromium steels can be hardened by heat-treatment and are used where resistance to wear, abrasion, and corrosion are required. Hardenable stainless steel is sometimes preferred for plastic injection molds. Here, the high-chromium content allows the steel to be highly polished and prevents deterioration of the cavity from heat and corrosion.

## Tool Steels

Tool steels are alloy steels produced primarily for use in cutting tools. Selecting the proper tool steels is complicated by their many special properties. The five principal properties of tool steels are heat resistance, abrasion resistance, shock resistance, resistance to movement or distortion in hardening, and cutting ability. Because no single steel can possess all of these properties to the optimum degree, hundreds of different tool steels have been developed to meet the total range of service demands.

The steels listed in *Table 2-2* will adequately serve 95% of all metal-stamping operations. The list contains 31 steels, nine of which are widely applied and readily available. The other steels represent slight variations for improved performance

Table 2-2. American Iron and Steel Institute (AISI) identification and classification of tool steels

| Steel Types* | Average Composition, % | | | | | | | |
|---|---|---|---|---|---|---|---|---|
| | C | Mn | Si | Cr | W | Mo | V | Other |
| W1 | 1.00 | | | | | | | |
| W2 | 1.00 | | | | | | 0.25 | |
| O1 | 0.90 | 1.00 | | 0.50 | 0.50 | | | |
| O2 | 0.90 | 1.60 | | | | | | |
| O7 | 1.20 | | | 0.75 | 1.75 | 0.25 | | |
| A2 | 1.00 | | | 5.00 | | 1.00 | | |
| A4 | 1.00 | 2.00 | | 1.00 | | 1.00 | | |
| A5 | 1.00 | 3.00 | | 1.00 | | 1.00 | | |
| A6 | 0.70 | 2.00 | | 1.00 | | 1.00 | | |
| D2 | 1.50 | | | 12.00 | | 1.00 | | |
| D3 | 2.25 | | | 12.00 | | | | |
| D4 | 2.25 | | | 12.00 | | 1.00 | | |
| D6 | 2.25 | | 1.00 | 12.00 | 1.00 | | | |
| S1 | 0.50 | | | 1.50 | 2.50 | | | |
| S2 | 0.50 | | 1.00 | | | 0.50 | | |
| S4 | 0.50 | 0.80 | 2.00 | | | | | |
| S5 | 0.50 | 0.80 | 2.00 | | | 0.40 | | |
| H11 | 0.35 | | | 5.00 | | 1.50 | | |
| H12 | 0.35 | | | 5.00 | 1.50 | 1.50 | 0.40 | |
| H13 | 0.35 | | | 5.00 | | 1.50 | 1.00 | |
| H21 | 0.35 | | | 3.50 | 9.00 | | | |
| H26 | 0.50 | | | 4.00 | 18.00 | | 1.00 | |
| T1 | 0.70 | | | 4.00 | 18.00 | | 1.00 | |
| T15 | 1.50 | | | 4.00 | 12.00 | | 5.00 | 5.00 Co |
| M2 | 0.85 | | | 4.00 | 6.25 | 5.00 | 2.00 | |
| M3 | 1.00 | | | 4.00 | 6.00 | 5.00 | 2.40 | |
| M4 | 1.30 | | | 4.00 | 5.50 | 4.50 | 4.00 | |
| L2 | 0.50 | | | 1.00 | | | 0.20 | |
| L3 | 1.00 | | | 1.50 | | | 0.20 | |
| L6 | 0.70 | | | 0.75 | | | | 1.50 Ni |
| F2 | 1.25 | | | | 3.50 | | | |

* W, water hardening; O, oil hardening, cold work; A, air hardening, medium alloy; D, high carbon, high chromium, cold work; S, shock resisting; H, hot work; T, tungsten base, high speed; M, molybdenum base, high speed; L, special purpose, low alloy; F, carbon tungsten, special purpose.

in certain instances. Their use is sometimes justified because of special considerations.

Tool steels are identified by letter and number symbols. All the steels listed, except those in the S and H groups, can be heat-treated to a hardness greater than Rockwell C 62 and, accordingly, are hard, strong, wear-resistant materials. Frequently, hardness is proportional to wear resistance, but this is not always the case, because wear resistance usually increases as the alloy content, and particularly the carbon content, increases.

The toughness of steels, on the other hand, is inversely proportional to the hardness, and increases markedly as the alloy or carbon content is lowered. *Table 2-3* lists basic characteristics of steels used for press tools, *Table 2-4* lists the hardening and tempering treatments, and *Table 2-5* lists typical applications of the various steels listed.

## Classes

The general nature and application of the various standard tool steel classes are as follows:

**W: water-hardening tool steels.** This group includes plain carbon (W1) and carbon vanadium (W2). Carbon steels were the original tool steels. Because of their low cost, abrasion- and shock-resisting qualities, ease of machinability, and ability to take a keen cutting edge, the carbon grades are widely applied. Both W1 and W2 steels are shallow hardening and are readily available.

**O: oil-hardening tool steels.** Types O1 and O2 are manganese oil-hardening tool steels. They are readily available and inexpensive. These steels have better stability than water-hardening steels, and are of equal toughness with water-hardening steels when the latter are hardened throughout. Wear resistance is slightly better than that of water-hardening steels of equal carbon content. Steel O7 has greater wear resistance because of its increased carbon and tungsten content.

**A: air-hardening die steels.** Type A2 is the principal air-hardening tool steel. It has minimum movement in hardening and higher toughness than the oil-hardening die steels, with equal or greater wear resistance. Steels A4, A5, and A6 can be hardened from lower temperatures. However, they have lower wear resistance and higher resistance against distortion.

**D: high-carbon, high-chromium die steels.** Type D2 is the principal steel in this class. It finds wide application for long-run dies. It is deep-hardening, fairly tough, and has good resistance to wear. Steels D3, D4, and D6, containing additional carbon, have very high wear

resistance and lower toughness. Steels D2 and D4 are air-hardened.

**S: shock-resisting tool steels.** These steels contain less carbon and have higher toughness. They are applied where heavy cutting or forming operations are required, and where breakage is a serious problem. Steels S1, S4, and S5 are readily available. Steels S4 and S5 are more economical than S1.

**H: hot-work die steels.** These steels combine red hardness with good wear-resistance and shock-resistance. They are air-hardening and used on occasion for cold-work applications. They have relatively low carbon content and intermediate to high alloy content.

**T and M: tungsten and molybdenum high-speed steels.** Steels T1 and M2 are equivalent in performance and have good red hardness and abrasion resistance. They have higher toughness than many of the other die steels. They may be hardened by conventional methods or surface-hardened by carburizing. Steels M3, M4, and T15 have greater cutting ability and resistance to wear. They are more difficult to machine and grind because of their increased carbon and alloy contents.

**L: low-alloy tool steels.** Steels L3 and L6 are used for special die applications. Other L steels find application where fatigue and toughness are important considerations, such as in coining or impression dies.

**F: finishing steels.** Steel F2 has limited use, but is occasionally applied where extremely high wear resistance in a shallow-hardening steel is desired.

## Cast Iron

Cast iron is essentially an alloy of iron and carbon, containing from 2–4% carbon, 0.5 to about 3.00% silicon, 0.4 to approximately 1% manganese, plus phosphorus and sulfur. Other alloys may be added depending on the properties desired.

The high compressive strength and ease of casting gray irons are utilized in large forming and drawing dies to produce such items as automobile panels, refrigerator cabinets, bath tubs,

Table 2-3. Comparison of basic characteristics of steels used for press tools

| AISI Steel Number | Non-deforming Properties | Safety in Hardening | Toughness | Resistance to Softening Effect of Heat | Wear Resistance | Machinability |
|---|---|---|---|---|---|---|
| W1 | Poor | Fair | Good | Poor | Fair | Best |
| W2 | Poor | Fair | Good | Poor | Fair | Best |
| O1 | Good | Good | Fair | Poor | Fair | Good |
| O2 | Good | Good | Fair | Poor | Fair | Good |
| O7 | Good | Good | Fair | Poor | Fair | Good |
| A2 | Best | Best | Fair | Fair | Good | Fair |
| A4 | Best | Best | Fair | Poor | Fair | Fair |
| A5 | Best | Best | Fair | Poor | Fair | Fair |
| A6 | Best | Best | Fair | Poor | Fair | Fair |
| D2 | Best | Best | Fair | Fair | Good | Poor |
| D3 | Good | Good | Poor | Fair | Best | Poor |
| D4 | Best | Best | Poor | Fair | Best | Poor |
| D6 | Good | Good | Poor | Fair | Best | Poor |
| S1 | Fair | Good | Good | Fair | Fair | Fair |
| S2 | Poor | Fair | Best | Fair | Fair | Fair |
| S4 | Poor | Fair | Best | Fair | Fair | Fair |
| S5 | Fair | Good | Best | Fair | Fair | Fair |
| H11 | Best | Best | Best | Good | Fair | Fair |
| H12 | Best | Best | Best | Good | Fair | Fair |
| H13 | Best | Best | Best | Good | Fair | Fair |
| H21 | Good | Good | Good | Good | Fair | Fair |
| H26 | Good | Good | Good | Best | Good | Fair |
| T1 | Good | Good | Fair | Best | Good | Fair |
| T15 | Good | Fair | Poor | Best | Best | Poor |
| M2 | Good | Fair | Fair | Best | Good | Fair |
| M3 | Good | Fair | Fair | Best | Good | Fair |
| M4 | Good | Fair | Fair | Best | Best | Poor |
| L2 | Fair | Fair | Good | Poor | Fair | Fair |
| L3 | Fair | Poor | Fair | Poor | Fair | Good |
| L6 | Good | Good | Good | Poor | Fair | Fair |
| F2 | Poor | Poor | Poor | Fair | Best | Fair |

and other large articles. Conventional methods of hardening result in little distortion.

Alloying elements are added to promote graphitization and improve mechanical properties or develop special characteristics.

## Invar®

Invar is an iron-nickel alloy with 35–36% nickel and smaller amounts of other elements added for machinability. It is strong, tough, and ductile, with a fair amount of corrosion resistance. The

## Table 2-4. Hardening and tempering treatments for press tools

| AISI Steel Number | Preheat Temperature, °F (°C) | Rate of Heating for Hardening | Hardening Temperature, °F (°C) | Time at Temperature, min | Quenching Medium | Tempering Temperature, °F (°C) | Depth of Hardening | Resistance to Decarburizing |
|---|---|---|---|---|---|---|---|---|
| W1 | | Slow | 1,425–1,500 (774–816) | 10–30 | Brine or water | 325–550 (163–288) | Shallow | Best |
| W2 | | Slow | 1,425–1,550 (774–843) | 10–30 | Brine or water | 325–550 (163–288) | Shallow | Best |
| O1 | 1,200 (649) | Very slow | 1,450–1,500 (788–816) | 10–30 | Oil | 325–500 (163–260) | Medium | Good |
| O2 | 1,200 (649) | Very slow | 1,400–1,475 (760–802) | Do not soak | Oil | 325–600 (163–316) | Medium | Good |
| O7 | 1,200 (649) | Slow | 1,575–1,625 (857–885) | 10–30 | Oil | 350–550 (177–288) | Medium | Good |
| A2 | 1,450 (788) | Very slow | 1,700–1,800 (927–982) | 30 | Air | 350–700 (177–371) | Deep | Fair |
| A4 | 1,250 (677) | Slow | 1,450–1,550 (788–843) | 15–30 | Air | 300–500 (149–260) | Deep | Very good |
| A5 | 1,250 (677) | Slow | 1,450–1,550 (788–843) | 15–30 | Air | 300–500 (149–260) | Deep | Very good |
| A6 | 1,250 (677) | Slow | 1,500–1,600 (816–871) | 15–30 | Air | 300–500 (149–260) | Deep | Very good |
| S1 | | Slow to 1,400 (760) | 1,650–1,750 (899–954) | 10–30 | Oil | 500–600 (260–316) | Medium | Fair |
| S2 | | Slow | 1,525–1,575 (829–857) | 10–30 | Brine or water | 350–700 (177–371) | Fair | |
| S4 | | Slow | 1,550–1,650 (843–899) | 10–30 | Brine or water | 350–700 (177–371) | Poor | |
| S5 | | | 1,600–1,700 (871–927) | | Oil | 350–700 (177–371) | Medium | Poor |
| H11 | 1,400 (760) | Slow | 1,800–1,850 (982–1,010) | 15–60 | Air | 900–1,200 (482–649) | Deep | Good |
| H12 | 1,400 (760) | Slow | 1,800–1,850 (982–1,010) | 15–60 | Air | 900–1,200 (482–649) | Deep | Good |
| H13 | 1,400 (760) | Slow | 1,800–1,850 (982–1,010) | 15–60 | Air | 900–1,200 (482–649) | Deep | Good |
| H21 | 1,550 (843) | Medium | 2,000–2,200 (1,093–1,204) | 5–15 | Air, oil | 1,000–1,200 (538–649) | Deep | Good |
| H26 | 1,550 (843) | Medium | 2,000–2,200 (1,093–1,204) | 5–15 | Air, oil | 1,000–1,200 (538–649) | Deep | Good |
| T1 | 1,500–1,600 (816–871) | Rapid from | 2,150–2,300 (1,177–1,260) | Do not soak | Air, oil or salt | 1,025–1,200 (552–649) | Deep | Good |
| T15 | 1,500–1,600 (816–871) | preheat | 2,125–2,270 (1,163–1,243) | Do not soak | Air, oil or salt | 1,000–1,200 (538–649) | Deep | Fair |
| M2 | 1,500 (816) | Rapid | 2,125–2,225 (1,163–1,218) | Do not soak | Air, oil or salt | 1,025–1,200 (552–649) | Deep | Poor |
| M3 | 1,450–1,550 (788–843) | from | 2,125–2,225 (1,163–1,218) | Do not soak | Air, oil or salt | 1,025–1,200 (552–649) | Deep | Poor |
| M4 | 1,450–1,550 (788–843) | preheat | 2,125–2,225 (1,163–1,218) | Do not soak | Air, oil or salt | 1,025–1,200 (552–649) | Deep | Poor |

Table 2-4. (continued)

| AISI Steel Number | Preheat Temperature, °F (°C) | Rate of Heating for Hardening | Hardening Temperature, °F (°C) | Time at Temperature, min | Quenching Medium | Tempering Temperature, °F (°C) | Depth of Hardening | Resistance to Decarburizing |
|---|---|---|---|---|---|---|---|---|
| L2 | | Slow | 1,550–1,700 (843–927) | 15–30 | Oil | 350–600 (177–316) | Medium | Good |
| L3 | | Slow | 1,425–1,500 (774–816) | 10–30 | Brine or water | 300–800 (149–427) | Medium | Good |
| L3 | | Slow | 1,500–1,600 (816–871) | 10–30 | Oil | 300–800 (149–427) | Medium | Good |
| L6 | | Slow | 1,450–1,550 (788–843) | 10–30 | Oil | 300–1,000 (149–538) | Medium | Fair |
| F2 | 1,200 (649) | Slow | 1,525–1,625 (829–885) | 15–30 | Brine or water | 300–500 (149–260) | Shallow | Good |

main feature of Invar is that it has a coefficient of thermal expansion of almost zero, thus providing the dimensional stability required for length standards and optical platforms. One problem with Invar is that it is difficult to machine.

## NONFERROUS TOOL MATERIALS

Nonferrous tool materials are used to some degree as die materials in special applications, and generally for limited-production requirements. On the other hand, in jig and fixture design, some nonferrous materials are used extensively where magnetism or tool weight is an important factor. Another area where nonferrous materials are finding increased use is for cutting tools. Alloys and compositions of nonferrous materials are used extensively to machine the newer, exotic, high-strength metals.

### Aluminum

Aluminum has been used for special tooling for a long time. The principal advantages to using aluminum are its high strength-to-weight ratio, nonmagnetic properties, and relative ease in machining and forming. Pure aluminum is corrosion resistant, but not well-suited for use as a tooling material except in very limited, low-strength applications. Aluminum alloys, while not as corrosion resistant as pure aluminum, are much stronger and well-suited for many special

tooling applications. Aluminum/copper (2000 series) and aluminum/zinc (7000 series) are the alloys most frequently used for tooling applications. Depending on composition, some aluminum alloys are weldable and some can be heat-treated. One form of aluminum alloy finding increased use is aluminum tooling plate. This material is available in sheets and bars made to very close tolerances. Aluminum tooling plate is very useful for a wide variety of tooling applications. From supports and locators to base plates and tool bodies, aluminum tooling plate provides a lightweight alternative to steel. Other variations of aluminum frequently used for tooling are aluminum extrusions and cast bracket materials. In most cases these materials can be used as is, with little or no machining required.

### Magnesium

Magnesium, like aluminum, is a lightweight yet strong tooling material. Magnesium is lighter than aluminum and has a very good strength-to-weight ratio. Magnesium is commercially available in sheets, bars, and extruded forms. The only disadvantage in using magnesium is its potential fire hazard. When specifying magnesium as a tooling material, make sure those who are to make the various parts are well-acquainted with the precautions that must be observed when machining this material.

Table 2-5. Applications of tool steels

| Application | Suggested AISI Tool Steels | Rockwell C Hardness Range |
|---|---|---|
| Arbors | L6, L2 | 47–54 |
| Axle burnishing tools | M2, M3 | 63–67 |
| Boring bars | L6, L2 | 47–54 |
| Broaches | M2, M3 | 63–67 |
| Bushings (drill jig) | M2, D2 | 62–64 |
| Cams | A4, O1 | 59–62 |
| Centers, lathe | D2, M2 | 60–63 |
| Chasers | M2 | 62–65 |
| Cutting tools | M2 | 62–65 |
| Dies, blanking | O1, A2, D2 | 58–62 |
| Dies, bending | S1, A2, D2 | 52–62 |
| Dies, coining | S1, A2, D2 | 52–62 |
| Dies, cold heading: Solid | W1, W2 | 56–62 |
| Insert | D2, M2 | 57–62 |
| Dies, hot heading | H12, H13 | 42–48 |
| Dies, lamination | D2, D3 | 60–63 |
| Dies, shaving | D2, M2 | 62–64 |
| Dies, thread rolling | D2, A2 | 58–62 |
| Die casting: Aluminum | H13 | 42–48 |
| Form tools | M2, M3 | 63–67 |
| Lathe tools | M2, T1 | 63–65 |
| Reamers | M2 | 63–65 |
| Shear blades: Light stock | D2, A2 | 58–61 |
| Heavy stock | S1, S4 | 52–56 |
| Rolls | A2, D2 | 58–62 |
| Taps | M2 | 62–65 |
| Vise jaws | L2, S4 | 48–54 |
| Wrenches | L2, S1 | 40–50 |

## Bismuth Alloys

Bismuth alloys have several different uses in special tools. One of the principal advantages of bismuth alloys is their comparatively low melting temperature. Many alloy compositions will melt in boiling water. In addition to acting as a reusable nesting material, it can be applied as a matrix material for securing punch and die parts in a die assembly, and as cast punches and dies for short-run forming and drawing operations. Another frequent application of these alloys is for cast workholders. In this case, the material is melted and poured around the part and, once cool, the part is removed and the cast nest is used to hold subsequent parts for machining.

Low-melt alloys also are useful when machining parts with very thin cross sections, such as turbine blades. In these applications, the material is cast around the thin sections and acts as a support during machining. Once the machining is complete, the material is melted off the part and can be reused.

## Carbides

Carbides are a family of tool materials made from the carbides of tungsten, titanium, tantalum, or a combination of these elements. They are powder metals consisting of the carbide with a binder, usually cobalt, hot-pressed and then sintered into desired shapes. The most common carbide material used for special tools is tungsten carbide. All carbides are characterized by high hardness values and resistance to wear. This makes them an excellent choice for cutting tools. The specific grades and characteristics commonly used to classify carbides are shown in *Tables 2-6* and *2-7*. *Table 2-6* shows the unofficial C-classification system, and *Table 2-7* contains the International Organization for Standardization (ISO) system. These classification systems are application-based and only roughly tied to material properties. Generally, there is a tradeoff between hardness and toughness, but micrograin carbides provide greater hardness and toughness together.

## Cermets

Cermets are similar to carbides, but they consist of titanium carbide (TiC) or titanium nitride (TiN) powder in a nickel or cobalt binder. Cermets are harder and stronger than carbides, and provide better cutting performance on ferrous materials.

### Table 2-6. Joint Industry Council (JIC) carbide-classification code

| Code | Application | Carbide Characteristics |
|---|---|---|
| C-1 | Roughing | Medium-high shock resistance<br>Medium-low wear resistance |
| C-2 | General purpose | Medium shock resistance<br>Medium wear resistance |
| C-3 | Finishing | Medium-low shock resistance<br>Medium-high wear resistance |
| C-4 | Precision finishing | Low shock resistance<br>High wear resistance |
| C-5 | Roughing | Excellent resistance to cutting temperature<br>Shock and cutting load<br>Medium wear resistance |
| C-50 | Roughing and heavy feeds | Same as above |
| C-6 | General purpose | Medium-high shock resistance<br>Medium wear resistance |
| C-7 | Finishing | Medium shock resistance<br>Medium wear resistance |
| C-70 | Semifinishing and finishing | High cutting-temperature resistance |
|  |  | Medium wear resistance |
| C-8 | Precision finishing | Very high wear resistance |
|  |  | Low shock resistance |

Toughness increases from bottom to top, hardness from top to bottom.

## NONMETALLIC TOOL MATERIALS

Nonmetallic tool materials are chiefly used where the production of parts is limited and where the cost of using tool steels or similar materials would not be economically practical. In many cases, where nonmetallic tool materials are used for special tools, they are used in conjunction with steel parts such as bushings or blades rather than by themselves. However, in other applications, nonmetallic materials may be used alone. The principal nonmetallic materials used for special tooling are wood, composite materials, plastics, epoxy resins, rubber, urethane, ceramics, and diamonds.

## Wood

Wood is frequently used for low-cost, limited-production tooling. Typical applications include jig plates with inserted steel bushings and backing, and support parts for steel-rule dies. When work-ing with wood, the designer must anticipate the problems inherent in this material. For example, wood has a tendency to swell and warp. However, by selecting a relatively stable type of wood and properly positioning the parts, these problems can be minimized or eliminated (see *Figure 2-4*).

Some variations of wood products often used in tooling applications include hardboard, densi-fied woods, plywood, and particleboard.

### Hardboard

Hardboard is basically a material made of com-pressed wood fiber. Typical uses for this mate-rial include forming punches and dies, blocks for rubber forming, and stretch dies.

### Densified Woods

Densified woods are impregnated with a phe-nolic resin and laminated. After lamination, the assembled parts are compressed to about 50% of

Table 2-7. ISO carbide-classification system

| Main Machining Group | Color Marking | Application Group | Operations and Working Conditions |
|---|---|---|---|
| P: steel, cast steel, long-chipping malleable | Blue | P01 | High-precision turning and boring, high cutting speeds, small chip cross section, dimensional accuracy, good surface finish, and vibration-free machining |
| | | P10 | Turning, copy turning, thread cutting and milling, high cutting speeds, and small-to-medium chip cross section |
| | | P20 | Turning, copy turning, milling, medium cutting speeds, and medium chip cross section; planing with small chip cross section |
| | | P30 | Turning, milling, planing, medium-to-low cutting speeds, medium-to-large chip cross section, also under unfavorable conditions |
| | | P40 | Turning, planing, milling, shaping, low cutting speeds, large chip cross section, high rake angles, unfavorable conditions; also automatic turning |
| | | P50 | Where highest demands are made on toughness of carbide: turning, planing and shaping, low cutting speeds, large chip cross section, and high rakes under unfavorable conditions, automatic turning |
| M: steel, cast steel, austenitic manganese steel, cast-iron alloys, austenitic steels, malleable and spheroidal cast iron, free-cutting mild steel | Yellow | M10 | Turning, medium-high cutting speeds, small-to-medium chip cross sections |
| | | M20 | Turning, milling, medium cutting speeds, and medium chip cross section |
| | | M30 | Turning, milling, planing, medium cutting speeds, medium to large chip cross sections |
| | | M40 | Turning, form turning, parting off and recessing, particularly for automatics |
| K: cast iron, chilled cast iron, short-chipping malleable cast iron, hardened steel, nonferrous metals, non-metallic materials | Red | K01 | Turning, precision turning and precision boring, finish milling, and scraping |
| | | K10 | Turning, milling, boring, countersinking, reaming, scraping, and broaching |
| | | K20 | Turning, milling, planing, countersinking, scraping, reaming, and broaching under tougher conditions than K10 |
| | | K30 | Turning, milling, planing, shaping under unfavorable conditions, high rakes |
| | | K40 | Turning, milling, planing, shaping under unfavorable conditions, high rakes |

Cutting speed and wear resistance increase from bottom to top, feed and carbide toughness from top to bottom.

the original thickness of the wood layers. Densified woods are used for basically the same applications as hardboard.

## Plywood

Plywood may be used for special tooling in either its natural condition or as clad plywood. Clad plywood has a thin sheet of metal applied to help it resist wear and damage during use. Plywood may be used for tumble jigs, steel-rule dies, milling fixtures, and stretch-forming dies. When using plywood for any tooling application, it is best to use an exterior plywood grade to prevent the wood plies from separating.

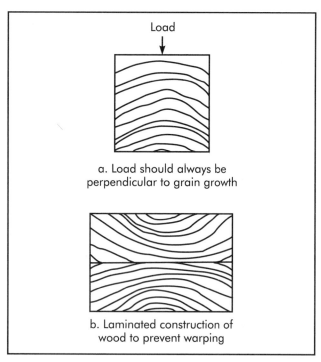

a. Load should always be
perpendicular to grain growth

b. Laminated construction of
wood to prevent warping

Figure 2-4. *Using wood properly.*

## Particleboard

Particleboard is a composition material made from wood chips and epoxy resins. These materials are available in several grades and serve basically the same functions as plywood. When using any type of wood or wood byproduct, it is always a good precaution to coat the tool with a lacquer or shellac to preserve the wood and help reduce swelling and warpage.

## Composite Materials

Composite materials normally are made from a filler, or base material, and some type of resin acting as a binder. Two typical composite materials used for special tooling are phenolic and Bakelite®. Both materials are used for the same applications as wood, but are more stable and less susceptible to moisture. Fiberglass and graphite composites are sometimes used to make tools that are lightweight without sacrificing strength.

## Plastics

Plastics are used where the operations are not severe and the production run is short to medium.

Plastics are resistant to chemicals, moisture, and temperature. They are inexpensive and facilitate tool repair and modification. In most cases, plastics can be machined with the same tools and equipment as metals, and can be easily adapted to toolroom uses. Plastics have been recently developed to withstand high heat and abrasion. Some have a nonstick surface that makes for excellent sliding. Newer plastics have tensile and shear strengths equal to some steels. With constant research and development in this industry, plastics will play an increasingly important role as a tool material.

## Epoxy Resins

Epoxy resins are mainly used for casting and laminating. Castable resins are used for jig plates, workholders, silkscreen fixtures, duplicating patterns, and large forming dies. In addition to the resin, a filler material is added to the mixture to increase its strength and provide better dimensional stability. Typical filler materials include glass and metal beads and metal filings. Cast resins are strong and relatively lightweight. When properly cast, they require little or no machining. Laminated resins are used for large, stretch-forming dies and checking fixtures. These materials are generally laminated over a wooden frame.

Epoxy resins combine the advantages of low cost, ease of modification, and shortened lead times into a single tooling material. They work well with intricate or complex part shapes, and depending on the filler material, normally will last a long time.

## Rubber

Rubber is used less now than it was in the past. This is due in large part to newer and better materials that have been developed. However, rubber is still used for specialized drawing, blanking, and bulging die operations, as well as protective elements for other special tools.

## Urethane

Urethane is becoming widely used for special tooling. It is available in solid bars that can be machined to suit a specific application or cast into almost any desired form. Urethane is not compressible and acts as a liquid when force is applied. That is, when force is applied, urethane displaces it equally in all directions. By containing and redirecting these displaced forces, urethane can be used to form complex shapes without marring the workpiece material. When used as a clamp pad, urethane transfers all the clamping forces without damaging the workpiece surfaces. Urethane is also used as a stripper in some larger, low-production, blanking dies.

Urethane does not shrink an appreciable amount and can be used to duplicate parts exactly. Lack of shrinkage makes urethane ideal for nests for ultrasonic fixtures or molds for model parts. Urethane is also used for embossing or shallow forming dies. When used in forming, only the die is made from urethane; the punch is normally made from steel or a similar material.

## Ceramics

Ceramics, or oxide cutting tools, are basically aluminum-oxide materials. They have a high compressive strength, high red hardness, high abrasive resistance, low heat conductivity, and good resistance to galling and welding. Ceramics are harder than carbides, but will chip or break easily when bending or twisting loads are applied. They should not be used for interrupted cuts since they have a low resistance to shock loads. For this reason, the machine selected to use ceramic cutting tools must be extremely rigid. Ceramic cutting tools are used to machine cast iron, carbon steels, low-alloy steels, and for finishing hard steels (Rockwell C 60 and 65) at high speeds. Also, they may be used for machining carbon, graphite, fiberglass, and other highly abrasive materials.

## Diamonds

Diamond is the hardest substance known, but has only limited use as a tool material. Industrial diamonds are either synthetic or natural. They are used for turning tools, grinding wheels, and grinding-wheel dressers. Diamonds are frequently used for turning plastics, precious met-

als, nonferrous metals, and general finishing operations with light cuts, fine feeds, and high cutting speeds. Diamond powder is used for lapping and polishing. These powders, or flours, create a very smooth finish and high luster, especially on ferrous materials.

Polycrystalline diamond (PCD) blanks consist of diamond powder bonded together under high pressure and temperature, generally on a tungsten-carbide substrate. The blank can be used either as an indexable insert or brazed into a cutting-tool body. PCD is generally used to machine nonferrous and nonmetallic materials. It is not recommended for machining ferrous materials because it has a high chemical affinity to the carbon in the PCD.

## Other Materials

Cubic boron nitride (CBN) is a product similar to diamond. It is used for grinding wheels and cutting tools. CBN can be used on ferrous materials because there is a low chemical affinity between this material and ferrous materials. Thus, it can be used at higher cutting speeds and for hardened ferrous materials. Polycrystalline CBN (PCBN) is a granular material compacted in a binder. It is used to make cutting tools in the same way PCD is used.

## HEAT-TREATING

The purpose of heat-treatment is to modify and control the properties of a metal or alloy. This is done by altering the structure of the metal or alloy by heating it to definite temperatures and cooling at various rates. This combination of heating and controlled cooling determines not only the nature and distribution of the microconstituents, which in turn determine the properties, but also the grain size.

Heat-treating should improve the alloy or metal for the service intended. It can:

- remove strains after cold-working;
- remove internal stresses, such as those produced by drawing, bending, or welding;
- increase the hardness of the material;
- improve the machinability of the material;

- improve the cutting capabilities of tools;
- increase the wear-resisting properties of the material;
- soften the material, as in annealing, and
- improve or change the properties of a material such as corrosion resistance, heat resistance, magnetic properties, or others as required.

## Hardening

*Hardening* is the process of heating to a temperature above the critical range, then cooling rapidly enough through the critical range to appreciably harden the steel. (See *Table 2-4* for specific treatment.)

What happens during the heat-treatment of die steels is represented graphically in *Figure 2-5*. Starting in the annealed condition at *A*, the steel is soft, consisting of an aggregate of ferrite and carbide. Upon heating above the critical temperature to *B*, the crystal structure of ferrite changes, becomes austenite, and dissolves a large portion of the carbide. The new structure, austenite, is always a prerequisite for hardening. By quenching it (cooling rapidly to room temperature), the carbon is retained in solution, and the structure known as martensite (*C*) results. This is the hard matrix structure in steels. It is initially highly stressed since the change from austenite involves some volumetric expansion against the natural stiffness of the steel. It must be reheated to an intermediate temperature (*D*) to soften it slightly and relieve those internal stresses that unduly embrittle the steel. This process is called *tempering*.

If quenching is not rapid enough, the austenite reverts to ferrite and carbide (*E*) and high hardness is not obtained. The rate at which quenching is required to produce martensite depends primarily on the alloy content. Low-alloy die steels are water- or oil-hardened, while highly alloyed steels usually can be hardened in air (that is, quenched at a much slower rate). Highly alloyed steels make the reaction more sluggish.

## Normalizing

*Normalizing* involves heating the material to a temperature between 100–200° F (56–111° C) above the critical range and cooling in still air. This is about 100° F (56° C) above the regular hardening temperature.

The purpose of normalizing is usually to refine grain structures that have been altered during forging. With most of the medium-carbon forging steels (alloyed and unalloyed), normalizing is highly recommended after forging and before machining to produce more homogeneous structures, and, in most cases, improved machinability.

High-alloy, air-hardened steels are not normalized, since to do so would cause them to harden and defeat their primary purpose. Increasing

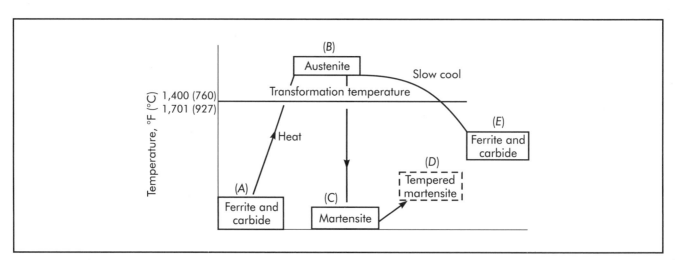

Figure 2-5. *Transformations in the hardening of steel.*

temperatures can cause grain growth to occur, which would further reduce hardness.

## Stress Relieving

*Stress relieving* is a method of relieving the internal stresses set up in steel during forming, cold-working, and cooling after welding or machining. It is the simplest heat-treatment and is accomplished merely by heating the steel to 1,200–1,350° F (649–732° C) followed by air or furnace cooling.

Large dies are usually roughed out, then stress-relieved and finish-machined. This will minimize the changing of a shape not only during machining, but during subsequent heat-treating as well. Welded sections also will have locked-in stresses due to a combination of differential heating and cooling cycles, as well as changes in the cross-section. Such stresses can cause considerable movement in machining operations.

## Annealing

The process of annealing consists of heating the steel to an elevated temperature for a defined period of time and cooling it slowly. Annealing is done to produce homogenization and establish normal equilibrium conditions with corresponding characteristic properties.

Tool steel is generally purchased in the annealed condition. Sometimes it is necessary to rework a tool that has been hardened, and the tool must then be annealed. For this type of anneal, the steel is heated slightly above its critical range, then cooled very slowly.

Finished parts may be annealed without surface deterioration by placing them in a closed pot and covering with compounds that will combine with the air present to form a reducing atmosphere. Partially spent carburizing compound is widely used for annealing, as well as cast-iron chips, charcoal, and commercial neutral compounds.

## Spheroidizing

Spheroidizing is a form of annealing that, in the process of heating and cooling steel, produces a rounded or globular form of carbide. Carbide is the hard constituent in steel.

Tool steels are normally spheroidized to improve machinability. This is accomplished by heating to a temperature of 1,380–1,400° F (749–760° C) for carbon steels and higher for many alloy tool steels, holding at heat for one to four hours, and cooling slowly in the furnace.

## Tempering

*Tempering* is the process of heating quenched and hardened steels and alloys to some temperature below the lower critical temperature to reduce the internal stresses set up in hardening. Thus, the hard martensite resulting from the quenching operation is changed in tempering in the direction of the equilibrium properties. The degree of change is dependent on the tempering temperature and rate of cooling. (See *Table 2-4* for specific treatment.)

## Case Hardening

The addition of carbon or nitrogen to the surface of steel parts and the subsequent hardening operations are important phases in heat-treating. The process may involve the use of molten sodium cyanide mixtures, pack carburizing with activated solid material such as charcoal or coke, gas or oil carburizing, and dry cyaniding.

Whether a solid charcoal or coke packing material is used, or a liquid gas, the objective is to produce a hard, wear-resistant surface with a tough center core. The carbon content of the surface is raised to 0.80–1.20% and the case depth can be closely controlled by the time, temperature, and carburizing medium used. Pack carburizing is generally done at 1,700° F (927° C) for eight hours to produce a case depth of 0.06 in. (1.5 mm). Light cases up to 0.005 in. (0.13 mm) can be obtained in liquid cyanide baths. Case depths to 0.03 in. (0.8 mm) are economically practical in liquid carburizing baths.

Usually, low-carbon and low-carbon alloy steels are carburized. The normal carbon range is 0.10–0.30%, though higher carbon-content steels may be carburized as well.

## Nonferrous Materials

Heat-treatment of nonferrous metals and alloys closely approximates that of steel, except that the temperature ranges used are lower, and hardening is accomplished by the precipitation of hard metallic compounds or particles.

Nonferrous metals and alloys that are not heat-treatable harden by cold work only.

For the heat-treatable alloys of aluminum, hardening is accomplished by precipitation. When an alloy is water-quenched from the hardening heat it is very soft; this is known as the solution treatment. Hardness is accomplished by aging, which follows the quenching operation. The aging temperature for some aluminum alloys is room temperature; others may require an elevated temperature of 290–360° F (143–182° C), depending on the alloy. As a rule, the lower the aging temperature, the longer the time required for the alloy to reach full hardness.

Beryllium copper is a precipitation-hardening alloy and is usually furnished by the manufacturer in the very soft, solution-treated condition. It has excellent forming properties in this condition. Formed parts are hardened by aging at 560–620° F (293–327° C) for two hours at heat. A hardness of 38–42 Rockwell C can be expected.

All other brass and bronze alloys are hardenable only by cold-working. They may be softened to various degrees by stress relieving or annealing.

## REVIEW QUESTIONS

1. What are the three principal categories of tool materials?
2. List three physical properties of materials.
3. List five mechanical properties.
4. What is resistance to penetration called?
5. Which hardness test system often uses a diamond, cone-shaped penetrator?
6. In what three conditions are steels normally purchased?
7. What are the grades of carbon steel?
8. What is another name for nickel-chromium steel?
9. List five classes of tool steel.
10. What advantage does Invar® alloy have? Disadvantage?
11. Which nonferrous metal is also known as "low-melt alloy?"
12. What are the main heat-treating processes?

# ◄ 3 ►

# Cutting Tool Design

The physics of metal cutting provide the theoretical framework by which all other elements of cutting tool design must be examined. Workpiece materials range from a very soft, buttery consistency to very hard and shear resistant. Each workpiece material must be handled by itself; the amount of broad information applicable to each material is reduced as the distinctions between workpiece characteristics increase. Not only is there a vast diversity of workpiece materials, but a large variety of tool shapes and compositions as well.

The tool designer must match many variables to provide the best possible cutting geometry. Trial and error once was normal for this decision, but today, with the ever-increasing variety of tools, it is far too expensive. The designer must develop expertise in applying data and making comparisons on the basis of others' experience.

## FORM AND DIMENSION

The primary method of imparting form and dimension to a workpiece is the removal of material using edged cutting tools. An oversize mass is literally carved into its intended shape. The removal of material from a workpiece is termed generation of form by machining, or simply *machining*.

Form and dimension also may be achieved using a number of alternative processes, such as hot or cold extrusion, and sand, die, and preci-

sion casting. Sheet metal can be formed or drawn by the application of pressure. In addition to machining, metal removal can be accomplished by chemical or electrical methods. A great variety of workpieces may be produced without resorting to a machining operation. Economic considerations, however, usually dictate form generation by machining, either as the complete process or in conjunction with another process.

## Elements of Machining

Material removal by machining involves the interaction of five elements: cutting tool, toolholding and guiding device, workholder, workpiece, and machine. The cutting tool may have a single or many cutting edges. It may be designed for linear or rotary motion. The geometry of the cutting tool depends on its intended function. The toolholding device may or may not be used for guiding or locating. Toolholder selection is governed by tool design and the intended function.

The physical composition of the workpiece greatly influences the selection of machining method, tool composition and geometry, and rate of material removal. The intended shape of the workpiece also influences the selection of machining method and choice of linear or rotary tool travel. To a great extent, the composition and geometry of the workpiece determines workholder requirements. Workholder selection also depends on forces produced by the tool on the workpiece. The workholder must hold, locate, and support the workpiece. Tool guidance may be incorporated into the workholding function.

Successful tool design for material removal requires, above all else, a complete understanding of cutting tool function and geometry. This knowledge enables the designer to specify the correct tool for a given task. The tool, in turn, governs the selection of toolholding and guidance methods. Tool forces govern selection of the workholding device. Although the process involves the interaction of five elements, everything begins with and is based on what happens at the point of contact between the workpiece and cutting tool.

## BASICS OF METAL CUTTING

Cutting tools are designed with sharp edges to shear the workpiece and minimize rubbing contact. Variations in the shape of the cutting tool influence tool life, surface finish of the workpiece, and the amount of force required to shear a chip from the parent metal. Various angles on a tool comprise what is often termed the tool geometry:

- the surface the chip flows across is called the *face* or rake face;
- the surface that forms the other boundary of the wedge is called the *flank*;
- the *rake angle* is the angle between the tool face and a line perpendicular to the cut workpiece surface, and
- the relief or *clearance angle* is the angle between the tool flank and the cut workpiece surface.

### Orthogonal and Oblique Cutting

*Orthogonal cutting (Figure 3-1)* is defined as two-dimensional cutting in which the cutting edge is perpendicular to the direction of motion relative to the workpiece, and the cutting edge is wider than the chip. *Oblique cutting* is used if the direction of cutting is not perpendicular to the cutting edge (*Figure 3-2*). This is a three-dimensional case requiring more cutting angles to define the tool. The effects on three-dimensional cutting must be described on the basis of the effective rake angle in the direction of chip flow.

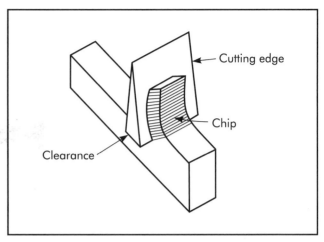

Figure 3-1. *Orthogonal cutting.*

## CHIP FORMATION

The majority of metal-cutting operations involve the creation of chips, or waste, from the workpiece. Chip formation involves three basic requirements. There must be:

- a cutting tool harder and more wear-resistant than the workpiece material;
- interference between the tool and workpiece as designated by the feed and depth of cut, and
- a relative motion or cutting velocity between the tool and workpiece with sufficient force to overcome the resistance of the workpiece material.

As long as these three conditions exist, the portion of material being machined that interferes with free passage of the tool will be displaced to create a chip. Many possibilities and combinations exist that may fulfill such requirements. Variations in tool material and tool geometry, feed and depth of cut, cutting velocity, and workpiece material affect not only chip formation, but also cutting force, horsepower, and temperatures; tool wear and life; dimensional stability; and the quality of the newly created surface. The interrelationship and interdependence among these manipulating factors constitute the basis for the study of machinability—a study that has been popularly defined as the response of a material to machining (Boston 1951).

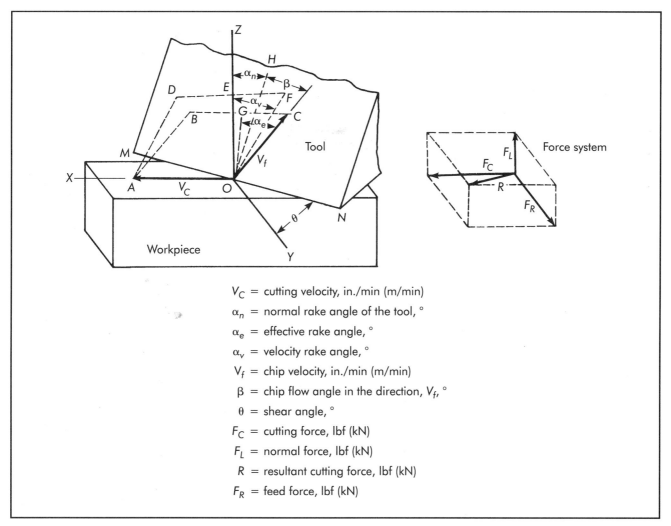

Figure 3-2. *Oblique cutting (Schrader and Elshennawy 2000).*

$V_C$ = cutting velocity, in./min (m/min)

$\alpha_n$ = normal rake angle of the tool, °

$\alpha_e$ = effective rake angle, °

$\alpha_v$ = velocity rake angle, °

$V_f$ = chip velocity, in./min (m/min)

$\beta$ = chip flow angle in the direction, $V_f$, °

$\theta$ = shear angle, °

$F_C$ = cutting force, lbf (kN)

$F_L$ = normal force, lbf (kN)

$R$ = resultant cutting force, lbf (kN)

$F_R$ = feed force, lbf (kN)

## Types of Chips

*Figure 3-3* illustrates the necessary relationship between cutting tool and workpiece for chip formation in several common machining processes. Although it is apparent that different shapes and chip sizes may be produced by each of the basic processes, all chips, regardless of process, are usually classified according to their general behavior during formation.

When a brittle material like cast iron or bronze is cut, it is broken along the shear plane. The same may happen if the material is ductile and the friction between the chip and tool is very high. The chips come off in small pieces or segments, and are pushed away by the tool. A chip formed in this way is called a *type I, discontinuous,* or *segmental chip.* A ductile material, cut optimally, is not broken up but comes off like a ribbon, as shown in *Figure 3-4a.* This is known as a *type II* or *continuous chip.* An evident line of demarcation separates the highly distorted crystals in the chip from the undistorted parent material.

### Built-up Edge

Consideration of chip flow along the face of the tool, in the formation of continuous chips, is of prime importance. If the friction force resisting the passage of the chip along the tool face is less than the force necessary to shear the chip material, the entire chip will pass off cleanly (*Figure*

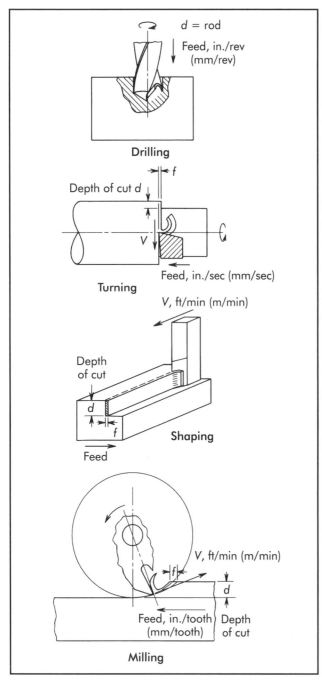

Figure 3-3. *Examples of feed depth and velocity relationships for several chip-formation processes.*

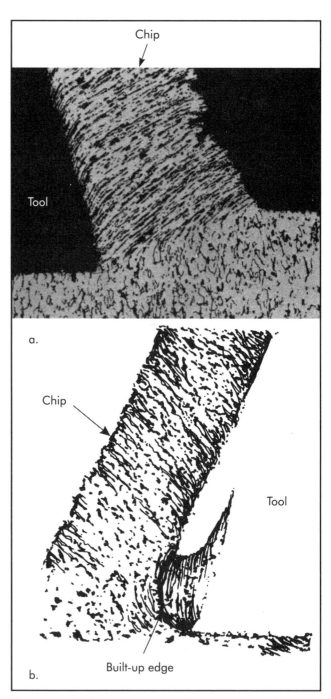

Figure 3-4. (a) *Photomicrograph of a partially formed continuous chip. AISI 1015 steel cut at 24 ft/min (7.3 m/min) with cutting fluid composed of water and 1% rust inhibitor. (b) Drawing of a photomicrograph of a continuous chip and tool with a built-up edge (Schrader and Elshennawy 2000).*

3-4a). This ideal case of chip formation may be approached, but is seldom realized. It is generally associated with materials of high strength and low work-hardening capacity, and with low coefficients of friction—factors that lead to large shear angles. The mechanism for built-up edge

formation is not completely understood, but there seems to be a correlation with temperature and, thus, a correlation with cutting speed. Often, a built-up edge will not appear at very low cutting speeds, then start to appear as the cutting speed is increased, peaking and starting to decrease as the cutting speed is further increased. Higher cutting speeds are favorable if there are no adverse effects.

In most cases, it is nearly impossible to prevent some amount of seizure between the chip and the tool face. Unless surfaces are perfectly flat, contact is made along the high spots over only a fraction of the total area. As the chip passes over the tool face, cutting forces give rise to extremely high unit pressures sufficient to form pressure welds. If these welds are stronger than the ultimate shear strength of the material, that portion of the chip welded to the tool shears off as the chip is displaced and becomes what is called a built-up edge. Continuous chips with built-up edges are illustrated in *Figure 3-4b*.

The built-up edge is common to most metal-cutting operations and particularly evident in machining aluminum and some stainless steels. The edge builds up to a point where it eventually breaks down, part of it going off with the chip, and part of it being deposited on the work surface. This characteristic occurs at rapid intervals. When steel is cut, a continuous chip usually is formed, but the pressure against the tool is high and the severe action of the chip quickly rubs the natural film from the tool face. The freshly cut chip and the newly exposed material on the face of the tool have an affinity for each other, and a layer of highly compressed material adheres to the face of the tool. Any change in cutting conditions that reduces or eliminates the built-up edge will usually improve surface quality. Built-up edge affords some protection to the cutting edge to reduce wear, and a small amount may be desirable. The problem then becomes one of size control through the effects of the various manipulating factors.

## Mechanism of Chip Formation

Observations during metal cutting reveal several important characteristics of chip formation (Shaw 1954; Ernst 1938). Those observations include:

- the cutting process generates heat;
- the thickness of the chip is greater than the thickness of the layer being cut;
- the hardness of the chip usually is much greater than the hardness of the parent material, and
- the above relative values are affected by changes in cutting conditions and in properties of the material to be machined to produce chips that range from small lumps to long continuous ribbons.

These observations indicate that the process of chip formation is one of deformation or plastic flow of the material, with the degree of deformation dictating the type of chip that will be produced.

Plastic flow takes place by means of a phenomenon called slip, along what are referred to as slip planes. The capacity for plastic flow depends on the number of slip planes available. The number of planes, in turn, depends on the crystal lattice structure of the material and prior treatment. When the resisting stresses in a material exceed its elastic limit, a permanent relative motion occurs between the adjacent slip planes most favorably oriented in the direction of the applied force. Once this motion or slip takes place, these particular planes are strengthened and resist further deformation in preference to other, now weaker, planes that are available. This strengthening is called work or *strain hardening*, and is characteristic of all steels but most dramatically exhibited in stainless steels.

As the tool advances into the workpiece, frictional resistance to flow along the tool face and resultant work hardening cause deformation to take place ahead of the tool along a shear plane. This plane extends from the vicinity of the cutting edge toward the free surface of the workpiece at some shear angle θ, as illustrated in *Figure 3-5*. If the workpiece material is brittle and has little capacity for deformation before fracture, when the fracture shear stress is reached separation will take place along the shear plane to form a segmental chip. Ductile materials, however, contain sufficient plastic flow capacity to deform along the shear plane without rupture. Strain

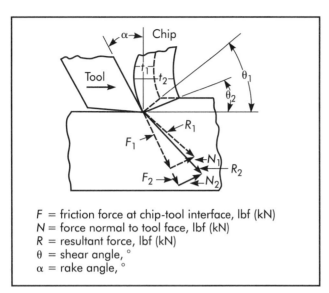

F = friction force at chip-tool interface, lbf (kN)
N = force normal to tool face, lbf (kN)
R = resultant force, lbf (kN)
θ = shear angle, °
α = rake angle, °

Figure 3-5. *Effect of friction force on shear angle and amount of chip distortion. Force polygons show effect of friction force on magnitude and direction of resultant force.*

hardening permits a transfer of slip to successive shear planes, and the chip tends to flow in a continuous ribbon along the face of the tool and away from the work surface. The chip is highly worked, and much harder than the material from which it is taken. *Figure 3-6* illustrates the relationship of all forces represented by the force $R$ acting at the edge of the tool with the components in their respective positions.

The effect of a change in shear is shown in *Figure 3-5*. For a given depth of cut, a smaller shear angle $\theta_2$ causes greater chip cross section and, therefore, greater distortion than in the case of the larger shear angle $\theta_1$. For a given material, the shear angle is a function of the tool rake angle and the coefficient of friction along the tool face. Materials that have fewer slip planes and low work-hardening capacity generally show higher shear angles than the more ductile materials, and have a lower ratio of chip thickness to undeformed chip thickness. Under ideal conditions, this ratio will approach 1.5 and can be checked to a rough extent by measuring the thickness of the chip and comparing it to the feed rate.

## Effect of Manipulating Factors

Certain manipulating factors provide some control of the metal-cutting characteristics. When studying various examples, keep in mind the relationship between tool, chip, and surface appearance. The size and the degree of the chip's brittleness may be a good indication of the severity of the cutting operation. The back of the chip, which is in contact with the tool face, gives a fairly good indication of the built-up edge condition. In the absence of built-up edge, the back of the chip should be clean, smooth, and highly burnished. The workpiece surface should be correspondingly good. As the size of the built-up edge increases, more and more markings are evident on the chip. Generally, the workpiece is affected in the same manner.

### Velocity

Velocity refers to the speed at which the cutting edge moves through workpiece material. This is slightly different than speed in terms of revolutions per minute (rpm) because the velocity is a function of the rate at which the workpiece (or cutting tool) is turning (rpm) and the radius at which the cutting edge is located. Velocity affects temperature, which in turn affects the cutting process. At low velocities, the temperature at the tool point is below the recrystallization temperature of the material. As a result, work hardening in the chip is retained and the workpiece material is not softened due to failure to reach the yield strength temperature of the material. If the velocity increases to the point where the cutting temperature is above the yield strength temperature of the material, chip material at the interface tends to soften and machine much more efficiently. Higher shear angles occur at higher velocities, and an ideal chip thickness of 1.5 times the feed can be approached. As mentioned, built-up edge formation is less at higher velocities. High-speed machining takes advantage of all of these factors. However, excessive velocity that results in excessive heat generation and tool wear will cause the tool to fail rapidly. Of all the manipulating factors, speed generally has the greatest effect on tool life.

The form or shape of chips at high velocities can be troublesome on ductile materials. The reduced resistance to chip flow and resultant increase in shear angle gives a thinner, less

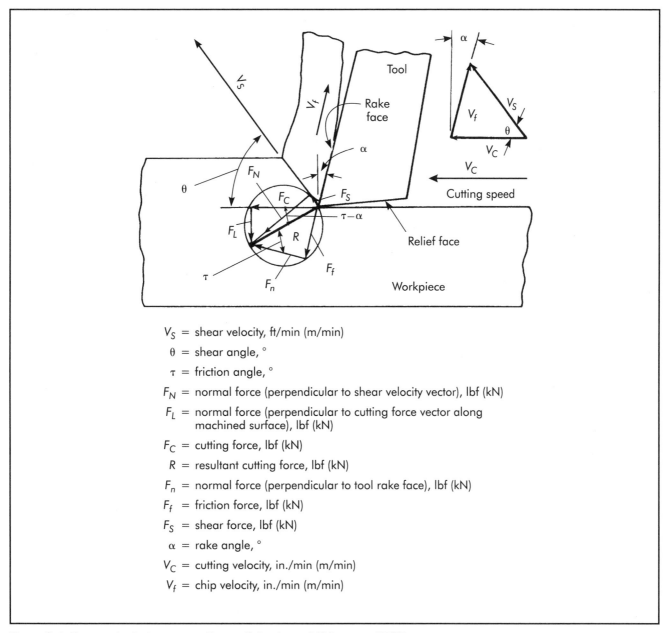

Figure 3-6. *Force and velocity system of a cut (Schrader and Elshennawy 2000).*

distorted chip, but one that becomes longer and straighter as velocity increases. Chip grooves specifically designed for thin chips can be employed to deal with this problem.

## Size of Cut

Changes in the size of cut effectively change the cross-sectional area of chip contact (*Figures 3-7a* and *b*). How this area is changed determines the effect on the cutting process. An increase in depth of cut for a constant feed merely lengthens the contact, but does not change the thickness, so the force per unit length remains the same. However, an increase in feed for a given depth widens the area of contact and changes the force per unit length. This results in greater chip distortion and reduced tool life, although increased feed reduces the machining cycle.

Several factors affect surface quality to a greater degree than may be predicted. Lack of

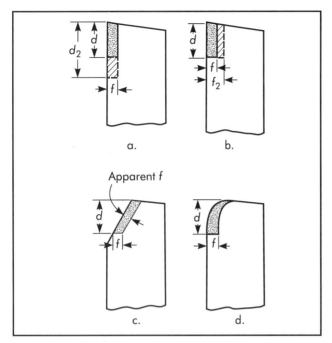

Figure 3-7. *Effect of size of cut and changes in tool geometry on chip thickness: (a) change in depth, (b) change in feed, (c) effect of side-cutting-edge angle, and (d) effect of nose radius. Crosshatched portions represent increase in contact area.*

rigidity will permit greater deflections as a result of higher forces. Increases in feed and depth of cut then may cause chatter, poor surface quality, and loss of dimensional stability. Deep turning cuts on relatively small diameters have a greater percentage of change in velocity along the length of the cutting edge. This might result in erratic tool life behavior with poor surface quality.

The effect upon the chip is also much more pronounced with increases in feed than in depth. Because of the greater distortion at high feed rates, chips tend to break up more readily. Chip control is essential for operator safety, since long, continuous chips can wrap around the rotating part and become dangerous.

## Tool Geometry

For specified cutting conditions, changes in tool geometry have two direct effects on chip formation:

- effect upon shear angle, and
- effect upon chip thickness.

The two are related in that a change in one usually affects the other. The effects of changes in

rake angles are shown in *Figure 3-8*. Lower rake angles decrease the shear angle, cause greater chip distortion, and increase the resistance to chip flow. Lower rake angles (negative) produce rougher and more work-hardened surfaces. At low or negative rake angles, the chip is so highly distorted that it facilitates chip control by breaking the chip into six short lengths.

## Tool Material

One difference between cutting tool materials is in the ability to sustain high cutting velocities. For example, high-speed-steel (HSS) is less able to sustain the forces and temperatures generated at high cutting velocities than carbide. The effect of high velocity on the cutting process and temperature generation has already been described. Another factor when selecting a cutting tool material is the coefficient of friction between chip and tool material. Usually, this is of little consequence with HSS tools because the coefficient of friction does not change appreciably among the various grades. Sintered carbide, cermet, and ceramic tools are made with different compositions and may behave quite differently for similar cutting conditions.

## Cutting Fluids

Ideally, if a cutting fluid provides lubrication between the chip and the tool, the coefficient of friction will be reduced and the shear angle increased. However, effective lubrication in this region may be difficult to achieve, except possibly at very low cutting speeds or higher application pressures. The effects of lubricant vary with cutting conditions and work materials. At high speed, fluids act principally as coolants but may effectively lubricate the tool-chip interfacial zone. This provides more efficient machining, which often results in increased tool life and improved surface finish. Constant, even flow is essential when cutting fluids are applied with carbide tools to prevent thermal shocking and resultant fracture.

## Workpiece Materials

Brittle materials form discontinuous chips, and can be machined with neutral rake (for example,

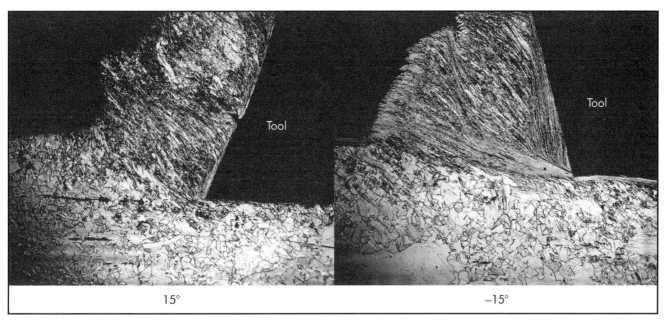

| 15° | −15° |

Figure 3-8. *Photomicrographs showing effect of rake angle on shear angle, chip distortion, built-up edge, and work hardening of machined surface. Shaper tools set for some depth of cut. Apparent difference in depth is due to higher separating force and greater tool deflection with a negative rake tool. Work material is 304 stainless steel.*

rake flat top inserts with no chip groove pressed in). Cutting forces are usually lower and tool life is longer than for a ductile material of corresponding strength because of generally large shear angles and lower resistance along the tool face.

Ductile materials produce continuous chips. Continuous chips can be dangerous to operators and machines, so the problem has to be solved. With low friction and high cutting velocities, particularly with materials of low work-hardening capacity, a thinner, less distorted chip is produced. For inserts, pressed-in chip grooves are required for low feed rates to break up the chip. High frictional resistance to flow, low shear angles, and materials of high work-hardening capacity are associated with large distortions during cutting, and breaking up chips is not such a problem. The addition of lead, sulfur, and phosphorus to low-carbon steels helps break up chips, reduces built-up edge, and improves surface quality.

## CUTTING FORCES

The forces arising from orthogonal cutting are shown in *Figure 3-9*. $F_c$ and $F_N$ are the cutting and normal force, respectively. These components

of the resultant force $R$ can be measured by means of a dynamometer,

$$\mu = \tan\beta = \frac{F}{N} = \frac{F_N - F_c \tan\alpha}{F_c - F_N \tan\alpha} \qquad (3\text{-}1)$$

Then,

$$F_T = F_c \cos\theta - F_N \sin\theta$$

where:

    $\mu$ = coefficient of friction
    $\beta$ = friction angle, °
    $F$ = force required to overcome friction between chip and face of tool, lbf (kN)
    $F_N$ = normal force, lbf (kN)
    $F_C$ = cutting force, lbf (kN)
    $\alpha$ = rake angle, °
    $F_T$ = shear force, lbf (kN)
    $\theta$ = shear angle, °

$F_T$ cannot be calculated from the static mechanical properties of the material cut, since the strain rate encountered in metal cutting is much greater than any conventional test. The shear angle $\theta$ must be determined before the shearing force and shear flow stress can be determined

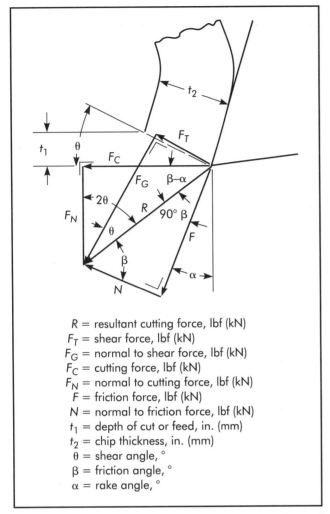

$R$ = resultant cutting force, lbf (kN)
$F_T$ = shear force, lbf (kN)
$F_G$ = normal to shear force, lbf (kN)
$F_C$ = cutting force, lbf (kN)
$F_N$ = normal to cutting force, lbf (kN)
$F$ = friction force, lbf (kN)
$N$ = normal to friction force, lbf (kN)
$t_1$ = depth of cut or feed, in. (mm)
$t_2$ = chip thickness, in. (mm)
$\theta$ = shear angle, °
$\beta$ = friction angle, °
$\alpha$ = rake angle, °

Figure 3-9. *Forces acting on a continuous chip in orthogonal cutting.*

from $F_c$ and $F_N$. One method of approximating the shear angle is by measuring the thickness of the chip. Optimum shear angles occur when $\mu$ is in the range of 1.5.

$$r = \frac{t_1}{t_2} \qquad (3\text{-}2)$$

where:

$r$ = cutting ratio
$t_1$ = depth of cut or feed, in. (mm)
$t_2$ = thickness of chip, in. (mm)

$$\tan\theta = \frac{r\cos\alpha}{1 - r\sin\alpha} \qquad (3\text{-}3)$$

The shearing force can be determined from $F_c$ and $F_N$ by the geometry of the force system in *Figure 3-9*.

*Figure 3-10* shows the resultant force acting on the cutting tool in a three-dimensional cut, and the components of this force that can be measured with a three-component dynamometer. A typical relationship of the magnitude of these forces is shown in *Figure 3-11*.

Although some variation of tangential cutting force with respect to changes in speed may occur at low cutting speeds, the cutting force can be considered to be independent of cutting speed within the practical ranges of cutting speeds normally used. The effect of feed and the depth of cut on cutting force is illustrated in *Figure 3-12*. The following equation is obtained from *Figure 3-12*:

$$F_T = cf^{n_3} d^{n_4} \qquad (3\text{-}4)$$

where:

$F_T$ = tangential cutting force, lbf (kN)
$c$ = constant of proportionality
$f$ = feed, in./rev (mm/rev)
$d$ = depth of cut, in. (mm)
$n_3$ = slope of $F_T$ versus $f$ graph (typical values = 0.05 to 0.98)
$n_4$ = slope of $F_T$ versus $d$ graph (typical values = 0.90 to 1.4)

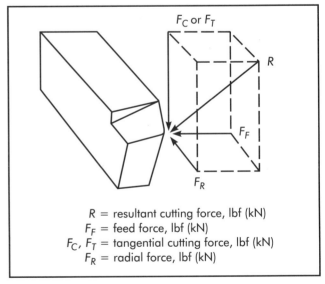

$R$ = resultant cutting force, lbf (kN)
$F_F$ = feed force, lbf (kN)
$F_C$, $F_T$ = tangential cutting force, lbf (kN)
$F_R$ = radial force, lbf (kN)

Figure 3-10. *Three components of cutting force.*

Figure 3-11. *Three components of cutting force through tool life (Boston 1951).*

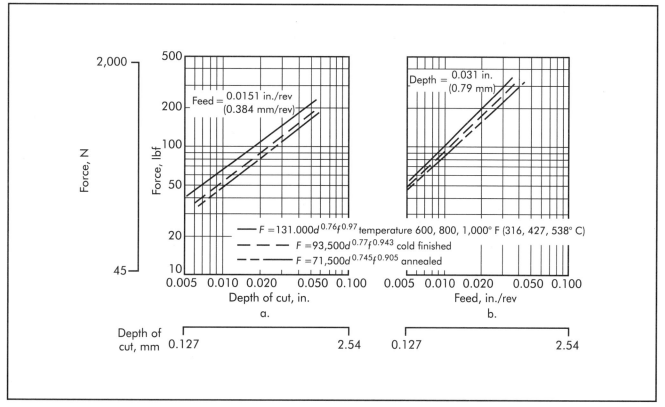

Figure 3-12. (a) *Cutting force vs. depth of cut. Tool material—high-speed steel, geometry—6, 11, 6, 6, 6, 15, 0.01, work material—SAE 1045 steel.* (b) *Cutting force vs. feed, depth—0.031 in.*

The cutting-tool geometry has a considerable effect upon the cutting forces. The most pronounced is due to variations in the rake angle, shown in *Figure 3-13*, for orthogonal cutting conditions. In three-dimensional cutting, the true rake angle is the particular rake angle on the face of the tool along which the chip is sliding.

The true rake angle can be closely approximated by the following (*Figure 3-14*):

$$\tan\phi = \frac{d}{N_R + (d - N_R)\tan\varepsilon} \qquad (3-5)$$

where:

$\phi$ = chip flow angle, °
$N_R$ = nose radius, in. (mm)
$\varepsilon$ = side-cutting-edge angle, °

This equation is correct for tools having a 0° rake angle. For most normal rake angles, it will be correct to within a few degrees of error. For very large rake angles, significant error can be introduced. *Equation 3-5* can be used for most practical applications.

*Equation 3-6* expresses the true rake angle corresponding to the chip flow angle.

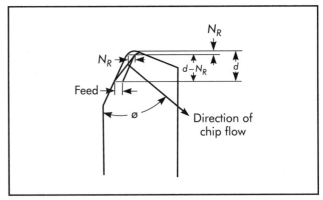

Figure 3-14. *Approximate chip flow direction.*

$$\tan\rho = \tan\gamma \sin\theta + \tan\phi \cos\theta \qquad (3-6)$$

where:

$\rho$ = true rake angle, °
$\gamma$ = side rake angle, °
$\theta$ = shear angle, °
$\phi$ = chip flow angle, °

The location of the maximum rake angle is found by:

$$\tan\phi_{max} = \frac{\tan\gamma}{\tan\phi} \qquad (3-7)$$

Figure 3-13. *Effect of feed on tangential and normal cutting forces for a range of rake angles when turning nickel steel.*

*Figure 3-15* is a graphical representation of the effect of feed on several forces for various depths of cut. The effects of cutting speed on cutting force (*Figure 3-16*) and tool forces (*Figure 3-17*) also are illustrated.

## POWER REQUIREMENTS

The power required at the cutting tool is defined by:

$$P_c = \frac{F_T V}{33,000} \qquad (3-8)$$

where:

$P_c$ = power at the cutting tool, hp (kW)
$F_T$ = tangential cutting force, lbf (kN)
$V$ = cutting speed, ft/min (m/min)

The radial cutting force does not contribute to power requirements. Although the feed force can be considerable, the feeding velocity generally is so low that the horsepower required to feed the tool can be neglected.

Substituting *Equation 3-4* in *Equation 3-8*,

$$P_c = \frac{c f^{n_3} d^{n_4} V}{33,000} \qquad (3-9)$$

Figure 3-16. *Effect of cutting speed on cutting force for various feeds when cutting bronze with a tungsten carbide tool. Depth of cut—0.040 in. (1.02 mm).*

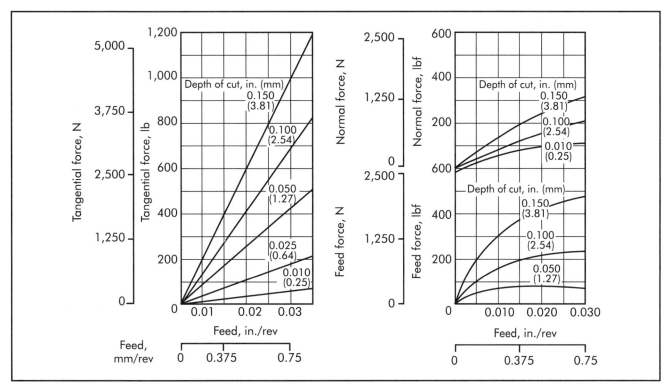

Figure 3-15. *Effect of feed on tangential feed and normal forces for various depths of cut. Tool material—high-speed steel, tool shape—8, 14, 6, 6, 6, 15, 3/64, work material—SAE 3135 steel, annealed, cutting speed—50 ft/min (15.2 m/min) (dry).*

Figure 3-17. Effect of cutting speed on tool forces. Tool material—tungsten carbide, tool shape—10, 10, 8, 8, 7, 0, 0.015, work material—AISI C 1118 steel, depth of cut—0.150 in. (0.38 mm).

There are losses in the machine that must be considered when estimating the size of the electric motor required:

$$P_G = \frac{P_c}{M_e} + P_t \qquad (3\text{-}10)$$

where:

$P_G$ = gross or motor power required, hp (kW)
$M_e$ = mechanical efficiency of the machine
$P_t$ = power required to run the machine at no-load conditions, hp (kW)

## Specific Power Consumption or Unit Horsepower

Specific power consumption, $W_p$ or unit horsepower, is defined as the horsepower required to cut a material at a rate of 1 in.$^3$/min:

$$W_p = \frac{P_c}{C} \qquad (3\text{-}11)$$

where:

$W_p$ = specific power consumption, hp (kW)
$P_c$ = power at the cutting tool, hp (kW)
$C$ = material removal by cut, in.$^3$/min (m$^3$/min)

for turning,

$$W_p = \frac{P_c}{12Vfd} \qquad (3\text{-}12)$$

$$= \frac{F_T}{396,000fd} \qquad (3\text{-}13)$$

where:

$V$ = cutting speed, ft/min (m/min)
$f$ = feed, in./rev (m/rev)
$d$ = depth of cut, in. (mm)
$F_T$ = tangential cutting force, lbf (kN)

Note that $W_p$ is independent of the cutting speed.

$$W_p = \frac{cf^{n_3}d^{n_4}}{396,000fd} = \frac{c}{396,000f^{1-n_3}d^{1-n_4}} \qquad (3\text{-}14)$$

from which,

$$\frac{c}{396,000} = c^1 = W_{p_1}(f_1)^{1-n_3}(d_1)^{1-n_4} = W_{p_2}(f_2)^{1-n_3}(d_2)^{1-n_4}$$

or,

$$W_{p_2} = W_{p_1}\left(\frac{f_1}{f_2}\right)^{1-n_3}\left(\frac{d_1}{d_2}\right)^{1-n_4}$$

This defines the effect of feed and depth of cut on the specific power consumption. The equation may be reduced to the following form, since the value of $n_4$ is often nearly equal to one:

$$W_{p_2} = W_{p_1}\left(\frac{f_1}{f_2}\right)^{1-n_3}$$

Typical values of specific power consumption are given in Table 3-1.

## TOOL WEAR

For the sake of recognition and understanding the fundamentals of metal cutting, the effects of changes in manipulating factors have been described without regard to their influence on such criteria as tool wear and tool life. Yet, there is no known tool material that can completely resist contact and rubbing at high tem-

Table 3-1. Power consumption values for various materials

| Dry Cutting; Depth, 1/8 in. (3.2 mm); Feed, 1/64 in./rev (0.4 mm/rev) | | | |
|---|---|---|---|
| Material Cut | Tool Shape | Brinell Hardness Number | $P_c$, in.$^3$/min (cm$^3$/sec) |
| Plain carbon steel | | 126 | 0.59–0.66 (1.6–1.8) |
| | | 179 | 0.70–0.79 (1.9–2.2) |
| | | 262 | 0.85–0.95 (2.3–2.6) |
| Free-cutting steel | | 118 | 0.36–0.39 (1.0–1.1) |
| | | 179 | 0.44–0.48 (1.2–1.3) |
| | 8, 14, 6, 6, 6, 0, 1/16 | 229 | 0.50–0.54 (1.4–1.5) |
| | | 131 | 0.46–0.57 (1.3–1.6) |
| Alloy steel | | 179 | 0.55–0.68 (1.5–1.9) |
| | | 269 | 0.67–0.83 (1.8–2.3) |
| | | 429 | 1.10–1.90 (3.0–5.2) |
| | | 140 | 0.22–0.32 (0.6–0.9) |
| Cast iron | | 179 | 0.45–0.68 (1.2–1.9) |
| | | 256 | 0.85–1.30 (2.3–3.6) |
| Leaded brass | | 33 | 0.18–0.27 (0.5–0.7) |
| | | 76 | 0.22–0.31 (0.6–0.8) |
| | | 131 | 0.25–0.35 (0.7–1.0) |
| Unleaded brass | 8, 14, 6, 6, 6, 15, 0 | 50.9 | 0.54 (1.5) |
| Pure copper | | 40.4 | 0.88 (2.4) |
| Magnesium alloys | | 32 | 0.084–0.10 (0.2–0.3) |
| | | 49 | 0.094–0.11 (0.26–0.3) |
| | | 68 | 0.10–0.12 (0.3–0.33) |
| | 8, 14, 6, 6, 6, 15, 0 | 55 | 0.28 (0.8) |
| | | 159 | 0.26 (0.7) |
| | | 32 | 0.12 (0.33) |
| Aluminum alloys | 20, 40, 10, 10, 10, 15, 0 | 94 | 0.15 (0.4) |
| | | 115 | 0.17–0.21 (0.5–0.6) |
| | | 153 | 0.200 (0.55) |
| Monel® metal | 8, 14, 6, 6, 6, 15, 3/64 | 147 | 0.58–0.75 (1.6–2.0) |
| | | 160 | 1.35 (3.7) |

peratures and high pressures without some changes from its original contours over time. It becomes necessary, therefore, to think of the effect of the manipulating factors not only on the cutting process itself, but also on the performance of the cutting tool, which may, in turn, itself affect the cutting process.

## Tool Failure

Failure of the cutting tool occurs when it is no longer capable of producing parts within required specifications. The point of failure, together with the amount of wear that determines it, is a function of the machining objective. Surface quality, dimensional stability, cutting forces, cutting horsepower (kW), and production rates may alone, or in combination, be used as criteria for tool failure. It may, for example, take very little wear to affect surface quality, although the tool itself could continue to remove metal with little if any loss of efficiency. In contrast, only a few thousandths of an inch of wear on a wide-form tool might cause such a large increase in thrust or feeding forces that it would result in a loss of dimensional stability, or require excessive power in addition to a loss of surface quality.

## Types of Tool Wear

Tool failure is associated with some form of breakdown of the cutting edge. Under proper operating conditions, this breakdown takes place gradually over a period of time. In the absence of rigidity, or because of improper tool geometry that gives inadequate support to the cutting edge, the tool may fail by mechanical fracture or chipping under the load of the cutting forces. This is not truly a wear phenomenon—it can be eliminated or at least minimized by proper design and application.

As a result of direct contact with the work material, there are three major regions on the tool where wear can take place: face, flank, and nose (*Figure 3-18*).

Figure 3-18. *Representative wear patterns on face, flank, and nose of cutting tool, typical of chip-removal process on ductile materials. Crater on face of tool in (a) started well back of cutting edge. In (b) crater wear had progressed to point where the weak cutting edge broke down under cutting forces.*

## Face Wear

The *face* of the tool is the surface over which the chip passes during its formation. Wear takes the form of a cavity or crater, originating not along the cutting edge but at some distance away and within the chip contact area. As wear progresses with time, the crater gets wider, longer, and deeper, and approaches the edges of the tool.

Crater wear usually is associated with ductile materials that give rise to continuous chips. If crater wear is allowed to proceed too far, the cutting edge becomes weak as it thins out and suddenly breaks down. Usually, there is some preliminary breakthrough of the crater at the nose and periphery prior to total failure of the cutting edge. These preliminary breaks serve as focal points for the development of notches along the flank. In general, crater wear develops faster than flank wear on ductile materials, and is the limiting factor in determination of tool failure.

## Flank Wear

Although crater wear is most prominent in the machining of ductile materials, flank wear always is present, regardless of work and tool material, or even cutting conditions. The *flank* is the clearance face of the cutting tool, along which the major cutting edge is located. It is the portion of the tool in contact with the work at the chip separation point, and it resists the feeding forces. Because of the clearance, initial contact is made along the cutting edge. Flank wear begins at the cutting edge and develops into a wider and wider flat of increasing contact area called a *wear land*.

Materials that do not form continuous chips promote little if any crater wear, and flank wear becomes the dominant factor in tool failure. In the case of most form tools and certain milling cutters, the wear land is in direct contact with the finished surface, and usually becomes the basis for failure even on ductile materials, particularly if surface-finish specifications are the controlling factors in the process. Quite often, flank wear is accompanied by a rounding of the cutting edge, particularly when machining abrasive materials. This results in large increases of cutting and feeding forces that, if carried too far, could lead to tool fracture.

## Nose Wear

Nose wear is similar to and often considered a part of flank wear, but can be considered separately. Nose wear sometimes proceeds at a faster rate than flank wear, particularly when working on rather abrasive materials and using small nose radii. In finish turning operations, for example, excessive wear will affect finished part dimensions as well as surface roughness. Where sharp corners are specified on the part drawing, the rounding or flattening of the nose can cause out-of-tolerance conditions long before flank wear itself becomes a factor.

## Mechanism of Tool Wear

Evidence indicates that wear is a complex phenomenon influenced by many factors. The causes of wear do not always behave in the same manner, nor do they always affect wear to the same degree under similar cutting conditions. The causes of wear are not fully understood, although various researchers have made great strides. Even though there is some disagreement regarding the true mechanisms by which wear actually takes place, most investigators feel that there are at least five basic causes of wear:

1. abrasive action of hard particles contained in the work material;
2. plastic deformation of the cutting edge;
3. chemical decomposition of the cutting-tool contact surfaces;
4. diffusion between work and tool material, and
5. welding of asperities between work and tool.

The relative effects of these causes are a function of cutting velocity or cutting temperatures and are shown in *Figure 3-19*. Investigations have been made on other possible causes, such as oxidation and electrochemical reactions in the tool-work contact zone.

The most important factor influencing tool wear is cutting temperature. Of the five basic causes of wear, temperature has a considerable effect in all but one. Cutting temperatures are important for two basic reasons: (1) most tool materials show rapid loss of strength, hardness,

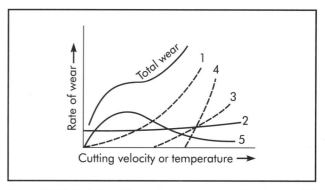

Figure 3-19. *Relative effects of various causes of tool wear: (1) abrasive wear, (2) plastic deformation of cutting edge, (3) chemical decomposition, (4) diffusion, (5) welding of asperities.*

and resistance to abrasion above some critical temperature, and (2) the rate of diffusion between work and tool materials rises very rapidly as temperature increases past the critical mark.

Analytical and experimental methods have been used to show that the average peak temperatures at the tool-chip interface occur near the point where the chip leaves the tool surface (*Figure 3-20*) (Chao and Trigger 1955, 1958). Crater wear appears greatest at this point. The relationship between crater-wear rate and average tool-chip interface temperature is shown in *Figure 3-21*. The rate of wear increases very rapidly beyond a critical temperature. Flank temperatures were found to be maximum near the tool point as shown in *Figure 3-22*.

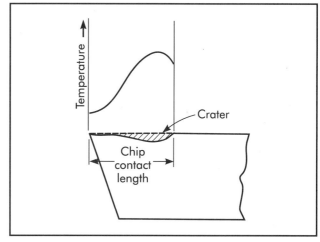

Figure 3-20. *Temperature distribution along tool-chip contact length (Chao and Trigger 1955).*

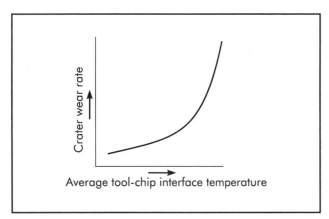

Figure 3-21. *Relationship between rate of crater wear and average tool-chip interface temperature (Chao and Trigger 1955).*

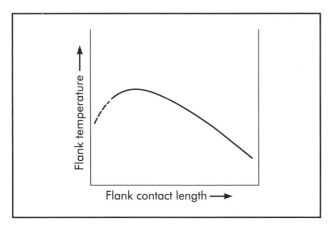

Figure 3-22. *Distribution of flank temperature along flank-work contact length (Chao and Trigger 1955).*

The significance of temperature on wear is associated with tool-material properties. High-speed steel (HSS) tools begin to lose their properties very rapidly at approximately 1,100° F (593° C). Carbides show less drastic sensitivity to temperature up to about 1,600° F (871° C). Chemical decomposition and diffusion will not occur at any appreciable extent until critical temperatures are reached.

## Abrasive Action

Wear by abrasive action may be partly explained by the fact that hard particles (sand inclusions, carbides, etc.) in the workpiece material literally gouge or dislodge particles from the tool, causing continuous wear under any cutting condition. The rate of wear is dependent on the num-

ber, size, distribution, and hardness of the particles in the work material, as well as the hardness of the cutting tool and the workpiece. At higher cutting speeds, even some of the softer constituents may contribute to the gouging action as a result of higher impact values and reduced tool resistance to abrasion.

## Plastic Deformation of the Cutting Edge

Plastic deformation of the cutting edge is believed to take place at all ranges of cutting temperatures; it arises from high unit pressures imposed on the tool. This results in a slight depression and bulging of the edge, similar to that shown in *Figure 3-23*. The net effect is greater tool pressure and increased cutting temperature, resulting in further deformation and concluding in edge wipe out. This mode of failure is common when machining hardened materials at high speeds.

## Chemical Distortion of the Cutting Tool

Chemical distortion of the cutting tool occurs through localized chemical reactions at the tool-workpiece interface. These reactions are temperature-dependent and result in weakening the bond between minute tool segments and the segments surrounding them. This may occur either through formation of weaker compounds or, in the case of carbide tools, by a dissolving action of

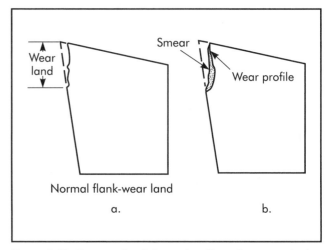

Figure 3-23. *Section of tool shows the effect of smear and other causes on flank wear.*

the bond between the binder and individual carbide particles. As a result of this weakening effect, particles are pulled out from the main body of the tool by the chip or work as it moves past the contact surfaces. Once the critical temperature for this chemical action is reached, the rate of wear is relatively rapid.

### Diffusion

Diffusion, a complex wear phenomenon between the work and tool, results in a rapid breakdown of tool material once the critical temperature is reached. Much has yet to be learned about this cause of wear, but basically it involves a change of composition at the tool-chip interface. There is an alloying effect that weakens the bond of tool particles and permits them to be pulled out by the chip as it sloughs off. Carbon transfer from the tool material to the workpiece is enhanced at higher temperatures and this greatly contributes to premature tool failure. Selection of a cutting tool material that is less susceptible to this transfer would allow machining at higher temperatures and speeds. For example, cubic boron nitride (CBN) is used for machining hard ferrous materials for this very reason.

### Welding of Asperities

Wear through the welding of asperities parallels that of the built-up edge. As shown in *Figure 3-19*, the greatest rate of wear by built-up edge occurs at lower cutting velocities or temperatures. A built-up edge forms because of a high resistance to chip flow along the tool face, which causes a portion of the chip to shear off as it moves past the tool. This action is most prominent when cutting temperatures are below the recrystallization temperature for the material. Work hardening is retained, and the built-up edge is harder and stronger than the rest of the chip. This same situation exists for wear through the welding of asperities.

The asperities on a tool are brittle and relatively weak in bending or tension. If welding takes place between the chip and the asperities because of the extremely high unit pressures, the work-hardened chip material is strong enough to pull these asperities off. However, if the temperature is near or beyond the recrystallization temperature, then the bond between the chip or built-up edge and tool is no weaker than the material adjacent to it, because work hardening has not been retained. Therefore, the rate at which these asperities are pulled out diminishes.

If cutting conditions are such that the resulting temperatures approach or surpass the critical temperature for a given tool material, the reduced resistance of the tool, and the increasing tendency for alloying between work and tool material, cause a high rate of wear and rapid failures. When cutting temperatures are low, the processes of wear by abrasion and welding at the asperities become most prominent.

### Effects of Manipulating Factors on Tool Wear

The effects of manipulating factors on tool wear are concerned either with modifications that influence the cutting process directly for given tool and workpiece materials, or inherent material properties that resist or promote wear. For a given tool and workpiece material combination, cutting temperatures are influenced mostly by cutting speed and, to a lesser extent, feed and depth of cut (*Figure 3-24*). Adjustments in speed

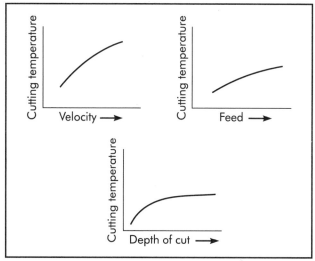

Figure 3-24. *Relative effects of velocity, feed, and depth of cut on cutting temperatures for a given work material and tool geometry.*

or feed, or both, will affect tool wear. It may be possible to use another material that inherently has better temperature-resistant properties to maintain original or even higher production rates with less sensitivity to temperature failure. The cost of the second material may be higher than that of the first, but it may be more than justified by higher production rates at increased operating temperatures.

Changes in tool geometry that result in higher shear angles, less chip distortion, lower frictional resistance, and thinner chips will lower cutting forces and decrease cutting temperatures. They contribute to a reduction in the rate of tool wear for given cutting conditions. Within practical and design limitations, rake, relief, and other angles should be matched to the application providing the most free-cutting strong geometry that directs cutting forces in the workpiece's most rigid section. Heat transfer characteristics also may be adversely affected if the point of the tool is too thin as a result of high relief and rake angles. The heat at the point does not dissipate as rapidly, and higher temperatures prevail.

Workpiece materials with relatively high hardness, shear strength, coefficient of friction, and work-hardening capacities, and containing hard constituents, promote more rapid wear for certain cutting conditions. Materials such as titanium or stainless steels, which have poor thermal conductivity, do not dissipate heat from the cutting zone as rapidly as others, and temperature failures are more common.

## Effect of Wear on Machinability Criteria

The study of machinability involves certain criteria that play a prominent part in the evaluation of the cutting process. Machinability ratings for a given material are entirely relative in that one material is used as a base. The ratings can vary, not only among the machining processes, but with the criteria used in the evaluation for a given process. Though much data is available about numerous materials, erroneous conclusions may result if it is not interpreted or applied properly.

Much of the available data, particularly with respect to cutting forces, specific power require-

ments, and surface quality, is based on the results of sharp-tool investigations. These investigations serve a valuable purpose when analyzing the cutting process and determining initial levels of performance. However, they give no indication as to how long the initial level of performance will be maintained when tools begin to wear. Some materials that are given very high machinability ratings with sharp tools are extremely sensitive to tool wear. Performance may drop off rapidly with time. On the other hand, similar materials may have lower initial levels of performance but are less sensitive to tool wear, maintain the original levels for longer periods of time, and actually receive higher production performance ratings.

Another sort of data that warrants some caution in direct application is the tool-life curve. These curves (see *Figure 3-25*) usually are based on accelerated-wear tests. They are plots of cutting velocity versus cutting time, or cubic inches (millimeters) of metal removal before failure for a given size of cut and otherwise constant cutting conditions. Failure can be identified in several ways. A preselected amount of flank wear (for example, 0.03 in. [0.8 mm]) may indicate failure in tests with carbide and ceramic tools. Also, the quality of a machined surface may be used to denote failure of any tool material. Whenever wear is used in some form as the criterion of fail-

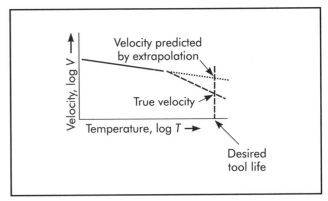

Figure 3-25. *Tool-life curve showing error that can be introduced in predicting velocity for longer tool lives by extrapolation of typical accelerated-wear results (solid line). Lower velocities affect the rate and type of tool wear to change the slope of the curve. Slopes approaching 0 indicate greater sensitivity to temperature while slopes approaching 1 indicate effect of abrasive wear.*

ure, the results can be plotted as a generally acceptable straight line on log-log coordinates.

For various reasons (mostly economic), points used to establish the tool-life curve come from comparatively short tool lives, usually less than one hour of cutting time. If the velocity for longer tool life is desired for a practical application, these curves are usually extended and the velocity for the desired tool life is extrapolated as indicated by the dotted line in *Figure 3-25*. This practice can sometimes lead to very unsatisfactory results.

High cutting velocities result in high cutting temperatures. When short tool lives are encountered, the cutting temperature is in or near critical temperature ranges, which promote rapid tool wear by diffusion and chemical decomposition. In this region, small changes in velocity or temperature may have relatively large effects on wear rate and, therefore, on tool life. The solid line in *Figure 3-25* represents a typical plot.

The further the cutting temperature is removed from the critical temperature, the less effective wear by diffusion and chemical decomposition becomes, and wear by abrasion and welding at the asperities becomes more prominent. Since the total wear rate decreases at this point, there should be a lesser effect upon tool life for the same increment change in cutting velocity. The absolute slope of the curve should increase, as represented by the dashed line in *Figure 3-25*. The actual velocity for a specified tool life may be considerably lower than the predicted velocity by extrapolation from a curve based on short time tests. Whether all workpiece and tool materials exhibit this kind of behavior to any predictable degree is not fully known. There is considerable evidence that accelerated-wear tests can give misleading information and care should be given to make proper use of this information. Some materials are much more sensitive to tool wear than others. The general wear trends are similar, although not to the same degree.

Numerous changes may take place during the life of a cutting tool. Whether these changes are important depends on the machining objective. When cutting ductile materials, increases in crater and flank wear literally cause a change in the true tool geometry, which usually affects chip formation. Changes in chip form can be very sudden, particularly when the crater starts to break through the cutting edge.

One of the most notable effects of tool wear is the change in surface quality. During a test on AISI 1045 steel at 190 ft/min (57.9 m/min), there were five distinct changes in surface appearance before total tool failure occurred at 57 minutes. *Table 3-2* lists these changes. The changes in surface quality reflect the unstable character of the cutting edge as wear progresses. With materials that are notorious for formation of built-up edge, surface roughness usually reaches unsatisfactory proportions long before actual cutting tool failure. In the previous example, if surface finish was the criterion of failure, tool life would have been about 11 minutes rather than the 57 minutes of actual cutting time.

Tool wear has a definite effect on cutting forces; the ratio of increase can be beyond expectation. In production operations, feeding or thrust forces can increase to as low as two and as high as 40 times the sharp-tool values, depending on the type of flank wear encountered. In one test, the feeding force rose from a sharp-tool value of 11–505 lbf (49–2,246 N) for a flank wear land of 0.0084 in. (0.213 mm) (University of Michigan). In another test on a material of the same commercial grade but from another source, the feeding force rose from an initial value of 16 lbf (71 N) to only 180 lbf (801 N) for the same width of wear land. The cutting conditions were exactly

Table 3-2. Effects of tool wear on surface quality

| Cutting Time, min | Surface Appearance |
|---|---|
| 0–2 | Clean, only minor traces of built-up edge |
| 2–11 | Dull and streaky |
| 11–25 | Numerous large deposits of built-up edge; unsatisfactory |
| 24–45 | Numerous small deposits of built-up edge; very streaky |
| 45–57 (failure) | Partially burnished at 45 minutes to very highly burnished at failure |

the same in each case, but there were sufficient differences in material behavior (even though both materials were within commercial specifications for the grade) to cause flank wear patterns as illustrated in *Figure 3-23a*.

On the basis of the condition of the flanks, one might suspect that tangential cutting forces or cutting power would, or should, show the same general characteristics as the feeding forces. Actually, in the tests cited, there was not only comparatively little difference in power requirements for the operation in spite of the large difference in feeding force, but power requirements increased by a factor of only one-half in each case.

Although a large increase in feeding force may not appreciably affect power requirements, the effect on tool or workpiece deflections can be quite pronounced. In the example cited above, the formed diameters increased by 0.050 in. (1.27 mm) and 0.028 in. (0.71 mm) respectively, with the higher and lower forces. Because of clearances in machine-tool assemblies, the typical force-deflection characteristics of a tool mounted on a slide controlled through a series of links can be represented by the curve shown in *Figure 3-26*. Initially, a small change in force can result in a rather large deflection. As the play between parts is taken up and elastic resistance increases, a comparable change in force results in a smaller deflection of the tool. Thus, it is apparent that a range of feeding force from a very low to a high value would absorb the greatest deflection range.

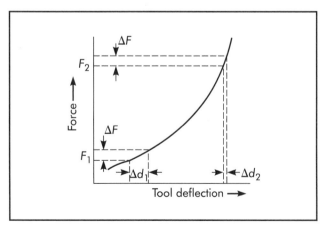

Figure 3-26. *Effect of forces on machine-tool component deflections.*

If the initial feeding force is high, most of the tool deflection occurs prior to any effects created by tool wear, and the change in dimensions is lower in magnitude.

The previous examples illustrate certain effects of tool wear, but more than that, they illustrate the ever-present variations and difficult-to-explain results that complicate metal-cutting practice.

## TOOL LIFE

The types and mechanisms of tool failure have been previously described. It was shown that excessive cutting speeds cause a rapid failure of the cutting edge; thus, the tool can be declared to have had a short life. Other criteria are sometimes used to evaluate tool life, including:

- change in quality of the machined surface;
- change in the magnitude of the cutting force, resulting in changes in machine and workpiece deflections, which lead workpiece dimensions to change;
- change in the cutting temperature, and/or
- costs, including labor, tool, and for tool changing time, etc.

The selection of the correct cutting speed has an important bearing on the economics of all metal-cutting operations. Fortunately, the correct cutting speed can be estimated with reasonable accuracy from tool-life graphs, provided that necessary data are obtainable.

The tool-life graph is shown in *Figure 3-27*. The logarithm of tool life in minutes is plotted against the logarithm of cutting speed. The resulting curve is very nearly a straight line in most instances. For practical purposes, it can be considered a straight line. This curve is expressed by the following equation (unless otherwise mentioned, equations and figures are for U.S. customary units only):

$$VT^n = C \qquad (3\text{-}15)$$

where:

$V$ = cutting speed, ft/min (m/min)
$T$ = tool life, min

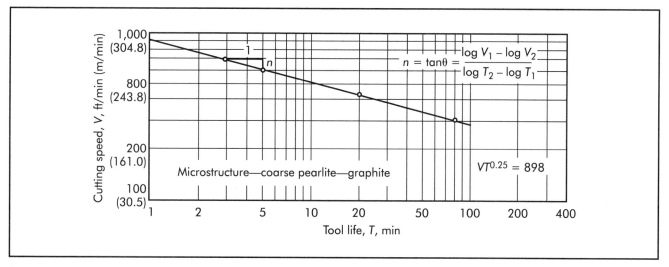

Figure 3-27. *Tool life versus cutting speed. Tool material—Kennametal carbide, tool geometry—0, 6, 6, 6, 6, 0, 0.050, work material—gray cast iron, 195 Bhn.*

$n$ = slope of the curve

$$\left( n = \tan\theta = \frac{\log V_1 - \log V_2}{\log T_2 - \log T_1} \right)$$

$C$ = constant equal to the intercept of the curve and the ordinate or cutting speed—actually it is the cutting speed for a one-minute tool life

## Example 1

A 2-in. (50.8-mm) diameter bar of steel was turned at 284 rpm and tool failure occurred in 10 minutes. The speed was changed to 232 rpm and the tool failed in 60 minutes of cutting time.

Assuming a straight-line relationship exists, what cutting speed should be used to obtain a 30-minute tool life ($V_{30}$)?

## Solution

Calculating the cutting speed

$$V_1 = \frac{\pi DN}{12} = \frac{\pi (2)(284)}{12} = 149 \text{ ft/min}$$

$$V_2 = \frac{\pi (2)(232)}{12} = 122 \text{ ft/min}$$

The slope $n$ of the tool-life curve can be determined from the previous equation. If a log-log

calculator is available, a faster method of solution is:

$$V_1 T_1^n = V_2 T_2^n = C$$

$$\frac{V_1}{V_1} = \left( \frac{T_2}{T_1} \right)^n$$

$$\frac{149}{122} = \left( \frac{60}{10} \right)^n$$

$$1.22 = 6^n$$

$$n = 0.11$$

From *Equation 3-15*,

$$C = 149(10)^{0.11} = 149(1.2888) = 192$$

thus,

$$VT^{0.11} = 192$$

$$V_{30} = \frac{192}{(30)^{0.11}} = \frac{192}{1.455} = 132 \text{ ft/min}$$

These values are for the particular feed, depth of cut, and tool geometry shown. Significant changes in the tool geometry, depth of cut, and feed will change the value of the constant $C$ and may cause a slight change in the exponent $n$. In general, $n$ is more a function of the cutting tool material. The value of $n$ for common cutting tool materials is:

$$\text{HSS: } n \cong 0.1 \text{ to } 0.15$$

$$\text{Carbides: } n \cong 0.2 \text{ to } 0.25$$

$$\text{Ceramics: } n \cong 0.6 \text{ to } 1.0$$

*Equation 3-16* incorporates the effect of the size of cut:

$$K = VT^n f^{n_1} d^{n_2} \qquad (3\text{-}16)$$

where:

$K =$ constant of proportionality
$V =$ cutting speed, ft/min (m/min)
$T =$ tool life, min
$f =$ feed, in./rev (mm/rev)
$n =$ slope of the curve

$$\left( n = \tan \theta = \frac{\log V_1 - \log V_2}{\log T_2 - \log T_1} \right)$$

$d =$ depth of cut, in. (mm)
$n_1 =$ exponent of feed (average value = 0.5 to 0.8)
$n_2 =$ exponent of depth of cut (average value = 0.2 to 0.4)

The optimum cutting speed for constant tool life is more sensitive to changes in feed than to depth of cut. Tool life is most sensitive to changes in the cutting speed, less sensitive to changes in the feed, and least sensitive to changes in the depth of cut. This relationship is shown by *Figures 3-28* and *3-29*.

## Example 2

The following equation was derived by machining AISI 2340 steel with HSS cutting tools having a 8, 22, 6, 6, 6, 15, 3/64 tool signature (see definition in section on single-point tools):

$$2.035 = VT^{0.13} f^{0.77} d^{0.37}$$

A 100-minute tool life was obtained using the following cutting condition:

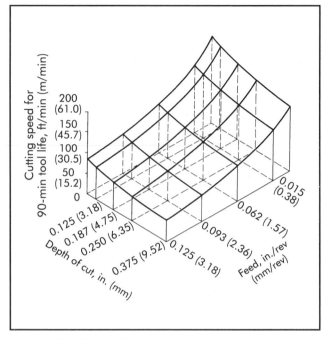

Figure 3-28. *Effect of feed and depth of cut on cutting speed for 90-min tool life. Workpiece material—gray cast iron. Tool material—HSS (Boston 1951).*

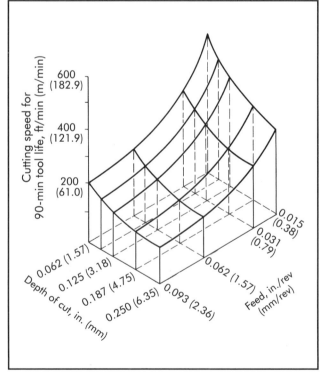

Figure 3-29. *Effect of feed and depth of cut with increased cutting speed for 90-min tool life. Workpiece material—gray cast iron. Tool material—HSS (Boston 1951).*

$V$ = 75 ft/min, $f$ = 0.0125 in./rev, $d$ = 0.100 in.

Calculate the effect on tool life for a 20% increase in the cutting speed, feed, and depth of cut, taking each separately. Calculate the effect of a 20% increase in each of the previous parameters taken together:

a. When:

$V = 1.2 \times 75 = 90$ ft/min

$$T^{0.13} = \frac{K}{Vf^n d^{n_2}} = \frac{2.035}{90(0.0125)^{0.77}(0.100)^{0.37}}$$

$$= \frac{2.035}{90(0.034)(0.426)} = 1.56$$

$T = 1.56^{1/0.13} = 1.56^{7.7} = 31$ min

b. When:

$f = 0.0125 \times 1.2 = 0.015$ in./rev

$$T^{0.13} = \frac{2.035}{75(0.015)^{0.77}(0.100)^{0.37}} = \frac{2.035}{75(0.039)(0.426)}$$

$T = (1.63)^{7.7} = 43$ min

c. When:

$d = 1.2 \times 0.100 = 0.120$ in.

$$T^{0.13} = \frac{2.035}{75(0.0125)^{0.77}(0.120)^{0.37}} = \frac{2.035}{75(0.034)(0.456)}$$

$T = (1.745)^{7.7} = 74$ min

d. When:

$V = 90$ ft/min

$f = 0.015$ in./rev

$d = 0.120$ in.

$$T^{0.13} = \frac{2.035}{90(0.039)(0.456)}$$

$T = (1.3)^{7.7} = 7.5$ min

Tool life is very sensitive to changes in cutting tool geometry. However, tests varying the geometry generally do not yield curves consistent enough to interpret in general mathematical terms, such as those already given. Tool life is also sensitive to the microstructure and hardness of the workpiece. An approximate equation relating tool life to Brinell hardness number (Bhn) is:

$$K = VT^n f^{n_1} d^{n_2} Bhn^{1.25} \qquad (3\text{-}17)$$

where:

$K$ = constant of proportionality
$V$ = cutting speed, ft/min (mm/min)
$T$ = tool life, min
$n$ = slope of the curve

$$\left( n = \tan\theta = \frac{\log V_1 - \log V_2}{\log T_2 - \log T_1} \right)$$

$f$ = feed, in./rev (mm/rev)
$d$ = depth of cut, in. (mm)
$n_1$ = exponent of feed (average value = 0.5 to 0.8)
$n_2$ = exponent of depth of cut (average value = 0.2 to 0.4)
$Bhn$ = Brinell hardness number

The microstructure of the metal has a more pronounced effect on tool life than hardness alone. It is possible to have two pieces of steel at the same hardness but with different microstructures. They will yield a different tool life and surface finish when machined under the same cutting conditions.

Tool life also is sensitive to the tool material and the use of cutting fluids. The following general equation and *Tables 3-3* and *3-4* take these factors into consideration:

$$V = \frac{K_1}{d^{0.37} f^{0.77}} \times \sqrt[6]{\frac{60}{T}} \times C_F \qquad (3\text{-}18)$$

where:

$K_1$ = constant of proportionality
$\sqrt[6]{\dfrac{60}{T}}$ = a factor that will correct the cutting speed from that obtained for a basic 60-minute tool life to the cutting speed for the desired tool life
$C_F$ = corrections factor for the tool material (for example, 18-4-1 HSS = 100)

## Table 3-3. Numerical values for $K_1$

| Metal to be Cut | 60 min Without Cutting Fluid, or 480 min With Cutting Fluid | 60 min With Cutting Fluid | 480 min Without Cutting Fluid |
|---|---|---|---|
| Light alloys | 25.0 | | |
| Brass (80–120 Bhn) | 6.7 | | |
| Cast brass | 4.2 | | |
| Cast steel | 1.5 | 2.1 | 1.1 |
| Carbon steel: | | | |
| SAE 1015 | 3.0 | 4.2 | 2.1 |
| SAE 1025 | 2.4 | 3.3 | 1.7 |
| SAE 1035 | 1.9 | 2.7 | 1.3 |
| SAE 1045 | 1.5 | 2.1 | 1.1 |
| SAE 1060 | 1.0 | 1.4 | 0.7 |
| Chrome-nickel steel | 1.6 | 2.3 | 1.1 |
| Cast iron | | | |
| 100 Bhn | 2.2 | 3.0 | 1.5 |
| 150 Bhn | 1.4 | 1.9 | 1.0 |
| 200 Bhn | 0.8 | 1.1 | 0.5 |

Column header note: $K_1$ For 18-4-1 High-speed-steel Tool and Tool Life of

## Table 3-4. Correction factors for compositions of tool material

| Type | W | Cr | V | C | Co | Mo | $C_F$ |
|---|---|---|---|---|---|---|---|
| 14-4-1 | 14 | 4 | 1 | 0.7–0.8 | — | — | 0.88 |
| 18-4-1 | 18 | 4 | 1 | 0.7–0.75 | — | — | 1.00 |
| 18-4-2 | 18 | 4 | 2 | 0.8–0.85 | — | 0.75 | 1.06 |
| 18-4-3 | 18 | 4 | 3 | 0.85–1.1 | — | — | 1.15 |
| 18-4-1+ 5% Co | 18 | 4 | 1 | 0.7–0.75 | 5 | 0.5 | 1.18 |
| 18-4-2+ 10% Co | 18 | 4 | 2 | 0.8–0.85 | 10 | 0.75 | 1.36 |
| 20-4-2+ 18% Co | 20 | 4 | 2 | 0.8–0.85 | 18 | 1.0 | 1.41 |
| Sintered carbide | — | — | — | — | — | — | Up to 5 |

Approximate Composition, %

*Table 3-3* lists values for $K_1$. *Table 3-4* lists correction factors ($C_F$) for different tool materials. *Table 3-5* lists values for $d^{0.37}$ and $f^{0.77}$.

## Example 3

A 5-in. (127-mm) diameter bar of SAE 1020 steel is to be turned to a 4-in. (102 mm) diameter in one cut, using cutting fluid. A feed of 0.020 in./rev (0.51 mm/rev) will provide a satisfactory surface finish and a tool life of 180 minutes is desired.

Determine the cutting speed that will accomplish the cut if an 18-4-3 HSS tool is to be used.

$K_1 = 3.75$ (interpolate from *Table 3-3*, 60 min with cutting fluids, basic)

## Table 3-5. Numerical values for $d^{0.37}$ and $f^{0.77}$

| $d$ | $d^{0.37}$ | $f$ | $f^{0.77}$ |
|---|---|---|---|
| 0.01 | 0.182 | 0.001 | 0.004 |
| 0.02 | 0.235 | 0.002 | 0.008 |
| 0.04 | 0.305 | 0.004 | 0.017 |
| 0.06 | 0.353 | 0.006 | 0.019 |
| 0.08 | 0.393 | 0.008 | 0.024 |
| 0.10 | 0.427 | 0.010 | 0.029 |
| 0.14 | 0.482 | 0.014 | 0.037 |
| 0.18 | 0.530 | 0.018 | 0.045 |
| 0.22 | 0.571 | 0.022 | 0.053 |
| 0.25 | 0.598 | 0.025 | 0.059 |
| 0.30 | 0.640 | 0.030 | 0.067 |
| 0.35 | 0.678 | 0.035 | 0.075 |
| 0.40 | 0.712 | 0.040 | 0.084 |
| 0.45 | 0.744 | 0.045 | 0.092 |
| 0.50 | 0.774 | 0.050 | 0.099 |
| 0.75 | 0.899 | 0.075 | 0.135 |
| 1.00 | 1.000 | 0.100 | 0.170 |

$d = 0.50$; $d^{0.37} = 0.774$ (from *Table 3-5*)

$f = 0.020$; $f^{0.77} = 0.049$ (interpolate from *Table 3-5*)

$T = 180$ min

$C_F = 1.15$ (from *Table 3-4*)

$$V = \frac{3.75}{0.774 \times 0.049} \times \sqrt[6]{\frac{60}{180}} \times 1.15$$

$$= 99 \times 0.33^{0.166} \times 1.15$$
$$= 99 \times 0.833 \times 1.15$$
$$= 95 \text{ ft/min}$$

$$N = \frac{12V}{\pi D} = \frac{12 \times 95}{\pi \times 5} = 72.5 \text{ rpm}$$

where:

$N$ = recommended lathe rpm

$D$ = diameter, in. (mm)

# GUIDELINES FOR CUTTING TOOL DESIGN

The design of cutting tools is not a pure science involving only computations to be carried out in functional isolation. Of itself, a cutting tool is only a piece of metal of special shape and construction, although frequently a very expensive piece of metal.

Good tool design is accomplished with consideration of many inseparably related factors: the composition, hardness, condition, and shape of the workpiece material; rate and volume of the specified production; type, motions, power, and speed of the machine tool to be used; toolholders and workholders available or to be designed; specified accuracy and surface of the finished workpiece; and many other factors, common or specific to the particular operation.

The following discussion is, therefore, very much the business of the good tool designer. He or she cannot escape responsibility by saying that some of the factors leading to undue tool wear, or to tool failure, are the responsibility of the process or methods functions. The designer should know, or anticipate, possible application difficulties that lie ahead for the tool about to be designed, then accommodate the difficulties through the design or consult with the other manufacturing functions as to changing certain conditions, or both.

## Rigidity

Setup rigidity is vital to maintaining dimensional accuracy of the cut surface since the tool shifts into or out of the cut with the accumulation of static deflections and take-up of loose fits. Rigidity also maintains surface-finish quality, avoiding the marks made by elastic vibration and free play of loose fits and backlash. In the control of vibration, rigidity of the part and cutting tool can make the difference between success or failure of the machining operation.

Increasing mass reduces vibration amplitude and resonant frequency, while dampening reduces amplitude by dissipating vibratory energy as frictional heat. Since each part of the cutting system (in other words, the machine, fixture, tool, and workpiece) can affect the mode and amount of vibration, most should be made oversize and broadly supported. This provides design latitude for those members having more severe cost and space limitations. Designers should be generous with rigidity, anticipating fast, efficient cuts.

## Strength

The strength of each member can be considered separately and related to the magnitude and application of force it will transmit. It should be sufficient to prevent breakage or deformation beyond the elastic limit when the operation is performed correctly. The designer also must consider overloads and damage that may be encountered, providing abundant strength wherever economically possible. In particular, generous size and material specification should give good working life to areas subject to abrasive wear and work hardening under impact loads. But chatter, packed chips, or binding due to setup misalignment can multiply normal operating forces many times. Tool failure, mechanical malfunctions, and operating errors threaten destructive casualties, even with costly over-design.

## Weak Links

A common practice is to protect the structural chain with weak links in anticipation of casualties, and confine or limit the possible damage. Suitable low strength with high rigidity is illustrated by the common soft shear pin. But these weak links must be strong enough to withstand normal operation and overload if possible. Permanent members are made unquestionably stronger by comparison and protected by location. Identical design criteria make weak links an ideal combination with wearing details. They should be comparatively cheap, with duplicates widely stocked or readily produced, and provide easy, accurate replacement mounting. Mass-produced, delicate workpieces are natural weak links, though cutting-tool inserts are the typical wear and breakaway members. Indexable insert cutting tools, using replaceable backing seats, are ideal examples of wear and damage protection.

## Force Limitations

Operating forces obviously may be limited by a weak-link member, as in the case of a delicate workpiece. A machine tool, such as a hydraulic-stroke planer or broaching machine, may be related in terms of force, with the ratings understated but subject to measurement and control. Some saws, mills, and grinders regulate feeding force instead of cutting force in the direction of cutting velocity. This causes the machine to stop feeding if an accident occurs, or if forces begin to exceed a safe level.

### Speed, Feed, and Size

A machine tool's speed and feed ranges, cutting tool adaptor capacity, and working clearance establish restrictions on tool design and production rate. The effect of forces is indirect but inevitable. A milling cutter with few teeth contacting a delicate workpiece exerts only a few times the cutting forces of one tooth, while high speed permits rapid completion of the cut. Limited cutter diameter and a low-speed range would require more teeth and more force or a longer cutting time. Variable infeed rates can be used to speed rough stock removal and then minimize distortion while finishing, as in the case of a grinding wheel spark-out. The important thing is to design around limitations and take full advantage of flexibility.

### Related Force Components

Total cutting force usually is resolved into three mutually perpendicular components. Force in the feeding direction of a turning, boring, facing, plunge forming, or parting tool corresponds to radial force on a peripheral-cutting milling cutter tooth or abrasive wheel or belt contact area. It is commonly taken with one-fourth up to three-fourths of the tangential force $F_T$ for sharp tools, the larger fraction being appropriate for extremely heavy feeds. Straight-line cutting tools, namely saw-blade teeth or broaches and planer or shaper tools, develop this force component in the direction of feed into the work. The radial force component for turning or boring tools is normal to the finished work surface in facing, planing, and shaping, and in an axial direction for peripheral milling. It may be negative, pulling into the work as a result of large positive back rake, but it is usually a pushing force of relatively low value.

The third and most significant force component is the tangential force $F_T$, which acts on the top of the tool tangent to the direction of rotation of the part or tool. Carbide turning tools typically have 1,000 lbf (4,448 N) of tangential force in general-purpose machining applications.

## Chip Disposal

One sure way to overload a cutting tooth is to block the path of the chip flowing across its face so that the chip is recut. Single-point tools cutting ductile work frequently employ a pressed-in chip breaker to curl an otherwise stringy chip so that it will break in the form of a figure nine and fall away. If groove design is too weak for the size of cut being taken, it can cause edge chipping or breakage. Small-diameter, coarse-pitch milling cutters commonly have ample chip spaces, provided the chips are thrown or washed out between successive passes through the cut. On the other hand, large-diameter cutters taking full-width cuts must carry the chip a half rotation before it can exit. These cuts require large chip slots. It is difficult to remove work materials like soft steel or copper alloys and titanium, whose chips tend to weld onto the tool face. Chip disposal in milling slots may demand high positive rake angles and climb milling instead of conventional cutter rotation to eject the chips. Complex selection and application methods have been developed for tapping and deep-hole drilling where chip clogging, misalignment, and runout can readily break tools. Tool design should provide space for chip flow and means of disposal, which may well be the solution to many problems of tool chipping.

## Uneven Motions

Another sure way to overload a cutting tooth is to increase the feed rate beyond its structural or chip-disposal capacity. Machine structural

deflection accomplishes this in the example of a drill breaking as it moves through the work. As the heavy thrust of the chisel edge is relieved, structural members spring back toward their unstressed shape, and the drill lips plunge into the work for an oversize bite. Feed mechanisms may employ air or hydraulic fluid, whose compression is elastic, or gearing and a leadscrew nut fit may introduce backlash. Machine way motion becomes jumpy at slow speeds (slip-stick motion), even with heavy lubrication. A milling cutter at slow feed may actually rub until pressure builds up. It then may dig into the work and surge ahead. Adding to the difficulty, the sudden change in cutting torque adds to the pounding caused by teeth entering the cut.

Torsional vibration and backlash tend to develop in a rotary drive train. Should cutter rotation become so erratic that it momentarily stops, carbide teeth will generally break at once by being bumped into the work. With some teeth gone, the entire cutter may fail progressively as each successive tooth is unable to carry the extra load left by the damaged teeth.

## Chatter

The rapid, elastic vibration that sometimes appears between tool and work is easily detected by marks on the work surface and by the sound that gives it the name "chatter." Chatter is the momentary separation of the tool and workpiece and the immediate banging back into contact at an audible frequency. It is a danger signal of the impending possibility of chipping or fracture. The remedy is to eliminate uneven motion and loose fits. Chatter is less likely with few teeth moving at high velocity taking thick chip loads, and having high rake and ample relief angles. A negative rake angle may prevent pulling into the cut. Changing the speed, even only a few percent, can reduce chatter. In grinding, harder action or broader contact helps withstand bumping. As an extreme simplification, chatter can be combated with lower cutting forces, while looseness and backlash cannot. Like all other problems in machining, chatter can be greatly reduced by proper tool design.

## SINGLE-POINT TOOLS

Single-point tools have one cutting edge. They are used for operations such as turning, boring, shaping, and threading.

## Basic Tool Angles

The tool signature or nomenclature for a single-point tool is a sequence of alpha and numeric characters representing the various angles, significant dimensions, special features, and the size of the nose radius. This method of identification has been standardized by the American National Standards Institute (ANSI) for carbide and HSS, and is illustrated in *Figure 3-30*, to-gether with the elements that make up the tool signature.

### Back-rake Angle

The *back-rake angle* is between the face of the tool and a line parallel to the base of the toolholder. It is measured in a plane parallel to the side cutting edge and perpendicular to the base. Variations in the back-rake angle affect the direction of chip flow and cutting force. As this angle is increased while other conditions remain constant, tool life will increase slightly and the required cutting force will decrease. Cutting-edge strength decreases dramatically as positive back-rake angles are increased above 5°. Similarly, cut-

Figure 3-30. A straight-shank, right-cut, single-point tool, illustrating the elements of the tool signature as designated by ANSI. Positive rake angles are shown.

ting-edge strength increases as back rake becomes negative and is optimized around –5°.

## Side-rake Angle

*Side-rake angle* is defined as the angle between the tool face and a plane parallel to the tool base. It is measured in a plane perpendicular to both the base of the holder and side-cutting edge. Variations in this angle have the largest effect on cutting force and, to some extent, affect direction of chip flow. As the angle is increased, forces are reduced about 1% for each degree of positive side rake as less tearing of the workpiece occurs. Negative side rake increases edge strength and is recommended for most steels.

## End-relief Angle

The *end-relief angle* is between the end flank and a line perpendicular to the base of the tool. The purpose of this angle is to prevent rubbing between the workpiece and the end flank of the tool. An excessive clearance or relief angle reduces the strength of the tool, so the angle should not be larger than necessary, typically in the 5–7° range.

## Side-relief Angle

The *side-relief angle* is between the side flank of the tool and a line drawn perpendicular to the base. Comments regarding end-relief angles are applicable to side clearance or relief angles as well. For turning operations, the side-relief angle must be large enough to prevent the tool from advancing into the workpiece before the material is machined away. Angles of 5–7° are sufficient for a feed ratio under 0.03 in. (0.8 mm) per revolution. Threading of low-pitch threads requires up to 25° clearance.

## End-cutting-edge Angle

The *end-cutting-edge angle* is between the edge on the end of the tool and a plane perpendicular to the side of the tool shank. The purpose of the angle is to avoid rubbing between the edge of the tool and the workpiece. As with end-relief angles,

excessive end-cutting-edge angles reduce tool strength with no added benefit.

## Lead Angle (Side-cutting-edge Angle)

The *lead angle* is between the straight cutting edge on the side of the tool and the side of the tool shank. This side edge provides the major cutting action and should be kept as sharp as possible. Increasing the lead angle tends to widen and thin the chip, and influences the direction of chip flow. An increase in the side-cutting-edge angle reduces the chip thickness for a given feed by a factor of the cosine of the angle. This, in effect, reduces the chip contact width to thin out the built-up edge. An excessive side-cutting-edge angle redirects feed forces in the radial direction, which may cause chatter. As the angle is increased from 0–45°, workpiece entry is moved away from the vulnerable tip (radius) of the tool to a stronger, more fully supported part of the tool, usually resulting in increased tool life. However, these benefits usually will be lost if chatter occurs, so an optimum maximum angle should be sought.

## Nose Radius

The nose radius connects the side- and end-cutting edges and dramatically affects tool life, radial force, and surface finish. Sharp, pointed tools have a nose radius of zero. Increasing the nose radius from zero avoids high heat concentration at a sharp point. Improvements in tool life and surface finish usually result as the nose radius is increased up to 0.06 in. (1.6 mm). An increase in nose radius has the same general effect as increasing the side-cutting-edge angle. The shape of the contact area changes, but at the point of contact between the machined surface and tool the chip is very thin. In comparison, the feed marks and resultant surface finish are much smoother than those left by a sharp-nosed tool. There is, however, a limit to radius size that must be considered. Chatter and poor surface finish will result if the nose radius is too large; an optimum maximum value should be sought. There is a correlation between ideal surface roughness and nose radius and feed that is given by:

$$R_a\,(\mu in.) = \frac{0.0321(f^2)}{r_e} \qquad (3\text{-}19)$$

where:

$R_a$ = surface roughness, $\mu$in. ($\mu$m)
$f$ = feed, in. (mm)
$r_e$ = nose radius, in. (mm)

## Tool Signature

A comparison of *Figures 3-30* and *3-31* illustrates the difference between a right- and left-hand tool. Right-hand tools are the most popular.

*Tables 3-6, 3-7*, and *3-8* give the recommended angles for single-point tools of HSS, carbide, and cast alloys, respectively.

## Chip Groove

Chip grooves can be ground into the cutting surface of HSS and brazed carbide tools. *Table 3-9* shows the typical dimensions used. Press technology has made the need to hand-grind carbide tools obsolete because indexable inserts have the grooves pressed in. Pressed-in chip grooves provide for chip control and force reductions with unique grooves designed for specific applications. Rapid advances in chip-groove design are the result of research into the effects of changes in groove geometry. *Figure 3-32* illustrates examples of standard chip-groove geometries.

Figure 3-31. *A left-cut tool. All other aspects are identical with Figure 3-30.*

## Forces and Power Requirements

The following is an example of how the equations for power given earlier are applied to turning: A steel part, shown in *Figure 3-33*, is to be machined between centers in a lathe equipped with an air-operated tailstock spindle. The maximum depth of cut is to be 0.100 in. (2.54 mm), the feed is 0.010 in./rev (0.25 mm/rev), and the cutting speed is 300 ft/min (1.5 m/sec). The cutting-tool geometry is 10, 10, 6, 6, 10, 15, 0.030. The feeding force is assumed to be two-thirds of the tangential cutting force. The following equation applies to this material.

$$F_T = cf^{0.8}d \qquad (3\text{-}20)$$

where:

$F_T$ = tangential cutting force, lbf (kN)
$c$ = constant of proportionality
$f$ = feed, in./rev (mm/rev)
$d$ = depth of cut, in. (mm)

Calculate the diameter $D_c$ of the air cylinder required to hold the work between centers against the cutting forces if the minimum pressure in the air cylinder may reach 60 psi (414 kPa). Both centers rotate; thus the friction between the work and the centers need not be considered.

Using *Equation 3-14* in its reduced form,

$W_p = 0.70$ (from *Table 3-1*)

$$W_{p_2} = W_{p_1}\left(\frac{f_1}{f_2}\right)^{1-0.8} = 0.70\left(\frac{0.0156}{0.010}\right)^{0.2}$$

$W_p = 0.765$ in.$^3$/min
$F_T = 396{,}000$
$P_c fd = 396{,}000(0.765)(0.010)(0.100)$
$\quad = 303$ lbf
$F_F = 2/3 \times 303 = 202$ lbf

where:

$W_p$ = specific power consumption, hp (kW)
$F_F$ = feeding force, lbf (kN)

Using *Equation 3-5*,

$$\tan\phi = \frac{d}{N_R + (d - N_R)\tan\varepsilon}$$

### Table 3-6. Recommended angles for high-speed steel, single-point tools

| Material | Side-relief Angle, ° | Front-relief Angle, ° | Back-rake Angle, ° | Side-rake Angle, ° |
|---|---|---|---|---|
| High-speed, alloy, and high-carbon tool steels and stainless steel | 7 to 9 | 6 to 8 | 5 to 7 | 8 to 10 |
| SAE steels: | | | | |
| 1020, 1035, 1040 | 8 to 10 | 8 to 10 | 10 to 12 | 10 to 12 |
| 1045, 1095 | 7 to 9 | 8 to 10 | 10 to 12 | 10 to 12 |
| 1112, 1120 | 7 to 9 | 7 to 9 | 12 to 14 | 12 to 14 |
| 1314, 1315 | 7 to 9 | 7 to 9 | 12 to 14 | 14 to 16 |
| 1385 | 7 to 9 | 7 to 9 | 12 to 14 | 14 to 16 |
| 2315, 2320 | 7 to 9 | 7 to 9 | 8 to 10 | 10 to 12 |
| 2330, 2335, 2340 | 7 to 9 | 7 to 9 | 8 to 10 | 10 to 12 |
| 2345, 2350 | 7 to 9 | 7 to 9 | 6 to 8 | 8 to 10 |
| 3115, 3120, 3130 | 7 to 9 | 7 to 9 | 8 to 10 | 10 to 12 |
| 3135, 3140 | 7 to 9 | 7 to 9 | 8 to 10 | 8 to 10 |
| 3250, 4140, 4340 | 7 to 9 | 7 to 9 | 6 to 8 | 8 to 10 |
| 6140, 6145 | 7 to 9 | 7 to 9 | 6 to 8 | 8 to 10 |
| Aluminum | 12 to 14 | 8 to 10 | 30 to 35 | 14 to 16 |
| Bakelite® | 10 to 12 | 8 to 10 | 0 | 0 |
| Brass, free-cutting | 10 to 12 | 8 to 10 | 0 | 1 to 3 |
| Red, yellow, bronze—cast, bronze—commercial | 8 to 10 | 8 to 10 | 0 | –2 to –4 |
| Bronze, free-cutting | 8 to 10 | 8 to 10 | 0 | 2 to 4 |
| Hard phosphor bronze | 8 to 10 | 6 to 8 | 0 | 0 |
| Cast iron, gray | 8 to 10 | 6 to 8 | 3 to 5 | 10 to 12 |
| Copper | 12 to 14 | 12 to 14 | 14 to 16 | 18 to 20 |
| Copper alloys: | | | | |
| Hard | 8 to 10 | 6 to 8 | 0 | 0 |
| Soft | 10 to 12 | 8 to 10 | 0 to 2 | 0 |
| Fiber | 14 to 16 | 12 to 14 | 0 to 2 | 0 |
| Formica® | 14 to 16 | 10 to 12 | 14 to 16 | 10 to 12 |
| Nickel iron | 14 to 16 | 10 to 12 | 6 to 8 | 12 to 14 |
| Micarta® | 14 to 16 | 10 to 12 | 14 to 16 | 10 to 12 |
| Monel® and nickel | 14 to 16 | 12 to 14 | 8 to 10 | 12 to 14 |
| Nickel silvers | 10 to 12 | 10 to 12 | 8 to 10 | 0 to –2 |
| Rubber, hard | 18 to 20 | 14 to 16 | 0 to –2 | 0 to –2 |

where:

$\varnothing$ = chip flow angle, °
$N_R$ = nose radius, in. (mm)
$\varepsilon$ = side-cutting-edge angle, °

and,

$\varnothing$ = 64°01'; rounded to 64°

$F_R = F_F \tan(90 - \varnothing) = 202 \tan 26°$
$\quad = 99 \text{ lbf}$

where:

$F_R$ = resultant force, lbf (kN)

Taking moments about the headstock center (*Figure 3-34a*),

$$\sum M = 0 = 6F_F - 20F_R + 20F_X$$
$$F_X = 38.4 \text{ lbf}$$

Table 3-7. Recommended angles for carbide, single-point tools

| Material | Normal End-relief Angle, ° | Normal Side-relief Angle, ° | Normal Back-rake Angle, ° | Normal Side-rake Angle, ° |
|---|---|---|---|---|
| Aluminum and magnesium alloys | 6 to 10 | 6 to 10 | 0 to 10 | 10 to 20 |
| Copper | 6 to 8 | 6 to 8 | 0 to 4 | 15 to 20 |
| Brass and bronze | 6 to 8 | 6 to 8 | 0 to –5 | +8 to –5 |
| Cast iron | 5 to 8 | 5 to 8 | 0 to –7 | +6 to –7 |
| Low-carbon steels up to SAE 1020 | 5 to 10 | 5 to 10 | 0 to –7 | +6 to –7 |
| Carbon steels SAE 1025 and above | 5 to 8 | 5 to 8 | 0 to –7 | +6 to –7 |
| Alloy steels | 5 to 8 | 5 to 8 | 0 to –7 | +6 to –7 |
| Free-machining steels SAE 1100 and 1800 series | 5 to 10 | 5 to 10 | 0 to –7 | +6 to –7 |
| Stainless steels, austenitic | 5 to 10 | 5 to 10 | 0 to –7 | +6 to –7 |
| Stainless steels, hardenable | 5 to 8 | 5 to 8 | 0 to –7 | +6 to –7 |
| High-nickel alloys (Monel®, Inconel®, etc.) | 5 to 10 | 5 to 10 | 0 to –3 | +6 to +10 |
| Titanium alloys | 5 to 8 | 5 to 8 | 0 to –5 | +6 to –5 |

Table 3-8. Recommended angles for cast alloy, single-point tools*

| Material | Back-rake Angle, ° | Side-rake Angle, ° | Side-relief Angle, ° | Front-relief Angle, ° | Side-cutting-edge Angle, ° | End-cutting-edge Angle, ° |
|---|---|---|---|---|---|---|
| Steel | 8–20[†] | 8–20[†] | 7 | 7 | 10 | 15 |
| Cast steel | 8 | 8 | 5 | 5 | 10 | 10 |
| Cast iron | 0 | 4 | 5 | 5 | 10 | 10 |
| Bronze | 4 | 4 | 5 | 5 | 10 | 10 |
| Stainless steel | 8–20[†] | 8–20[†] | 7 | 7 | 10 | 15 |

* Stellite® 98M2-turning tools.
[†] Angle depends on grade and type of steel. Boring tools use the same rake but greater relief to clear the work.

Table 3-9. Dimensions for parallel- and angular-type chip breakers, in. (mm)

| Depth of Cut | Feed | 0.006–0.012 (0.15–0.30) | 0.013–0.017 (0.33–0.43) | 0.018–0.025 (0.46–0.64) | 0.028–0.040 (0.71–1.02) | Over 0.040 (1.02) |
|---|---|---|---|---|---|---|
| | R | 0.010–0.025 (0.25–0.64) | 0.035–0.065 (0.89–1.65) | 0.035–0.065 (0.89–1.65) | 0.035–0.065 (0.89–1.65) | 0.035–0.065 (0.89–1.65) |
| | T | 0.010 (0.25) | 0.015 (0.38) | 0.020 (0.51) | 0.030 (0.76) | 0.030 (0.76) |
| 1/64–3/64 (0.4–1.2) | W | 1/16 (1.6) | 5/64 (2.0) | 7/64 (2.8) | 1/8 (3.2) | |
| 1/16–1/4 (1.6–6.4) | W | 3/32 (2.4) | 1/8 (3.2) | 5/32 (4.0) | 3/16 (4.8) | 3/16 (4.8) |
| 5/16–1/2 (7.9–12.7) | W | 1/8 (3.2) | 5/32 (4.0) | 3/16 (4.8) | 3/16 (4.8) | 3/16 (4.8) |
| 9/16–3/4 (14.3–19.0) | W | 5/32 (4.0) | 3/16 (4.8) | 3/16 (4.8) | 3/16 (4.8) | 3/16 (4.8) |
| Over 3/4 (19.0) | W | 3/16 (4.8) | 3/16 (4.8) | 3/16 (4.8) | 3/16 (4.8) | 1/4 (6.4) |

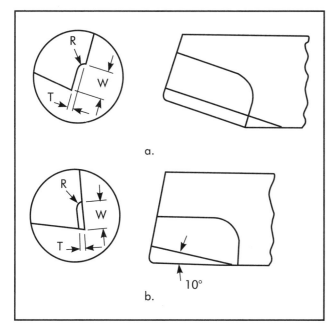

Figure 3-32. *Chip grooves ground into tools.*

Figure 3-33. *Steel part in a turning operation.*

where:

   $M$ = moment of inertia
   $F_X$ = force in $x$, lbf (N)

Resolving the forces at the cutting tool in a vertical plane normal to the work axis (*Figure 3-34b*):

$$R = (F_T{}^2 + F_X{}^2)^{1/2} = 307 \text{ lbf}$$

where:

   $R$ = resolved force, lbf (N)
   $F_T$ = tangential cutting force, lbf (N)

Resolving $R$ about the tailstock center (*Figure 3-34c*):

Figure 3-34. *Forces in turning operations.*

$$F_y = \frac{R}{\tan 60} = 177 \text{ lbf}$$

$$D_c = \frac{4F_y}{\pi 60} = 3.76 \text{ in.}$$

where:

   $F_y$ = force in $y$, lbf (N)
   $D_c$ = diameter of air cylinder required to hold the work, in. (mm)

## MULTIPLE-POINT TOOLS

Multiple-point cutting tools are a series of single-point tools mounted in or integral with a holder or body and operated in such a manner that all the teeth (tools) follow essentially the same path across the workpiece. The cutting edges may be straight or in the form of various contours to be reproduced on the workpiece. Multiple-point tools may be either linear-travel or rotary. With linear-travel tools, the relative motion between the tool and workpiece is along a straight-line path. The teeth of rotary cutting tools revolve about the tool axis. The relative motion between the workpiece and a rotary cutting tool may be either axial or in a plane normal to the tool axis. In some cases, a combination of the two motions is used. Certain form-generating tools involve a combination of linear travel and rotary motions.

*Figures 3-35* and *3-36* illustrate two types of milling cutters and indicate the difference in nomenclature between their angles and single-point tools. *Figure 3-36* shows the peripheral cutting edge angle as 0°, with positive and negative directions indicated.

Whether a cutting tool is single-point or one component of a milling cutter, various angles must provide the most efficient cutting action. Theoretical considerations may dictate larger angles, but actual cutting experience may dictate smaller angles for greater tool strength without chatter. Advantages from increasing any angle always must be considered together with the effect on tool strength.

## Cutting Processes

Cutting processes for multiple-point tools are similar to those for single-point tools. Linear-travel tools produce a series of chips similar to those produced by single-point tools on planing cuts. Milling cutters produce chips that vary in thickness because of the nature of the tooth path, as illustrated in *Figure 3-37*. The chips produced by axial-feed tools tend to be conical because varying diameters across the cutting edge cause different portions of the cutting edge to travel discrete distances. Aside from these differences, studies have shown no fundamental difference between the metal-formation process involved in forming chips using these tools or single-point tools on turning or planing cuts.

## Design Considerations

The single most important factor affecting the performance of any cutting tool is the attainment of a high degree of rigidity in the entire machining system. This includes the cutting tool, machine tool, fixture, and workpiece. A lack of rigidity in any of the system elements can largely nullify the benefits of high rigidity in other elements. This interrelationship is all too often overlooked by fixture designers—workpieces are not adequately supported at the point of cutting.

With adequate workpiece support and machine-tool rigidity, an increase in the rigidity of the cutting tool can enable large improvements to tool life. Such improvements are particularly evident in the case of the super-strength alloys used in aircraft and missile production. When drilling such alloys, improvements in tool life by a factor of 50 or more have been obtained by using short, heavy-web drills in place of conventional drills.

When designing multiple-point tools, there must always be some compromise with maximum rigidity. Most multiple-point tools are required to carry the chips generated for some distance before they can be ejected. Adequate chip space must be provided to avoid jamming, which can cause tool breakage. The amount and shape of the chip space provided depend on the material and the nature of the cut. If the chips are discontinuous, less chip room is required and closer tooth spacing can be used. More chip room and wider tooth spacing are required for high-tensile continuous chips. With some tools, it is possible to incorporate some form of chip breaker on the cutting face to produce smaller chips and thus improve the chip-conveying ability of the tool.

With tools such as milling cutters and broaches, tooth spacing can also affect the smoothness of operation and accuracy of work. If too many teeth are in contact with the work at the same time, the tool part or machine may be overloaded. Overloading may cause breakage of the tool or driving key. The deflection due to the increase in

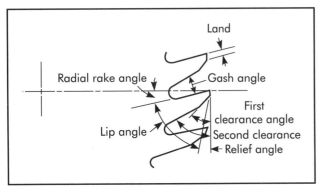

Figure 3-35. *A solid plain milling cutter showing the basic tooth angles and the tooth land.*

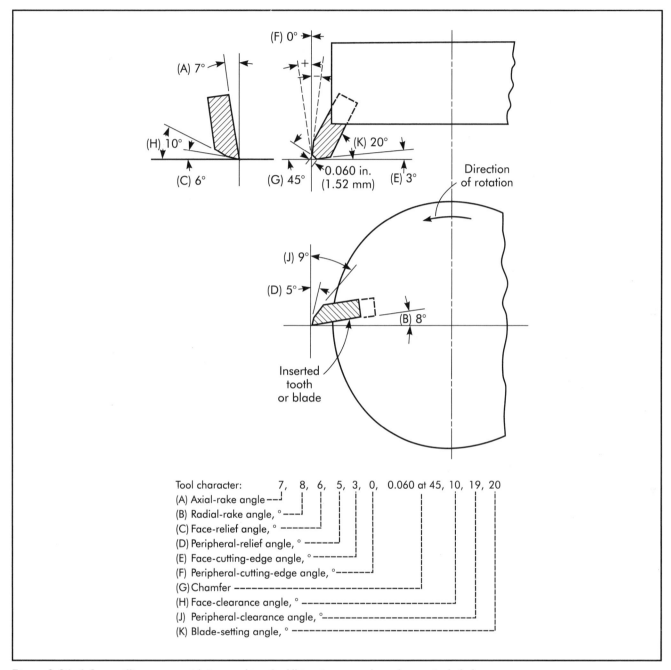

Figure 3-36. *A face milling cutter with inserted teeth. All pertinent tooth angles are included.*

loading may cause a decrease in accuracy. With milling cutters it is generally desirable that at least one tooth be in contact with the work at all times. This keeps the tool and workpiece under load at all times, avoiding the vibration that occurs from varying shock loads.

The rake and relief angles on multiple-point cutting tools affect both the tool performance and tooth strength. High rake angles usually make cutting freer and more efficient. High relief angles reduce the rubbing that occurs on the flank of the tool. These angles, however, cannot be increased without limit. As either the rake or relief angle is increased, cutting-edge strength is reduced. In addition, while high relief angles reduce the temperature resulting from flank friction, they do

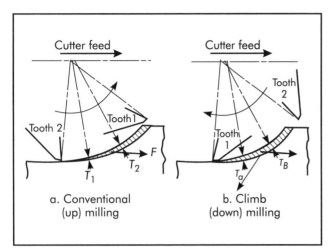

Figure 3-37. *A comparison of undeformed chip shapes.*

cause a greater change in workpiece size for the development of a wear land of given size. The selection of rake and relief angles, therefore, requires a compromise.

## Operating Considerations

In setting up the operating conditions for any multiple-point cutting tool, there are three important variables that can be adjusted—cutting speed, feed per cutting edge, and cutting fluid. Of these, the feed is the most important and should be established first. With some tools, such as taps and broaches, the feed per tooth is determined by the design of the tool and can be changed only by tool modification. With most other tools, the feed is determined by selecting an appropriate machine setting. It has long been demonstrated that it is more efficient to remove metal in the form of thick chips rather than thin ones, so the maximum possible feed per tooth should be used. The maximum feed per tooth is limited by the following factors: cutting-edge strength; rigidity and allowable deflection; the surface finish required; and tool chip space.

The feed cannot be increased without limit because some point will be reached at which either the tool or machine will be overloaded. Overload can cause breakage, either immediately or with the accumulative increase in cutting forces due to dulling. As the feed is increased, the cut-

ting forces increase, causing greater deflection between the tool and workpiece. Deflection can become so large that it is impossible to hold the accuracy required. The surface finish produced on the workpiece usually deteriorates as the feed is increased. This may necessitate the use of light feeds. With some metals, the greater volume of chips produced with heavy feeds may overload the chip space in the tool. In other cases, heavier feeds will produce broken chips that can be handled better than continuous chips resulting from light feeds.

In peripheral milling, the actual chip thickness is affected by the depth of cut. Maximum undeformed chip thickness equal to the feed per tooth is obtained when the width of cut equals one-half the cutter diameter. With shallow finishing cuts, the maximum undeformed chip thickness is much less than the cutter advance per tooth. In the extreme case of peripheral milling, with a shallow width of cut, the undeformed chip thickness may be so low that there is no chip load, which causes the tool to burnish or work harden the material rather than cut it. To prevent this on shallow cuts, feed rates of three to four times the normal should be used.

After the maximum allowable feed has been established, the cutting speed should be considered. At a constant feed per tooth, there is relatively little change in cutting forces as the cutting speed is increased. In the normal operating range, speed has relatively little effect on surface finish, but sometimes a large increase of speed (usually made possible by a change of tool material) will yield an improvement in surface finish by increasing the temperature to such that the workpiece is softened, resulting in less tearing. The principal effect of increasing cutting speed is to produce parts faster at a constant feed per tooth. This increase in production normally justifies the tool-life reduction resulting from operating at higher temperatures. The optimum cutting speed is what will permit parts production at the lowest cost per piece. This requires analysis of all the costs, including machining, tool changing, and cutting-tool acquisition and maintenance costs.

## Cutting Fluids

The use of the correct cutting fluid can make substantial improvements in a machining operation. The proper cutting fluid can permit higher feeds and increased speeds, as well as attain better surface finishes. Cutting fluids should be directed to the exact point where the cutting is being done and be applied in a constant, even flow. No machining operation should be set up without some consideration of cutting fluids. Dry or mimimal quantity lubrication machining is a consideration, however, to minimize cleanup and environmental considerations, such as waste disposal. Correct operating parameters, cutting tool design, and cutting material selection (including coatings) are critical for making dry or minimal quantity lubrication machining work. Furthermore, applicability depends greatly on the workpiece material.

## Forces and Power Requirements

Knowledge of cutting forces is essential to machine and fixture design, and for the determination of power requirements. Force and power predictions for multiple-point tools are more complex than those for a single-point tool. Varying numbers of teeth may be in contact with the work, chip size can vary in different parts of the cut, and orientation of the cutting teeth may not be constant with respect to the workpiece. The best methods to determine forces on such tools are by actual measurement on the machine to be used or a simulated setup in a metal-cutting laboratory. Since such measurements are sometimes inconvenient or impossible to make, methods for making reasonable force and power estimates may be of value.

One approach is to consider the multiple-point tool equivalent as a series of single-point tools, then estimate the contribution of each tool and sum these to arrive at the resultant forces. The following methods of calculation are preferable.

When it is not possible to estimate forces and power directly, a fairly good estimate still can be made by considering the cutting energy. The removal of 1 in.$^3$ (16.4 cm$^3$) of an alloy steel with a hardness of about 200 Bhn at normal feeds requires the expenditure of about 56,492 J (15.7 Wh) of energy. In terms of engineering units, this can be stated as 1.25 hp/in.$^3$/min. Thus, if the maximum rate of metal removal is computed in terms of cubic inches per minute, this value can be multiplied by 1.25 to give a reasonable estimate of the horsepower required at the cutting tool.

For SI metric conversion:

$$\text{Unit horsepower} = (\text{hp/in.}^3/\text{min}) \times 2.73$$
$$= \text{unit power (kW/cm}^3/\text{sec)}$$

If a rotary tool is involved, the tool torque $M$, in inch-pounds (Newton · meters [N/m]), can be estimated from the following equation:

$$M = \frac{63,025\, h_p}{N} \tag{3-21}$$

where:

$M$ = tool torque, in.-lb (N/m)
$h_p$ = tool horsepower
$N$ = tool rotational speed, rpm

To convert to SI units:

$$M \times 0.11298 = \text{Newton/meter (N/m)}$$

In the case of linear-travel tools, the force in the cutting direction can be estimated from the following formula:

$$F = \frac{33,000\, h_p}{V} \tag{3-22}$$

where:

$F$ = cutting force, lbf
$h_p$ = tool horsepower
$V$ = cutting speed, ft/min

To convert to SI units:

$$F \times 4.448 = \text{Newtons (N)}$$

*Equations 3-21* and *3-22* must be applied with caution, since they will be useful in estimating torque or cutting force only if an accurate estimate of the maximum rate of metal removal is made. If they are applied to the average rate of metal removal, they will indicate only average torque or average force. This could be considerably below the peak forces in the case of intermittent cutting, as might be encountered in milling or broaching. Further, all of the equations presented ignore the even higher peak forces resulting from impact or vibration during cutting. Because of these limitations, the equations should not be used as a basis for the design of critical elements in the machine or fixtures.

## LINEAR-TRAVEL TOOLS

Tools traveling in straight-line motions are referred to as linear-travel tools. These tools can be used in high production applications where close tolerance work is being performed. They normally produce good surface finishes and are versatile and economical.

### Broaches

The most common multiple-point linear-travel tool is the broach. Broaches are used for producing either external or internal surfaces. The surfaces produced may be flat, circular, or of an intricate profile, as viewed in a section normal to the tool travel. A broach is essentially a series of single-point tools following each other in the axial direction along a tool body or holder. Successive teeth vary in size or shape in such a manner that each following tooth will cut a chip of the proper thickness. The basic elements of broach construction are illustrated in *Figure 3-38*.

The spacing and shape of broach teeth are determined by the length of the workpiece and the chip thickness per tooth as well as by the type of chips formed. The chip space between the broach

teeth must be sufficient to take care of the volume of chips generated. Broach teeth are provided with rake and relief angles in the same manner as other cutting tools. Standard broaching nomenclature designates the rake angle as the face angle and the relief clearance as the back-off angle. Rake angles fall in the same range used for other tools, but back-off angles normally are quite low, in the range between 0.5° and 3.5°. Low back-off angles are used on broaches to minimize the loss of size in resharpening. Final finishing teeth are often provided with an unrelieved land behind the cutting edge to ensure proper sizing of the workpiece. Sometimes, noncutting burnishing teeth follow the final cutting teeth.

Internal broaches (*Figure 3-39*) are either pulled or pushed through the work. Strength considerations limit the design of such broaches.

Figure 3-39. *Typical broaches: (a) special form broach; (b) titanium slotting broach for compressor disks; (c) semifinish rock-gear form broach; (d) round broach used for half-round operation on connecting rods; (e) titanium dovetail roughing broach for compressor disks; (f) Inconel® pine-tree finish broach for turbine disks; (g) rebroach for hardened gears. (Courtesy Marbaix Lapointe)*

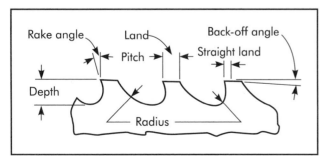

Figure 3-38. *Basic elements of broach construction.*

Surface broaches (*Figure 3-40*) ordinarily are carried on a large, guided ram; here, strength is not as critical since the cutting tool can be transferred to the ram at many points along the broach length.

Broaches are commonly made of HSS as solid units, but carbide-tipped and inserted-blade broaches are sometimes economical. This is particularly true in the case of surface broaches, which are better adapted to mounting carbide indexable inserts. Broaches can be used to cut helical internal forms if the broaching machine is equipped to rotate the broach at the proper lead rate as it passes through the work.

## Gear-shaper Cutters

A gear-shaper cutter is a tool that looks, to some extent, like a gear, with the teeth relieved to provide cutting edges. Typical gear-shaper cutters are illustrated in *Figure 3-41*. Gear-shaper cutters are reciprocated in the axial direction, but at the same time are rotated in timed relationship to the gear being generated. The rotational speed is low compared to the speed of

Figure 3-41. *Standard gear-shaper cutters.*

reciprocation, so the shaper-cutter principally is a linear-travel tool. The feed per tooth is determined by the rate at which the cutter is fed into depth in the gear blank, and to some extent by the speed of rotation. Gear-shaper cutters can, within limits, be designed for forms other than gears, such as splines, sprockets, etc.

## AXIAL-FEED ROTARY TOOLS

Tools that feed into a workpiece while either the tool or workpiece is rotating are considered axial feed rotary tools. These tools are old but

Figure 3-40. *Hole broach details.*

are still some of the most widely used of all machining processes.

## Drills and Related Tools

The production, enlargement, or recontouring of a hole takes place in drilling and its related processes. Either a rotating cutter and fixed workpiece, or a rotating workpiece and fixed cutter, can be used to produce a chip that is formed and removed from the workpiece.

## Twist Drills

In its most basic form, a twist drill (*Figure 3-42*) is made from a round bar of tool material. It has a pair of helical flutes that form the cutting surfaces and act as chip conveyors. Relief is provided behind the two cutting edges or lips. The intersection of the two relief surfaces across the web between the two flutes is known as the *chisel edge*. The lands between the flutes are cut away to a narrow margin to reduce the area of contact between the lands and the wall of the hole and/or guide bushing. The metal cut away forms the margin known as *body diameter clearance*.

The chisel edge of an ordinary drill indents as it is forced into the metal and, as it turns, partially cuts like a cutting tool with a large nega-

tive rake angle, as depicted in *Figure 3-43*. The chisel edge does not penetrate like a sharp point, and tends to make a drill start off center. Some drills are ground to reduce or eliminate the chisel edge; among a number of forms are the thinned point, crankshaft grind, and spiral point. Too much compensation may weaken the point, but optimum amounts have been found to reduce thrust force by one-third, increase tool life by as much as three times, and improve hole location precision as compared to the full chisel edge.

The cutting edges illustrated in *Figure 3-43* correspond to the cutting edges of a single-point tool. An average value for the helix angle is 30°, but angles from about 18° for hard materials to 45° for soft materials are used. The effective rake angle at the edge is larger at the periphery than toward the center of the drill. Some users grind drills at the end of the flutes to create a uniform rake angle along the edges.

The heel of the drill point must be backed off when ground to give relief behind the cutting edges, like the relief on a single-point tool.

Twist drills can be provided with a variety of shanks, but the straight shank and the Morse taper shank are most common. Twist drills ordinarily are made of HSS, but certain sizes and types are available, made from solid carbide or

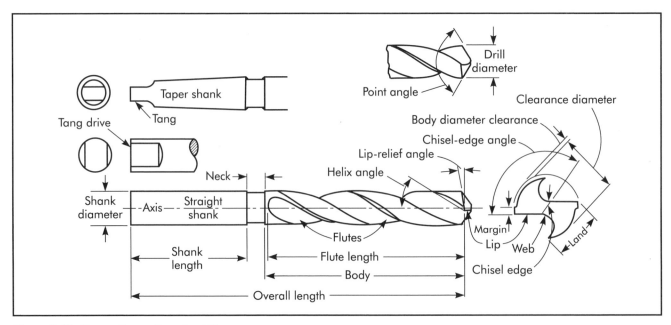

Figure 3-42. *Nomenclature for twist drills.*

Figure 3-43. *Sketches made from photomicrographs of chip formation at the two cutting zones of a twist drill.*

with carbide cutting tips. Indexable carbide inserted drills greatly improve metal removal rates for drilling holes from 1–3 in. (25.4–76.2 mm) in diameter.

Standard drills are available in a wide variety of designs; a few are shown in *Figure 3-44*, which may be used for various materials and services. The most common variations are to helix angle and web thickness. Tough materials usually require the use of rigid drills with heavy webs and reduced helix angles. Free-machining materials can be cut with drills having higher helix angles and lighter webs, which provide more efficient cutting and better chip ejection.

## Reamers and Core Drills

Twist drills produce holes; core drills and reamers are hole-enlarging and finishing tools. The cutting ends of these tools have relieved chamfers that extend to a small enough diameter to permit their entry into the hole being enlarged or finished.

Core drills are designed principally to enlarge existing holes, and provide a greater degree of accuracy and better finish than a two-flute twist drill. These tools usually have three or four moderately deep helical flutes. The lands between the flutes are given body diameter clearance to produce a margin similar to that on a twist drill. Typical core drills are shown in *Figure 3-44*. Core drills are usually made of HSS, but in some sizes they can be furnished with carbide tips. Sometimes they are made in a two-piece construction so that only the cutting end needs to be replaced.

Reamers are designed principally for hole sizing where only a moderate amount of stock is to be removed from the hole walls. The stock is removed by a larger number of flutes than are common with drills. Reamed holes usually have superior accuracy and finish. Because of the greater number of flutes and cutting edges, the margins on reamers are much narrower than those on twist drills and core drills, which minimizes galling of the margins. Reamers have little

Figure 3-44. *Conventional and special-purpose drills.*

or no starting taper, which improves their ability to accept guidance from bushings. The flutes on reamers may be either straight or helical, depending on the type of work to be finished.

Solid, HSS construction of reamers is most common, but carbide-tipped and solid carbide reamers are available in certain sizes and styles. *Figure 3-45* shows various commercial types of reamers.

Figure 3-45. *Commercial types of reamers.*

## Counterbores and Countersinks

Counterbores and countersinks are tools for modifying the ends of holes. They are often provided with pilots that engage the existing hole to improve alignment, and are commonly used to provide a seat in a plane normal to the hole axis. When the seating surface is flat and relatively shallow, tools called spot-facers are often used. Deeper seats, such as those used for recessing the heads of socket-head cap screws, are called counterbores. *Figure 3-46* illustrates various counterbores and spot-facers.

When a conical seat is required at the end of a hole, the operation is known as countersinking and the tools are called countersinks. Such seats are required to receive machine centers and permit the use of flat-headed screws with the heads flush to the drilled surface.

Except for the angle of the cutting edges, counterbores, spot-facers, and countersinks are similar in construction. They are commonly made with two or more flutes, which usually have a right-hand helix. The flutes are usually shorter than those on drills because the seats produced are relatively shallow. When pilots are provided, they may be either integral with the tool or removable. The tools themselves often are made quite short and used as readily replaceable tips in a holder mounted in the machine spindle. In some cases, one holder will drive a fairly wide range of cutter sizes.

The combined drill and countersink is a specialized tool, used for producing countersunk holes for flush-head fasteners and the center holes required for lathe and other machine centers. This tool combines a short two-flute drill with a two-fluted countersink so the entire drilling and countersinking operation can be performed in one pass on a single setup. Such tools are often made in double-end construction as shown in *Figure 3-47*. When one end becomes dull, the tool can be reversed in the chuck to provide another cutting end.

### *Multiple-diameter Tools*

Multiple-diameter holes can be produced or finished in a single operation. Common situations involve drilling and counterboring or drilling and chamfering a hole on the same setup. In high-production operations, this can result in lower machining costs.

The simplest way to make a multiple-diameter tool is to start with a tool of the size that will produce the largest diameter and then grind

Facing tools

High-speed, short-taper-shank counterbore and spot-facer with interchangeable pilot

Cap-screw counterbore, straight shank

High-speed, short-taper, straight-shank counterbore and spot-facer with interchangeable pilot

Counterbore with interchangeable blade and guide

Machine-screw counterbore

Step counterbores

Figure 3-46. *Types of counterbores and spot-facers.*

a. Regular type

b. Bell type

Figure 3-47. *Combined drills and countersinks.*

down the cutting end of the tool to the size required for the small diameter. The cutting portions of the tool are provided with appropriate relief. This is known as step construction.

Another style of multiple-diameter tool employs what is known as subland construction. With a multiple-diameter tool, separate tool lands are provided for each cutting diameter. Small-diameter lands run between large-diameter lands. Subland tools are more expensive because additional manufacturing operations are required when producing increased cutting surfaces.

Step and subland drills are included in *Figure 3-44*. Multiple-diameter reamers (*Figure 3-45*) and counterbores (*Figure 3-46*) also are commercially available. Where the length of the small-diameter portion of the tool is relatively long in terms of the usable tool length, the simpler step construction usually is preferred. Besides lower acquisition costs, such tools provide more chip space in the flutes.

The use of subland tools may be advantageous where the small-diameter step is relatively short. In step construction, the usable sharpening life is limited to something less than the length of the small-diameter step. After this is used up, it is necessary to cut off the tool and completely remanufacture the step end. In subland construction, the two diameters can be sharpened for the entire length of the tool and the length of the step is not critical. The restricted flute space due to the presence of the additional lands sometimes limits the depth of holes to which they may be applied. With multiple diameters, it is important that the difference in diameters be relatively small. With a large difference in diameter, it is not possible to maintain optimum cutting speeds on both diameters, and there is some difficulty in maintaining adequate strength in the tool.

## Power Requirements for Drilling and Reaming

For rotary axial-feed tools, such as twist drills, core drills, and reamers, reasonably accurate estimates of forces and power can be made through the use of a tried-and-true formulae developed some time ago (Shaw and Oxford 1957).

The torque and thrust for a twist drill operating in an alloy steel with a hardness of 200 Bhn can be represented by the following formulae:

$$M = 23,300 f^{0.8} d^{1.8} \left[ \frac{1 - \left(\dfrac{c}{d}\right)^2}{\left(1 + \dfrac{c}{d}\right)^{0.2}} + 3.2 \left(\dfrac{c}{d}\right)^{1.8} \right] \quad (3\text{-}23)$$

$$T = 42,600 f^{0.8} d^{1.8} \left[ \frac{1 - \left(\dfrac{c}{d}\right)}{1 + \left(\dfrac{c}{d}\right)^{0.2}} + 2.2 \left(\dfrac{c}{d}\right)^{0.8} \right] + 19,300 c^2$$

$$(3\text{-}24)$$

where:

$M$ = torque, in.-lb (N/m)
$f$ = feed, in./rev (mm/rev)
$c$ = chisel edge length, in. (mm) (approximately 1.15 times the web thickness for normal sharpening)
$d$ = drill diameter, in. (mm)
$T$ = thrust, lb (kg)

$K$ is a material constant (23,300 for an alloy steel with a hardness of 200 Bhn). For other materials, the values of $K$ are given in *Table 3-10*.

For drills of regular proportions, the ratio $c/d$ can be set equal to 0.18, and for split-point drills the ratio $c/d$ can be set to 0.03. For drills of regular proportions, drilling in 200 Bhn alloy steel, *Equations 3-23* and *3-24* simplify to:

$$M = 25,200 f^{0.8} d^{1.8}$$

$$T = 57,500 f^{0.8} d^{0.8} + 625 d^2$$

There is a difference in the two power requirements for drilling because of the configuration of the point. A reduced web area will take far less power to drive because of the rubbing action of the flat created by the chisel point.

For reamers or core drills, which are used for enlarging existing holes, the effects of the chisel-

Table 3-10. Work material factor, K,
for drilling with a sharp drill (Oberg et al. 1996)

| Work Material | Work Material Constant, K |
|---|---|
| AISI 1117 (Resulfurized free machining mild steel) | 12,000 |
| Steel, 200 Bhn | 24,000 |
| Steel, 300 Bhn | 31,000 |
| Steel, 400 Bhn | 34,000 |
| Cast iron, 150 Bhn | 14,000 |
| Most aluminum alloys | 7,000 |
| Most magnesium alloys | 4,000 |
| Most brasses | 14,000 |
| Leaded Brass | 7,000 |
| Austenitic stainless steel (Type 316) | 24,000* for torque 35,000* for thrust |
| Titanium alloy T16A 4V, 40 HRC | 18,000* for torque 29,000* for thrust |
| Rene 41, 40 HRC | 40,000*[†] minimum |
| Hastelloy-C | 30,000* for torque 37,000* for thrust |

* Values based upon a limited number of tests.
[†] Will increase with rapid wear.

edge region can be eliminated and the equations reduced to:

$$M = 23,300 k f^{0.8} d^{1.8} \left[ \frac{1 - \left(\dfrac{d_1}{d}\right)^2}{\left(1 + \dfrac{d_1}{d}\right)^{0.2}} \right] \quad (3\text{-}25)$$

$$T = 42,600 k f^{0.8} d^{0.8} \left[ \frac{1 - \dfrac{d_1}{d}}{\left(1 + \dfrac{d_1}{d}\right)^{0.2}} \right] \quad (3\text{-}26)$$

where:

$d_1$ = diameter of hole to be enlarged, in. (mm)
$k$ = constant depending on the number of flutes (see *Table 3-11*)

Table 3-11. Values of k
for different numbers of flutes*

| Number of flutes | Constant, k |
|---|---|
| 1 | 0.87 |
| 2 | 1.00 |
| 3 | 1.08 |
| 4 | 1.15 |
| 6 | 1.25 |
| 8 | 1.32 |
| 10 | 1.38 |
| 12 | 1.43 |
| 16 | 1.51 |
| 20 | 1.59 |

* U.S. customary units

The constant $k$ is necessary since, for a given feed per revolution, the number of flutes affects the feed per tooth. It has been pointed out that it is more efficient to remove metal in the form of thick chips than thin ones, so tools with a large number of flutes will require proportionately more energy because of the thinner chips.

Whereas the thrust forces can be substantial and thus can have a large influence on the required strength and rigidity, the power required in feeding the tool axially is very small (less than 2% of the total power requirements), and can usually be disregarded. The cutting power is a function of the torque and rotational speed and can be computed by:

$$P_c = \frac{MN}{63,025} \quad (3\text{-}27)$$

where:

$P_c$ = horsepower
$M$ = tool torque, in-lb (N/m)
$N$ = speed, rpm

To convert to SI Units:

$P_c \times 746$ = watt (W)

In using these formulae for estimating purposes, an allowance of at least 25% should be made for increases due to tool dulling, and further allowances should be made for the efficiency of the machine drive train. For other materials, the cutting forces vary in about the same ratio as noted for single-point tools.

## Taps

A tap essentially is a screw fluted to form cutting edges. The cutting end of the tap has a relieved chamfer that forms the cutting edges and permits entry to an untapped hole. If succeeding chamfered teeth are followed along the thread helix, it will be seen that the effect of the chamfer is to make each tooth have a slightly larger major diameter than the preceding one. Thus, the major feed involved in tapping is a radial feed, although the feed is built into the tool rather than controlled by the machine. This is the case with most other rotary tools. The actual feed per tooth depends on the chamfer angle and the number of flutes. A reduction in the chamfer angle or an increase in the number of flutes will reduce the feed per tooth.

The most common type of tap is the straight-fluted hand tap (*Figure 3-48*). This tap has a straight shank with a driving square at the shank end. In spite of the hand-tap designation they are generally used in tapping machines. Hand taps are usually available with three different chamfer lengths of nominally 1.25-, 4-, and 8-thread pitches, which are designated respectively as bottoming, plug, and taper taps. Taps also are available with helical flutes and other modifications to suit them to specific applications. Most taps used in production operations are made of HSS. Some solid carbide and carbide-tipped taps are used on work materials with high abrasiveness or hardness. However, these are more susceptible to breakage or chipping.

## Milling Cutters

Milling cutters are cylindrical cutting tools with cutting teeth spaced around the periphery (*Figure 3-49*). *Figure 3-35* showed the basic tooth angles of a solid plain milling cutter. A workpiece is traversed under the cutter in such a manner that the feed of the workpiece is measured in a plane perpendicular to the cutter axis. The workpiece is plunged radially into the cutter, and sometimes there is an axial feed of the cutter as well, which results in a generated surface on the workpiece.

Milling-cutter teeth intermittently engage the workpiece, with chip thickness determined by the motion of the workpiece, number of teeth in the cutter, rotational speed of the cutter, cutter lead angle, and overhang of the cutter on the workpiece.

There are two modes of operation for milling cutters. In conventional (up) milling, the workpiece motion opposes the rotation of the cutter (*Figure 3-50a*), while in climb (down) milling, the rotational and feed motions are in the same direction (*Figure 3-50b*). Climb milling is preferred wherever it can be used, since it provides a more favorable metal-cutting action and generally yields a better surface finish. Climb milling requires more rigid equipment, and there must be no looseness in the workpiece feeding mechanism since the cutter will tend to pull the workpiece.

Indexable milling cutters have precision-pressed or ground-carbide inserts positioned around the cutter body. They are held in by screws, pins, or wedges that can be released for indexing. Some milling cutters may either have profile-sharpened or form-relieved teeth. Profile-sharpened cutters are sharpened on the relief surface using a conventional cutter-grinding machine. Form-relieved cutters are made with uniform radial relief behind the cutting edge. They are sharpened by grinding the face of the teeth. The profile style provides greater flexibility in adjusting relief angles for the job to be done, but it is necessary to reproduce any form on the cutter during each resharpening. In the form-relieved style, the relief angle cannot be changed since it is fixed in the manufacture of the cutter. However, the form-relieved construction is well adapted to cutters with intricate profiles since the profile is not changed by resharpening.

Most large milling cutters are provided with an axial hole for mounting an adapter or arbor, and usually have a drive-key slot. Certain small-diameter cutters, and some cutters for specialized applications, are made using an integral shank construction where the cutting section is at the end of a straight or tapered shank that fits into the machine-tool spindle or adapter. Also,

Figure 3-48. *Tap nomenclature.*

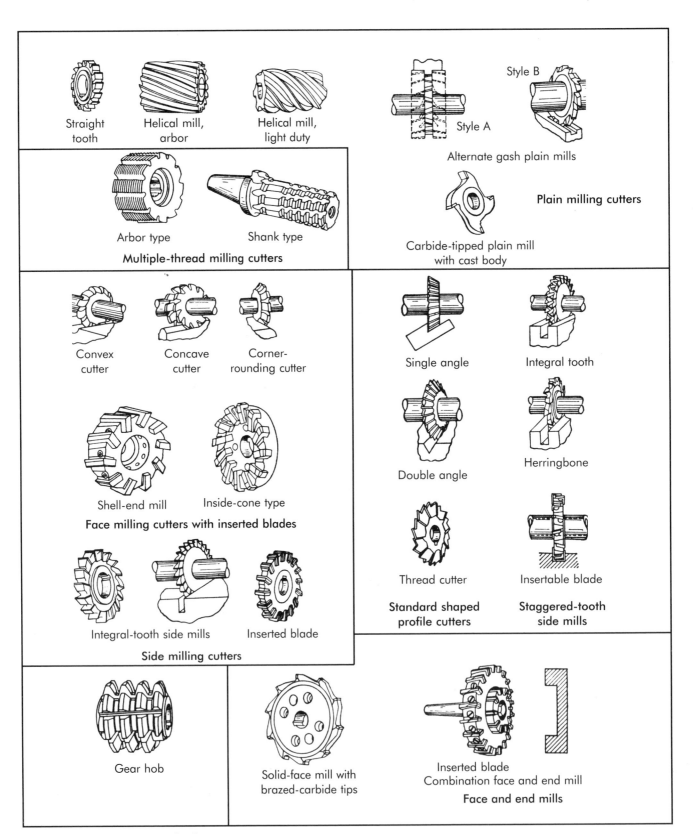

Figure 3-49. *Common types of milling cutters.*

Figure 3-50. (a) *"Out-cut," "conventional," or "up" milling; also called "feeding against the cutter."* (b) *"In-cut," "climb," or "down" milling; also called "feeding with the cutter."*

some large facing cutters are designed to mount directly on the machine-tool spindle nose.

## Number of Teeth in a Milling Cutter

The relationship between the number of teeth and the feed and speed may be expressed as (Drozda and Wick 1983):

$$n = \frac{F}{F_t N}$$  (3-28)

where:

$n$ = number of teeth
$F$ = feed rate, in./min (mm/min)
$F_t$ = feed per tooth, in. (mm) (chip thickness)
$N$ = cutter speed, rpm

*Equation 3-29* has been found satisfactory for HSS cutters such as plain, side, and end mills for both production and general-purpose use:

$$n = 19.5\sqrt{R} - 5.8$$  (3-29)

where:

$R$ = cutter radius, in. (mm)

Example: For a 4-in. (101.6-mm) diameter cutter

$$n = 19.5\sqrt{2} - 5.8 = 21.8 \text{ or } 22 \text{ teeth}$$

*Equation 3-30* was developed for calculating the number of teeth ($n$) when using carbide-tipped cutters (Cincinnati Milling Machine Co. 1951). Since it involves power available at the cutter, it is helpful in permitting full utilization of available power without overloading the motor (U.S. customary units only).

$$n = \frac{KP_c}{F_t N d w}$$  (3-30)

where:

$K$ = constant related to the workpiece material. It may be given the values of 0.650 for average steel, 1.5 for cast iron, and 2.5 for aluminum (these values are conservative, taking into account the dulling of the cutter in service).
$P_c$ = horsepower available at cutter
$d$ = depth of cut, in.
$w$ = width of cut, in.

## Power Requirements for Milling

Milling machines, like other machine tools, are not 100% efficient. That is, it takes more horsepower (kW) than that required at the cutting tool to run the milling machine in a machining operation. The reason is due to frictional losses, gear-train inefficiencies, spindle speeds that are too high for the particular machine, mechanical condition, etc. Consequently, the power required for milling must include machine power losses and power actually used at the cutter. Efficient use of power at the cutter is influenced by cutter speed, design, cutting-edge geometry, and workpiece material.

Using the power equations given earlier, the total horsepower required at the cutter ($P_c$) can be calculated. For example, an alloy steel having a hardness of 250 Bhn is to be machined in a milling machine. The depth of cut is to be 0.250 in. (6.35 mm), the feed is 0.005 in. (0.13 mm) per tooth, and the cutting speed is 300 ft/min (91.4 m/min). The milling cutter has 12 teeth and is 10 in. (25.4 cm) in diameter. The width of the cut is 5 in. (12.7 cm). The specific power consumption is 0.70 hp/in.$^3$/min.

$$P_c = W_p \times \text{in.}^3/\text{min}$$

$$= (W_p)\left(\frac{12V}{\pi D}\right)(n)(f)(w)(d) \qquad (3\text{-}31)$$

$$= (0.70)\left[\frac{12(300)}{\pi(10)}\right](12)(0.005)(5)(0.25)$$

$$= 6.03$$

Assume:

$$P_t = 0.5 \text{ and } E = 0.8$$

$$P_g = \frac{P_c}{E} + P_t = \frac{6.03}{0.8} + 0.5$$

$$= 8.05$$

Use a 10-hp motor.

where:

$W_p$ = specific power consumption, hp
$V$ = cutter speed, ft/min
$D$ = tool diameter, in. (mm)
$n$ = number of teeth
$f$ = feed, in./rev (mm/rev)
$P_t$ = power required to operate machine at no-load capacity, hp (kW)
$E$ = efficiency of power
$P_g$ = motor power required, hp (kW)

The total horsepower required at the cutter ($P_c$) also is given by the equation

$$P_c = \frac{M_R}{K} \qquad (3\text{-}26)$$

where:

$P_c$ = horsepower at the cutter
$M_R$ = metal removal rate, in.$^3$/min
$K$ = factor reflecting the efficiency of the metal-cutting operation

To convert to SI Units:

$$P_c \times 746 = \text{watt (W)}$$

The $K$ factor varies with type and hardness of material, and feed per tooth, increasing as the chip thickness increases. Time-consuming trials are required to determine the quantities involved, because in each case the $K$ factor represents a particular rate of metal removal and not a general or average rate.

To make available a quick approximation of the total power requirements, a milling-machine selector table has been devised (*Table 3-12*), which estimates the metal removed for the rated power of various machines under constant load conditions.

The metal removal rates in *Table 3-12* may be considered products of the $K$ factor (see *Table 3-13*). All $K$ constants given are for dull cutters; hence no allowance for increase in power due to dulling needs to be made. All values apply to average milling speeds and rake angles recommended for the various materials, and 0.010 in. (0.25 mm) feed per tooth (Shaw and Oxford 1957).

## End Mills

*End mills* are shank-type milling cutters usually designed with some form of relieved end teeth (*Figure 3-51*). This construction enables them to do some end cutting, but the majority of the cutting takes place on the periphery. Indexable end mills generally use radius corner inserts, but can be furnished with parallel land wiper inserts for generating finishes down to 50 μin. (1 μm). End mills are usually considered in a separate category from other milling cutters.

High-speed steel end mills have helical flutes with a flute helix angle between 20° and 40°. Flutes are usually several diameters long, but longer and shorter designs are available. Tools with two and four flutes are most common— larger sizes are available with more flutes. The

Table 3-12. Milling-machine selector

| Rated Power of Machine, hp (kW) | 3 (2.2) | 5 (3.7) | 7.5 (5.6) | 10 (7.5) | 15 (11.2) | 20 (14.9) | 25 (18.7) | 30 (22.4) | 40 (29.8) | 50 (37.3) |
|---|---|---|---|---|---|---|---|---|---|---|
| Overall Machine Efficiency, % | 40 | 48 | 52 | 52 | 52 | 60 | 65 | 70 | 75 | 80 |
| Material | Maximum Metal Removal, in.$^3$/min (cm$^3$/min) | | | | | | | | | |
| Aluminum | 2.7 (44) | 5.5 (90) | 8.7 (143) | 12 (197) | 18 (295) | 27 (442) | 37 (606) | 48 (787) | 69 (1,131) | 91 (1,491) |
| Brass, soft | 2.4 (39) | 4.7 (77) | 7.5 (123) | 10 (164) | 16 (262) | 24 (393) | 32 (524) | 41 (672) | 60 (983) | 79 (1,295) |
| Bronze, hard | 1.7 (28) | 3.3 (54) | 5.3 (87) | 7.3 (120) | 11 (180) | 17 (279) | 23 (377) | 30 (492) | 43 (705) | 56 (918) |
| Bronze, very hard | 0.78 (13) | 1.6 (26) | 2.5 (41) | 3.4 (56) | 5.3 (87) | 7.8 (128) | 11 (180) | 15 (246) | 20 (328) | 26 (426) |
| Cast iron, soft | 1.6 (26) | 3.2 (52) | 5.2 (85) | 7.1 (116) | 11 (180) | 16 (262) | 22 (361) | 28 (459) | 41 (672) | 54 (885) |
| Cast iron, hard | 1 (16) | 2 (33) | 3.3 (54) | 4.6 (75) | 7 (115) | 10 (164) | 14 (229) | 18 (295) | 26 (426) | 35 (574) |
| Cast iron, chilled | 0.78 (13) | 1.6 (26) | 2.5 (41) | 3.4 (56) | 5.3 (87) | 7.8 (128) | 10 (164) | 13 (213) | 19 (311) | 26 (426) |
| Malleable iron | 1 (16) | 2.1 (34) | 3.4 (56) | 4.7 (77) | 7.3 (120) | 11 (180) | 14 (229) | 18 (295) | 26 (426) | 36 (590) |
| Steel, soft | 1 (16) | 2 (33) | 3.3 (54) | 4.6 (75) | 7 (115) | 10 (164) | 14 (229) | 18 (295) | 26 (426) | 35 (574) |
| Steel, medium | 0.78 (13) | 1.6 (26) | 2.5 (41) | 3.4 (56) | 5.3 (87) | 7.8 (128) | 10 (164) | 13 (213) | 19 (311) | 26 (426) |
| Steel, hard | 0.56 (9) | 1.1 (18) | 1.8 (29) | 2.5 (41) | 3.9 (64) | 5.7 (93) | 7.7 (126) | 10 (164) | 14 (229) | 19 (311) |

(Courtesy Kearney & Trecker Corp.)

relieved end teeth permit their use in keywaying or pocketing operations with minimal contact with the surface at the bottom of the recess. Most two-flute end mills have end teeth that extend from the periphery to the center of the tool. This permits their use as drills to sink into depth before starting a cut. Some four-flute end mills are made with two diametrically opposed teeth cutting to the tool center. The corners at the intersection of the peripheral and end teeth can be provided with radii to minimize the formation of stress risers in the work. When this radius becomes equal to the cutter radius, the end mill is said to have a ball end. Such tools are widely used in forming recesses and cavities in dies. An operation of this sort often is called *die sinking*.

## Hobs

A *hob* is a generating, rotary cutting tool with its teeth arranged along a helical thread (*Figure 3-52*). The hob and workpiece are rotated in timed relationship to each other, while the hob is fed axially or tangentially across or radially into the workpiece (*Figure 3-53*). The cutting edges of the hob teeth lie in a helicoid, which usually is conjugate to the form produced on the workpiece. The teeth on a hob resemble those on a form-relieved milling cutter in that they are designed to be sharpened on the rake face. Most hobs are designed for generating gear teeth or splines. Certain other evenly spaced forms on the periphery of a cylindrical workpiece also can be hobbed.

Table 3-13. Value of *K* factor
for various materials (U.S. customary units)

| Material | *K* |
|---|---|
| Aluminum and magnesium | 2.5–4.0 |
| Bronze and brass, soft | 1.7–2.5 |
| Bronze and brass, medium | 1.0–1.4 |
| Bronze and brass, hard | 0.6–1.0 |
| Cast iron, soft | 1.5 |
| Cast iron, medium | 0.8–1.0 |
| Cast iron, hard | 0.6–0.8 |
| Malleable iron and 6140 cold-drawn steel, SAE | 0.9 |
| Cold-drawn steel, SAE 1112, 1120, and 1315 | 1.0 |
| Forged and alloy steel, SAE 3120, 1020, 2320, and 2345, 150–300 Bhn | 0.63–0.87 |
| Alloy steel, 300–400 Bhn | 0.5 |
| Stainless steel, AISI 416, free-machining | 1.1 |
| Stainless steel, austenitic, AISI 303, free-machining | 0.83 |
| Stainless steel, austenitic, AISI 304 | 0.72 |
| Monel® metal | 0.55 |
| Copper, annealed | 0.84 |
| Tool steel | 0.505 |
| Nickel | 0.525 |
| Titanium | 0.75 |

Figure 3-51. *End mills.*

There are some limitations on the shape of forms that can be generated by hobs.

Most hobs are manufactured from HSS. Carbide-tipped hobs are sometimes used for hobbing low-strength, abrasive, nonmetallic materials.

## REFERENCES

American Standard ASA B5.3—1959. "Milling Cutters, Nomenclature, Principal Dimensions, etc." New York: American Standard Association.

Boston, O. 1951. *Metal Processing,* 2nd Ed. New York: John Wiley & Sons, Inc.

Chao, B., and Trigger, K. 1955. "Temperature Distribution at the Tool-chip Interface in Metalcutting," *Trans. ASME,* 77. Fairfield, NJ: American Society of Mechanical Engineers.

——. 1958. "Temperature Distribution at the Tool-Chip and Tool-Work Interface in Metalcutting," *Trans. ASME,* 80. Fairfield, NJ: American Society of Mechanical Engineers.

Cincinnati Milling Machine Co. 1951. *A Treatise on Milling and Milling Machines,* 3rd Ed. Cincinnati, OH: Cincinnati Milling Machine Co.

Drozda, T. and Wick, J., eds. 1983. *Tool and Manufacturing Engineers Handbook,* Volume 1, Machining. Dearborn, MI: Society of Manufacturing Engineers, pp. 10-36.

Ernst, H. 1938. "Physics of Metalcutting." *Machining of Metals.* Materials Park, OH: ASM International.

Oberg, E., Jones, Franklin D., Horton, Holbrook L., and Ryffel, Henry H. 1996. *Machinery's Handbook,* 25th Ed. New York: Industrial Press, Inc.

Shaw, M. 1954. *Metalcutting Principles,* 3rd Ed. Boston, MA: Massachusetts Institute of Technology.

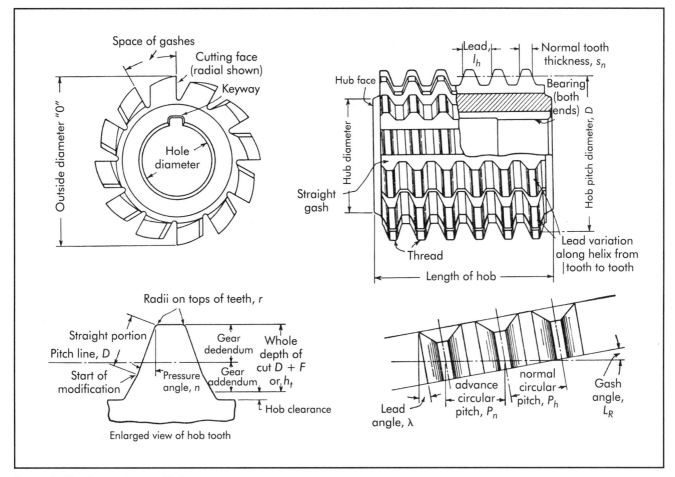

Figure 3-52. *Nomenclature of hobs.*

Schrader, G. and Elshennawy, A. 2000. *Manufacturing Processes and Materials,* 4th Ed. Dearborn, MI: Society of Manufacturing Engineers.

Shaw, M. and Oxford, Jr., C. 1957. "On the Drilling of Metals," and "The Torque and Thrust in Milling," *Trans. ASME,* 79:1. Fairfield, NJ: American Society of Mechanical Engineers.

University of Michigan Research Institute Project No. 2575, Reynolds Metals Company.

## REVIEW QUESTIONS

1. In general terms, when are positive back- and side-rake angles used?
2. What benefits can be realized from using negative side- and back-rake angles?
3. What are the advantages of a large side-cutting-edge angle?
4. What are the disadvantages of a large side-cutting-edge angle?
5. How can you tell if a tool that is failing prematurely has insufficient side or end clearance?
6. What is an acceptable ratio of actual chip thickness to undeformed chip thickness?
7. How do you measure the chip thickness ratio to undeformed chip thickness?
8. What is the only acceptable mode of tool failure?
9. Of speed, feed, and depth of cut, which has the least effect on tool life?
10. Of speed, feed, and depth of cut, which has the greatest effect on tool life?
11. What are common values for the tangential, feed, and radial forces for a single point turning application involving carbide tools and a steel workpiece?

Figure 3-53. *(a) Setup for bang-hobbing four helical gears on a hobbing machine. (b) Hob cutting the teeth on a set of four helical gears. (Courtesy Bourn & Koch Machine Tool Company)*

12. When should fine-pitch milling cutters be used?
13. How many teeth should be in the workpiece in a common milling application?
14. How should coolant be applied to a cutting application?

15. If a multiple-diameter rotation tool requires 1 in. difference in diameter, which tool would provide the most effective design? (a) large diameter 2 in., small diameter 1 in., or (b) large diameter 6 in., small diameter 5 in.

# 4

# Workholding Principles

Workholding is one of the most important elements of the machining process. The term *workholder* includes all devices that hold, grip, or chuck a workpiece to safely perform a manufacturing operation. The holding force may be applied mechanically, electrically, hydraulically, or pneumatically. This chapter considers workholders used in traditional material-removing operations.

## BASIC WORKHOLDERS

*Figure 4-1* illustrates almost all the basic elements present in a material-removing operation intended to shape a workpiece. The right hand is the toolholder, the left hand the workholder, the knife is the cutting tool, and the piece of wood is the workpiece. Both hands combine their motion to shape the piece of wood by removing material in the form of chips. The body of the person whose hands are shown may be considered a machine that imparts power, motion, position, and control to the elements shown. Except for the element of force multiplication, these basic elements may be found in all of the forms of manufacturing setups where toolholders and workholders are used.

*Figure 4-2* shows a pair of pliers or tongs used to hold a rod on which a point has to be ground or filed. This simple workholder illustrates the element of force multiplication by a lever action

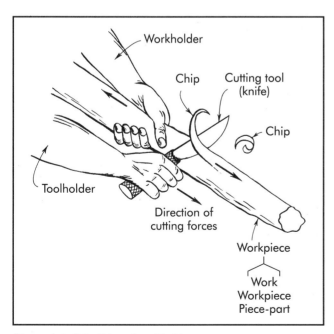

Figure 4-1. *Principles of workholders.*

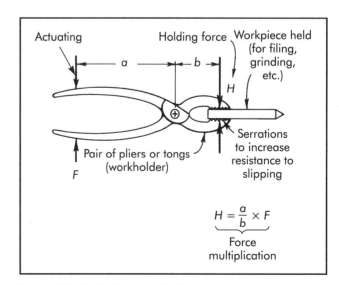

$$H = \frac{a}{b} \times F$$

Force multiplication

Figure 4-2. *Multiplication of holding force.*

and serrations on the parts contacting the rod to increase resistance against slippage.

*Figure 4-3* shows a widely used workholder, the screw-operated vise. The screw pushes the movable jaw and multiplies the applied force. The vise remains locked by the self-locking characteristic of the screw, provides means of attachment to a machine, and permits precise placement of the work.

A vise with a number of refinements often used in workholders is depicted in *Figure 4-4*. The main holding force is supplied by hydraulic power, the screw being used only to bring the jaws in contact with a workpiece. The jaws may be replaceable inserts profiled to locate and fit a specific workpiece. Other, more complex jaw forms are used to match complicated workpieces.

Figure 4-3. *Elementary workholder (vise).*

Figure 4-4. *Vise with hydraulic clamping.*

Another large group of workholders are chucks. They are attached to a variety of machine tools and used to hold a workpiece during turning, boring, drilling, grinding, and other rotary operations. Many types of chucks are available. Some are tightened manually with a wrench, others are power-operated by air or hydraulic methods or electric motors. On some chucks, each jaw is individually advanced and tightened, while others have all jaws advance in unison. *Figure 4-5* shows a workpiece clamped in a four-jaw independent chuck. The drill, which is removing material from the workpiece, is clamped in a universal chuck.

Figure 4-5. *Holding (chucking) a round workpiece.*

## WORKHOLDER PURPOSE AND FUNCTION

A workholder must position or locate a workpiece in a definite relation to the cutting tool. It must withstand holding and cutting forces while maintaining a precise location. A workholder consists of several elements, each performing a certain function. Locating elements position the workpiece; the structure, or tool body, withstands the forces; brackets attach the workholder to the machine; and clamps, screws, and jaws apply holding forces. Elements may have manual or power activation. All functions must be performed with the required firmness of holding, accuracy of positioning, and a high degree of safety for the operator and equipment.

## GENERAL CONSIDERATIONS

The design or selection of a workholder is governed by many factors, the first being the physical characteristics of the workpiece. The workholder must be strong enough to support the workpiece without deflection. Workholder material must be carefully selected with the workpiece in mind so neither will be damaged by abrupt contact (for example, damage to a soft copper workpiece by hard steel jaws).

Cutting forces imposed by machining operations vary in magnitude and direction. A drilling operation induces thrust and torque, while a shaping operation causes straight-line thrust. The workholder must support the workpiece in opposition to the cutting forces. Workholders are generally designed for a specific machining operation.

The workholder establishes the location of the workpiece relative to the cutting tool. If the operation is to be performed at a precise location on the workpiece, locating between the workpiece and workholder must be equally precise. If a cutting tool must engage the workpiece at a specified distance from a feature, such as a line or plane of the workpiece, then the workholder or workholding fixture must establish the line or plane at the specified distance. The degree of precision required in the workholder usually will exceed that of the workpiece because of cumulative error.

The strength and stiffness of the workpiece will determine to what extent it must be supported for the machining operation. If the workpiece design is such that it could be distorted or deflected by machining forces, the workholder must support the affected area. If the workpiece is sufficiently rigid to withstand machining forces, workholder support at the edge of the workpiece may be adequate. The strength of the workholder is determined by the direction and magnitude of the machining forces and weight of the workpiece.

Production requirements will greatly influence workholder design. If a large number of workpieces are to be processed, the cost of an elaborate workholder might well be offset by savings due to increased hourly production. High production rates and volume, therefore, can justify expensive fixturing. Conversely, if only one or two workpieces are to be machined, the operation usually will be performed with standard toolroom equipment since little to no fixturing costs can be justified. Production schedules may limit the time available for workholder acquisition and compel the use of standard equipment.

Safety requirements must always dictate workholder design or selection. A workholder must not only withstand normal cutting forces and workpiece weight, but also may have to withstand large momentary loads. In machining a cast workpiece, the cutting tool might strike an oxide inclusion causing instantaneous multiplication of force. The tool might cut through the inclusion, break, or stall. If the workholder breaks, however, the tool might impart motion to the workpiece. A workpiece in uncontrolled motion is a missile. The workholder also must be designed to protect workers from their own negligence. Where possible, a shield should be interposed between the worker and the tool.

A workholder should be designed to receive the workpiece in only one position. If a symmetric workpiece can be clamped in more than one position, it is probable that a percentage of workpieces will be incorrectly clamped and machined. Workholders should be designed to prevent incorrect placement and clamping.

It is advisable to use standard workholders and commercially available components whenever possible. Not only can these items be purchased for less than the cost of making them, but they are generally stronger and accurate enough.

Many workholders are used in other than material-removing operations. Workholders may be used for the inspection of workpieces, assembly, welding, and so on. There may be very little difference in their basic design and appearance. Quite often, a standard commercial design may be used in one application for a turning operation and for the same or another workpiece in an inspection operation.

## LOCATING PRINCIPLES

To ensure the successful operation of a workholding device, the workpiece must be accurately located to establish a definite relationship between the cutting tool and some points or surfaces of the workpiece. This relationship is established by locators in the workholding device that position and restrict the workpiece to prevent movement from its predetermined location. The workholder then will present the workpiece to the cutting tool in the required relationship. The locating device should be designed so each successive workpiece, when loaded and clamped, will occupy the same position in the workholder. Various methods have been devised to effectively restrict the movement of workpieces. The locating design selected for a given workholder will depend on the nature of the workpiece, requirements of the metal-removing operation to be performed, and other restrictions on the workholder.

## Workpiece Surfaces

One major consideration involved in selecting locators for a workpiece is the shape of the locating surface. All workpiece surfaces can be divided into three basic categories insofar as location is concerned: flat, cylindrical, and irregular. While most workpieces are a combination of different surfaces, the designer must identify which ones will be used to locate the part.

Flat surfaces are those which, regardless of their position, have a flat bearing area for the locators. Typical examples of flat surfaces include edges, flanges, steps, faces, shoulders, and slots.

Cylindrical surfaces are located on a circumference or diameter. Typical examples of cylindrical surfaces include internal (concave) surfaces of holes or external (convex) surfaces of turned cylinders.

Irregular surfaces provide neither a flat nor cylindrical locating surface. Typical examples of irregular surfaces include many cast or forged workpieces. Since the category of irregular locating surfaces is so inclusive, it is necessary that every locating surface be grouped into one, and only one, of the three surface categories.

## Location Types

Basic workpiece location can be divided into three fundamental categories: plane, concentric,

and radial. In many cases, more than one location category may be used for a particular workpiece. However, for the purpose of identification and explanation, each will be discussed individually.

### Plane

Plane location normally is considered the act, or process, of locating a flat surface. Many times, however, irregular surfaces may be located this way as well. Plane location is simply locating a workpiece with reference to a particular surface or plane (*Figure 4-6*).

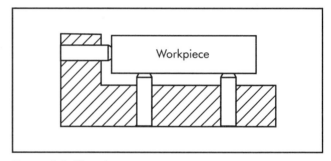

Figure 4-6. *Plane location.*

### Concentric

Concentric location is the process of locating a workpiece from an internal or external diameter (*Figure 4-7*).

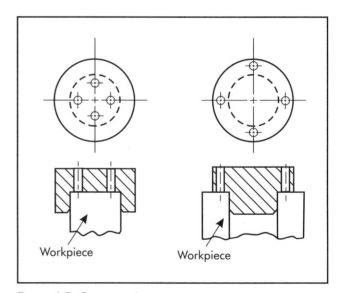

Figure 4-7. *Concentric location.*

## Radial

Radial location normally is a supplement to concentric location. With radial location (*Figure 4-8*), the workpiece is first located concentrically, then a specific point on the workpiece is located to provide a specific fixed relationship to the concentric locator.

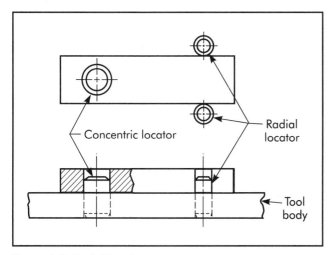

Figure 4-8. *Radial location.*

## Combined

Most workholders use a combination of locating methods to completely position a workpiece. The part shown in *Figure 4-9* is an example of all three basic types of location being used to reference a workpiece.

Figure 4-9. *Plane, concentric, and radial location.*

## Degrees of Freedom

A workpiece in space, free to move in any direction, is designed around three mutually perpendicular planes and has six degrees of freedom. It can move in either of two opposed directions (+ or −) along three mutually perpendicular axes, and can rotate in either of two opposed directions around each axis, clockwise (−) and counterclockwise (+). Each linear direction can be either positive or negative, and each rotation direction can be positive (counterclockwise) or negative (clockwise) as one looks at the origin (0, 0, 0). The six degrees of freedom as they apply to a rectangular prism are illustated in *Figure 4-10*.

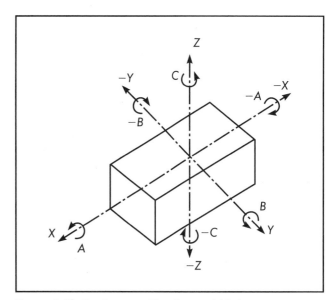

Figure 4-10. *Six degrees of freedom and 12 directions.*

## Location Methods

To accurately locate a workpiece, it must be confined to restrict movement in all of the six degrees of freedom (12 directions) except those called for by the operation. When this condition is satisfied, the workpiece is accurately and positively confined in the workholding device. It is important to indicate that the axis setup in *Figure 4-10* represents a device space orientation. Many computer-aided design (CAD) systems and horizontal machining centers transpose the $Y$ and $Z$ axes so that $Y$ values are up (+) or down (−) and $Z$ values are forward (+) and backward (−).

### 3-2-1 Method

A workpiece may be positively located by means of six pins positioned so they collectively restrict the workpiece in nine directions. This is known as the 3-2-1 method of location. *Figure 4-11* shows the prism resting on three pins, *A, B,* and *C.* The faces of the three pins supporting the prism form a plane parallel to the plane that contains the *X* and *Y* axes. The prism can neither rotate about the *X* and *Y* axes nor move downward in the *Z* direction. Therefore, directions (movements) one, two, three, four, and five have been restricted.

Figure 4-11. *Three pins arrest three directional movements.*

In *Figure 4-12,* two additional pins, *D* and *E,* whose faces are in a plane parallel to the plane containing the *X* and *Z* axes prevent rotation of the prism about the *Z* axis. It is not free to move to the left in direction nine. Therefore, directions six, seven, and nine are restricted, and the prism cannot rotate.

Figure 4-12. *Five pins arrest eight directional movements.*

Finally, with the addition of pin *F* as shown in *Figure 4-13,* direction eight is restricted. Thus, nine directional movements have been restricted by means of six locating points, three in a base plane, two in a vertical plane, and one in a plane perpendicular to the first two.

Figure 4-13. *Six pins arrest nine directional movements.*

Three directions, 10, 11, and 12, still remain unrestricted. This is necessary for loading the workpiece into the workholder. The remaining three directions may be restricted with clamps, which also serve to resist the forces generated by the metal-removing operation being performed on the workpiece. Any combination of three clamping devices and locating pins may be used if this is more suitable to the design of a particular workholder.

### Concentric and Radial Methods

A workpiece that is located concentrically and radially is restricted from moving in nine directions. *Figure 4-14* illustrates a typical workpiece located concentrically. The base and center pin restrict nine directional movements. The base restricts any downward movement and rotation around the *X* and *Y* axes. The center locating pin prevents any movement in either a traverse or longitudinal direction along the *X* and *Y* axes. Located in this manner, the part is only free to move vertically or radially around the *Z* axis.

Figure 4-14. *Base and center pin restrict nine directional movements.*

To prevent movement around the *Z* axis, a radial locator is positioned as illustrated in *Figure 4-15*. In this position, both rotational directions (+ and −) around the *Z* axis are restricted. The only possible movement this part can make is vertically, up the *Z* axis. This direction is restricted by the clamping device.

Figure 4-15. *Base, center pin, and radial locator restrict 11 directional movements. Clamping will restrict Z11 if required.*

## Basic Rules

To function properly, locators must be positioned correctly, designed properly, and sized accurately. To do all this and still permit easy

loading and unloading of the tool requires forethought when planning the locational elements of a workholder. The following are a few basic principles every designer should keep in mind when planning part location.

### Position and Number of Locators

Locators and part supports should always contact a workpiece on a solid, stable point. When possible, the surface should be machined to ensure accurate location. Locating points should be chosen as far apart as possible on any one workpiece surface. Thus, for a given displacement of one locating point from another, the resulting deviation decreases as the distance between the points increases.

The most satisfactory locating points are in mutually perpendicular planes. Other arrangements are possible but not desirable. Two disadvantages result when locating from other than perpendicular surfaces: (1) the consequent wedging action tends to lift the workpiece, (2) the displacement of a locating point or particle (chip or dirt) adhering to the workpiece introduces a correspondingly larger error. In *Figure 4-16* the introduced error, *T*, is projected to become the resulting error, *E*. The projection factor, *F*, is zero when the locating surfaces are perpendicular and increases as the angle between them becomes more acute.

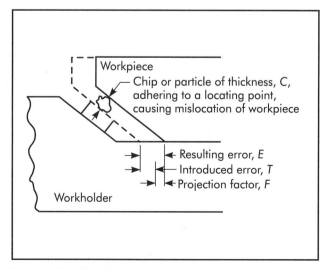

Figure 4-16. *Magnification and projection of error.*

# REDUNDANT LOCATORS

Always avoid redundant or duplicate locators on any part. Redundant location occurs when more than one locator is used to locate a particular surface or plane of a workpiece. *Figure 4-17* shows examples of redundant location. The principal objection to using more than one locator, or series of locators, to reference one location is the variance in part sizes. Any variation in the part size, even within the tolerance, can cause the part to be improperly located or bind between the duplicate locators. Besides these obvious problems, it is not cost-effective to use more locators than necessary.

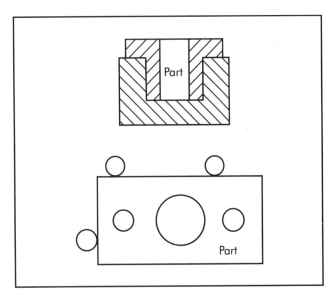

Figure 4-17. *Redundant locators.*

## Locational Tolerances

Locational tolerance is one point that must always be considered when specifying locators for any workholder. As a general rule, locational tolerance should be approximately 20–50% of the part tolerance. Making the tolerance excessively tight only increases costs. Likewise, overly large tolerances can shorten the life of a workholder. The designer must balance the cost against the expected life of the tool and required accuracy of parts to determine a locational tolerance that will provide the required number of parts without excessive tooling costs.

# FOOLPROOFING

Foolproofing is the process of positioning locators so a part will only fit in the workholder in its proper position. This is accomplished by a number of different means. The simplest and most cost-effective method is positioning a foolproofing pin to prevent incorrect loading (*Figure 4-18*). In any case where a part has the possibility of being loaded improperly, use a foolproofing device.

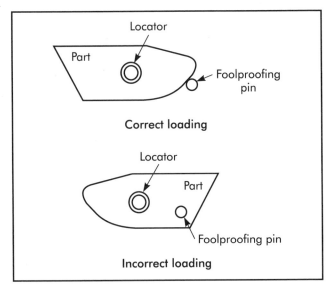

Figure 4-18. *Foolproofing.*

## Locator Types

Locators are made in a wide variety of shapes and sizes to accommodate the large range of workpiece configurations. In addition, commercial locators are available in many styles to suit their ever-increasing use. To properly design and specify an appropriate locator, the designer first must be familiar with the different types of locators commonly used in jig and fixture applications.

### External

External locators are used to position a part by its external surfaces. These locators normally are classified as either locators or supports. Locators are elements that prevent movement in a horizontal plane. Supports are locating devices positioned beneath the workpiece and prevent downward movement of the part and rotation around the horizontal axes. The two basic forms of external locators or supports are fixed or adjustable.

**Fixed locators.** These are solid locators that establish a fixed position for the workpiece. Typical examples of fixed locators include integral, assembled, pins, V-type, and locating nests.

*Integral.* Integral locators are machined into the body of the workholder (*Figure 4-19*). In most instances, this type of locating or supporting device is the least preferred. The principal objections to using integral locators are the added time required to machine the locator and the problem of replacing the locator if it wears or becomes damaged. Another drawback to using integral locators is the additional material required to allow for the machining of the locator.

Figure 4-19. *Integral locators.*

*Commercial Pin.* Commercial pin locators are made in two general styles—plain and shouldered. Their ends are made in either round, flat, or bullet shapes and facilitate easy loading and unloading of parts. These locating pins are normally made between 0.0005–0.0020 in. (0.013–0.051 mm) under size to prevent jamming and binding in the located hole. The installed end of these pins generally is 0.0625 in. (1.588 mm) smaller than the location end to prevent improper installation.

Commercial locating pins also are made in press-fit and slip-fit styles. Press-fit pins are installed directly into the tool body. Slip-fit pins are used along with liner bushings, which are installed into the tool body. A lock screw also is used to hold the pin in place.

*Assembled.* Assembled locators are similar to integral locators in that they both must be ma-

chined. However, these locators have the advantage of being replaceable. Assembled locators may be used as locators for supports (*Figure 4-20*). Since they are not part of the major tool body, using assembled locators does not require additional material for the tool body. Assembled locators frequently are made of tool steel and hardened to prevent wear.

Figure 4-20. *Assembled locators.*

*Pins.* A locating pin is the simplest and most basic form of locating element. These locators may be made in-house from steel drill rod or purchased commercially. Commercial locating pins are available in several styles and types (*Figure 4-21*). Standard hardened dowel pins are another form of commercial component frequently used for locating devices. Due to their simplicity, easy application, and replaceability, round pins are the most commonly used form of locating device.

The location and number of locating pins generally is determined by the size, shape, and configuration of the part, but in most cases the 3-2-1 principle is applied. In this principle there are three pins placed under the part to control five directional movements, and three more pins placed perpendicular to the base to control four more directional movements. The remaining two directional movements are controlled by the clamping element.

Consideration other than location of the workpiece will often affect the number of locating pins used. The workholding device must be designed to clamp the workpiece securely and support it to resist the forces generated by machining. If the operation performed applies considerable force, the workpiece may spring out of shape. Thus, the locating elements must be designed to provide

Figure 4-21. *Locating pins.*

adequate support for the workpiece against the forces acting upon it.

Many workpieces are essentially flat, or have a flat surface that can be used for locating purposes. These are commonly located by placing the workpiece on a plane surface to restrict it in five directional movements, as illustrated in *Figure 4-22*. The addition of six locating pins, *A, B, C, D, E,* and *F* will restrict it in six more directions. The workpiece can move only in an upward direction. This final movement is restrained by a suitable clamp having a plane surface parallel to the one on which the workpiece is placed. When this method of location is used, a great deal of planning must be done before specifying the locator positions. Since the workpiece is confined on all sides, the designer must specify locator positions that will not interfere with each other. Locators must be positioned to eliminate the possibility of duplicate location.

If the workpiece must be held in a vertical position, the same principle of clamping and supporting between two plane surfaces may be used. Again, this will restrict motion in six directions. Of the six remaining directions, only five must be restricted by locating pins (*Figure 4-23*). The four pins *A, B, C,* and *D* restrict motion downward, to the left and right, and both clockwise and counterclockwise around axis *X*. Gravity may be used to locate the workpiece and restrict movement during machining or other operations to be

Figure 4-22. *Simple workholder made of plane surface and pins.*

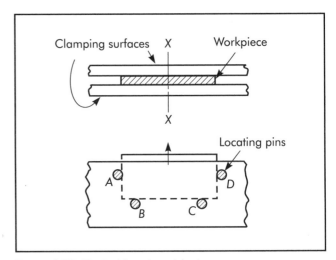

Figure 4-23. *Vertical locating with pins.*

performed. Here again, locators must be positioned to minimize the chance of duplicate location.

***V-type.*** A cylinder, like the prism, also has six degrees of freedom. The cylinder in *Figure 4-24* is free to move in two opposed directions along each axis and rotate both clockwise and counterclockwise around each axis. To accurately locate a cylindrical workpiece, it must be confined to restrict motion in each of its directions.

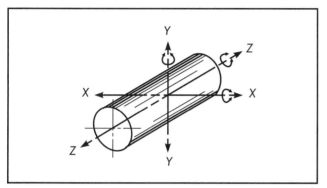

Figure 4-24. *Six degrees of freedom and 12 directions of a cylindrical workpiece.*

*Figure 4-25* shows a cylinder placed in the intersection of two perpendicular planes. The base plane is parallel to the *X* and *Z* axes, and the vertical plane is parallel to the *Y* and *Z* axes. The horizontal plane restricts movement in the two rotational freedoms around the *X* axis and the downward freedom along the *Y* axis. The vertical plane restricts the two rotational movements around the *Y* axis and the leftward movement

along the *X* axis. The pin that forms the end stop restricts one directional movement (forward movement along the *Z* axis). This corresponds to the basic 3-2-1 method of location used for the prism, but it restricts movement only in seven directions. The cylinder can move backward along the *Z* axis; in addition, it is free to rotate clockwise and counterclockwise around the *Z* axis.

Rotation around the *Z* axis can be restrained by clamping friction applied against the V formed by the two planes. This does not locate in a definite angular position about the *Z* axis and, therefore, cannot be considered true locating. No provision has been made to accurately locate a particular point on the cylindrical surface.

Locating a cylinder in a V places its longitudinal axis in true location. Often, this is sufficient for the operation to be performed. In addition, the basic principle of V location can be applied to workpieces that are not true cylinders but contain cylindrical segments.

A single V locator provides two points for locating where the cylindrical end of the workpiece is tangent to both sides of the V. The equivalent of three points in a base plane and a radial locator are required for complete location of the workpiece. In *Figure 4-26*, a workpiece with two cylindrical ends is confined by means of two V locators. The movable V locator serves only to locate one point, the center of the cylindrical portion.

The included angle between the two surfaces of a V locator governs the positions of circular sections of varying diameters (*Figure 4-27*). A V

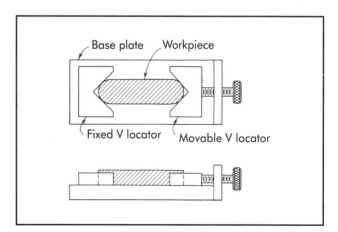

Figure 4-26. *Workholder with multiple V locators.*

Figure 4-25. *Seven directional movements arrested by V locator with stop pin.*

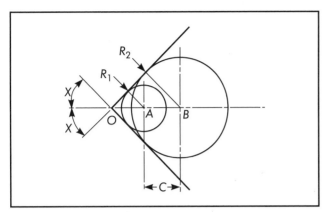

Figure 4-27. *Positions of circular sections of varying diameter in a V locator.*

with an included angle of $2X$ locates a circle of radius $R_1$ with center $A$ and a circle of radius $R_2$ with center $B$. The distance between centers $= C$.

By similar triangles:

$$\frac{R_1}{OA} = \frac{R_2}{OB} = \frac{R_2}{OA+C} \qquad (4\text{-}1)$$

but $OA = R_1$ cosecant $X$

$C = $ cosecant $X(R_2-R_1)$

and $D_2 = 2R_2$; $D_1 = 2R_1$

therefore,

$$C = \frac{\text{cosecant } X(D_2 - D_1)}{2} \qquad (4\text{-}2)$$

Consequently, the distance between positions of any two diameters in a V varies as one-half the included angle of the V. The smallest variation occurs when $X = 90°$ and where cosecant

$$X = \frac{0.5}{2}$$

However, as $X$ approaches 90°, there is less inclination for the circular section to seat positively in the V and more difficulty in retaining it. Note that with $X = 90°$, $2X = 180°$ or a straight line. The best compromise is achieved with $X = 45°$ and the included angle of the V is 90°.

Irregularities in the circular section of a workpiece, or a chip lodged between the workpiece and the V locating surface can introduce errors of lo-

cation. The included angle of the V has a definite influence on the effect of such displacement. In *Figure 4-28*, $E$ is the displacement caused by a rough surface or chip. The circular section of the workpiece may be considered to rest on another side of the V indicated by a dotted line. The displaced side of the V, shown by the solid line, forms a new V with the opposite side to define the displaced position of the circular section. The change in the position of the center of the workpiece is identical to the shift in the apex from the original to the new V. The original V is $BOC$ and the new V is $AO_1C$. The axis of the workpiece is displaced by the distance $OO_1$. For a constant displacement $E$, $OO_1$ is a minimum when $2X = 90°$.

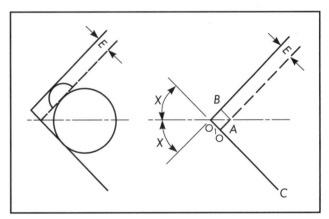

Figure 4-28. *Influence of the included angle or errors of location.*

Consideration must be given to the axis of the rotating tool and its relationship with the position of the V locator. The V locates the longitudinal axis of the cylindrical workpiece. When work is done perpendicular to this axis, the position of the V locator should be arranged to keep displacement of the workpiece to a minimum. In *Figure 4-29a*, a cylindrical workpiece is placed in a V locator so a hole can be drilled perpendicular to the longitudinal axis. Any variation in the diameter of the workpiece will cause a displacement in the location of the vertical axis. The drill bushing, however, remains in its original position and the drilled hole will deviate from its required position by the amount of the displacement. In *Figure 4-29b*, the V locator is positioned so its axis is parallel to that of the drill bushing. Varia-

Figure 4-29. *Minimizing error by proper placement of a V locator.*

tion in the diameter of the workpiece will cause no displacement of the vertical axis, and the drilled hole will not deviate from its required position.

***Locating nests.*** The nesting method of location features a cavity in the workholding device into which the workpiece is placed and located. If the cavity is the same size and shape as the workpiece, this is an effective means of locating. *Figure 4-30* illustrates a nest that encloses the workpiece on its bottom surface and around the entire periphery. The only degree of freedom remaining is in an upward direction. A similar nest can be used to locate cylindrical

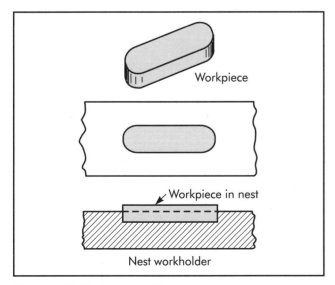

Figure 4-30. *Nest-type workholder.*

workpieces. Cavity nests are used to locate a wide variety of workpieces regardless of the complexity of their shape. All that is necessary is to provide a cavity of the required size and shape. No supplementary locating devices, such as pins, are normally required.

The cavity nest has some disadvantages. Since the workpiece is completely surrounded, often it is difficult to lift out of the nest. This is particularly true when no portion of the workpiece projects out of the nest to afford a good grip for unloading. The workholding device can, of course, be turned over, and the workpiece shaken out. When the workpiece tends to stick, an ejecting device, such as an ejector pin or pins, can be incorporated in the workholder. This, however, adds time to the processing. Another disadvantage is that the operation performed may produce burrs on the workpiece that tend to lock it into the nest. In this case, the workpiece must be pried out or an ejector must be provided. Chips from the cutting operation may lodge in the nest and must be removed before loading the next workpiece. Any remaining chips may interfere with the proper positioning of the next workpiece.

To avoid the disadvantage of a cavity-type nest, partial nests often are used for locating. Flat members, shaped to fit portions of the workpiece, are fastened to the workholder to confine the workpiece between them. *Figure 4-31* shows two partial nests, each confining one end of a bow-shaped workpiece. Each nest is fastened to the

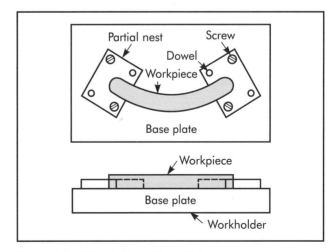

Figure 4-31. *Workholder with partial nests.*

flat supporting surface of the workholder with two screws. Accurate positioning of the nests is ensured by dowels that prevent each nest from shifting from its required position.

Partial nests eliminate many of the disadvantages of cavity nesting. Since they do not have the entire contour of the workpiece, they require less time to make. They do not completely confine the workpiece, so it is easily lifted out of the workholder. Normally, an ejecting device is not required. The cavity in each nest is open at one end to permit easy chip removal.

With the development of plastic casting materials, making cavities for locating workpieces has been simplified. The casting material can be poured around a workpiece prototype. After solidification and curing, the workpiece is removed. The resulting cavity becomes the nest for locating workpieces during the production operation. This method is much simpler than machining the cavity from a solid piece of material, and requires considerably less time, particularly when the workpiece is complex.

Many varieties of plastic materials are available, some with fillers of solid material, including steel or aluminum, to increase strength. Those with metallic fillers may be sawed, milled, drilled, and tapped after curing to a solid state. This machinability makes it easier to fasten the poured plastic cavity to the workholding device and eases alteration to accommodate future changes in the workpiece. Selection of the proper plastic depends on the forces generated by the operation to be performed, the quantity of parts to be made, and the effects of shrinkage of the material as it solidifies and cures.

Since plastic materials do not possess the strength of steels, plastic cavities often are reinforced with steel members. They increase resistance to tool forces, but do not materially increase the wear resistance of the plastic nesting surface. Consequently, plastic cavities generally are not used when large quantities of parts must be made. Also, if the plastic material used has a high degree of shrinkage, the resulting cavity may be too small to adequately locate the work. This, however, is not a critical factor since continuing improvements in the formulation of plastic compounds have consistently reduced inherent shrinkage.

Since the workpiece itself is required for the preparation of a nesting cavity, this type of workholder cannot be completed in advance of the production process. Of course, a prototype workpiece may be used, but making the prototype costs time and money so that the primary advantage of the cast locating cavity is lost. Therefore, the cavity is not usually made until production parts are available. This, however, does not usually cause a serious delay, since a cast cavity can be poured and cured in a few hours and be ready for use the following day.

A minimum-size workpiece must not be used for making the cavity since it will not accept parts that vary toward the maximum dimension. When the contours of the workpiece permit, its surfaces can be built up to increase vital locating dimensions to ensure casting a cavity large enough to adequately locate a maximum-sized part. Often, this can be done by applying strips of thin masking tape to the proper surfaces of the workpiece.

Experience has developed the casting process of making locating cavities to the extent that a complete workholding device can be poured in one piece. *Figure 4-32* illustrates a plate-type drill jig made this way. The workpiece containing the required holes is laid on a flat surface. Round pins are pressed into the holes and project upward to locate the drill bushings, whose outer periphery is serrated to ensure a firm grip in the

Figure 4-32. *Cast plastic, plate-type drill jig.*

cast plastic. A dam may be made from four rectangular bars placed around the workpiece. The pins are then removed, and the drill jig is ready for use.

Other materials frequently used for casting locating nests are low-melt alloys. These bismuth, lead, tin, and antimony alloys are ideal for many difficult workholding and locating problems. Quite often, low-melt alloys are used for applications where plastics cannot be used. One principal advantage to using low-melt alloys rather than plastic is their ability to be reused.

In use, the part is first positioned in a container. The container can be made of almost any material, such as sheet metal, low-carbon steel, or aluminum. Once positioned, the molten alloy is poured into the container and allowed to solidify. The part then is removed, the nest cleaned, and all burrs removed. If the part has a configuration that prevents its removal from the cast alloy, it is machined in the encasing alloy and, when finished, the alloy is melted off the part. Frequently, additional materials, such as metal filings, ball bearings, or similar materials, are added to the alloy to add wear resistance and permit the nest to hold up under longer production runs.

Both plastic and low-melt alloys provide cost-effective alternatives to expensive machined nests. The low initial costs and adaptability of these materials make them suitable for a wide range of workholding applications.

**Adjustable locators.** These are movable and frequently used for rough-cast or similar parts with surface irregularities. Examples of adjustable locators are threaded, spring pressure, and equalizing. Adjustable locators are used in conjunction with fixed locators to permit variations in part sizes while maintaining a fixed relative position of the part against the fixed locators.

Adjustable locators are widely used for applications where the workpiece surface is irregular or large variations between parts make solid locators impractical. The principal type of adjustable locator is the threaded style (*Figure 4-33*). In some cases, this type of locator also can be used as a clamping device rather than a locator (shown). However, this type mostly is used as a locator.

Figure 4-33. *Threaded adjustable locator.*

When adjustable locators are specified for a workholder, the position of the locator is not as critical as with solid locators, so the relative cost is greatly reduced. Frequently, adjustable locators are used as solid locators simply by adding a locknut, or screw, to secure the adjusting screw (*Figure 4-34*).

Figure 4-34. *Adjustable locators with locknut or screw.*

Adjustable supports simply are adjustable locators positioned beneath the workpiece. Threaded supports are used along with solid supports to permit easy leveling of irregular parts in the workholder (*Figure 4-35*). Spring supports also are used with solid supports to level the workpiece, but rather than using threads to elevate the locator a secondary threaded element, such as a thumbscrew, is used to lock the position of the spring support (*Figure 4-36*). Equalizing supports are used to insure constant contact of the supports and workpiece. These supports are normally self-adjusting. That is, as one is depressed, the other rises (*Figure 4-37*).

Figure 4-35. *Threaded type adjustable supports.*

Figure 4-36. *Spring-type adjustable supports.*

Figure 4-37. *Equalizing-type adjustable supports.*

## Sight

Sight locators are an effective means of locating sand castings and similar rough and irregular parts for first-operation machining. These elements, while not locators in the conventional sense, are well-suited to applications where machined details must be in an approximate area rather than at a specific point. Sight location uses lines, slots, and holes in the workholder body to position the workpiece in an approximate position for machining. *Figure 4-38* shows two examples where sight locators are used to position a workpiece. In most cases, the part is simply centered between the sight locators and clamped before machining.

## Internal

*Internal locators* are locating features, such as holes or bored diameters, used to locate a part by internal surfaces. The two basic forms of internal locators are fixed-size and compensating. Fixed-size locators are made to a specific size to suit a certain hole diameter, and include machined, commercial pin, and relieved locators. Compensating locators generally are used to centralize the location of a part or allow larger variations in hole sizes. The two typical forms of compensating locators are conical and self-adjusting.

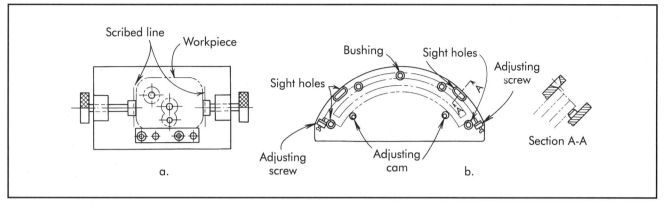

Figure 4-38. *Sight locators:* (a) *by scribed lines and* (b) *by sighting holes.*

**Machined internal.** Machined internal locators are made to suit special-size hole diameters. In most cases, machined locators are made for larger hole diameters. The exact form and shape of these locators normally are determined by the part to be located. They are generally machined to size and then attached to the tool using screws and dowels (*Figure 4-39*). In cases where small-diameter locating pins are required, materials such as drill rod or commercial drill blanks are frequently purchased in the desired diameter and then cut to the required length. Both drill rod and drill blanks normally are available in most standard sizes.

When round plugs are used in holes for locating, there is a tendency to stick when a close-fitting workpiece is applied. A plug of diameter $d$ extending from a faceplate (*Figure 4-40a*) will not stick in the hole of the diameter of a workpiece if the plug length:

Figure 4-39. *Machined internal locator.*

$$L = \sqrt{2WC}$$

where:

  $L$ = plug length, in. (mm)
  $W$ = outside diameter of the workpiece concentric with the hole
  $C = (D - d)$
  $D$ = hole diameter, in. (mm)
  $d$ = plug diameter, in. (mm)

A projection of diameter $d$ on a workpiece will not stick in a locating hole (*Figure 4-40b*) of diameter $D$ if the hole is relieved or countersunk so the length of engagement between hole and workpiece is

$$E = L - m \leq \sqrt{WC} \qquad (4\text{-}4)$$

where:

  $E$ = length of engagement, in. (mm)
  $m$ = depth of countersink, in. (mm)

A method of reducing the tendency for workpieces to stick on a locating plug extending from a faceplate is to relieve the plug by cutting away three equal segments (*Figure 4-40c*). A workable value for the angle, $\sigma$, is 15°, which results in $m = 0.35d$. A plug cut away in this manner will not stick if its length $L = 2.4 (2a + .085d)(D - d)$. One disadvantage is that a workpiece can be displaced in three directions on a relieved plug, and the extra error in inches (millimeters) of location as compared with a full round plug is $0.207 (D - d)$, or about a 20% loss in locating accuracy.

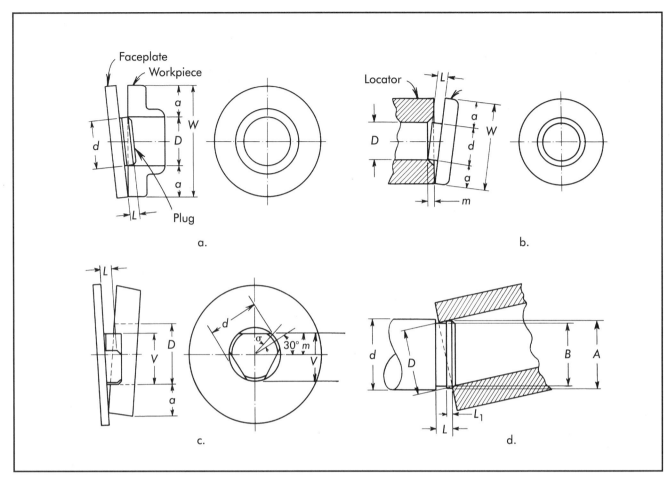

Figure 4-40. *Nonsticking locator design.*

An aligning groove may be put on the end of a plug of diameter $d$ of unlimited length $L$ to keep it from sticking when inserted in a hole of diameter $D$ (*Figure 4-40d*). Workable dimensions for such a design are:

$$L = \mu d$$
$$L_1 = 2AC_2 = \mu d$$
$$B = 0.95d$$

where:

$A$ = plug alignment (pilot) diameter, in. (mm)
$C_2$ = clearance between the plug pilot and hole
    = $(D - A)$ in. (mm)
$\mu$ = coefficient of friction between plug and hole surfaces, usually about 0.15 to 0.25 for steel

The pilot diameter may be made as convenient between size $d$ and $A = (2d_2/D) - d$.

## Relieved

Relieved locators, as their name implies, are designed to minimize the area of contact between the workpiece and locating pin. This reduces the chances of the locator sticking or jamming in the part. *Figure 4-41* shows several examples of relieved locators. The most commonly used form of relieved locator is the diamond pin.

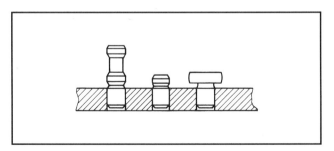

Figure 4-41. *Relieved locators.*

**Diamond pins.** Diamond pins are used for radial location in conjunction with round locating pins. It is possible to accurately locate a workpiece with two round pins, but allowances must be made for the variations encountered in hole sizes and locations. For instance, the distance between holes $A$ and $B$ (*Figure 4-42*) will vary to the extent of tolerance $X$. Similarly, the distance between pins $A$ and $B$ in the workholder has a tolerance $Y$. For accurate location, there should be an allowance between pin $A$ and hole $A$ of only a few ten-thousandths of an inch (micrometers). But if pin $B$ is a complete cylinder (same as pin $A$), the allowance between pin $B$ and hole $B$ must be at least as great as the sum of tolerances $X$ and $Y$. This is necessary for the pins to engage holes within the permissible tolerance $X$. The diameter of pin $B$ can be calculated by:

$$B = H_D - P_T \qquad (4\text{-}6)$$

where:

$B$ = diameter of pin $B$, in. (mm)
$H_D$ = minimum hole diameter, in. (mm)
$P_T$ = sum of the center-to-center pin tolerance

Extreme cases occur when both hole and pin center-to-center dimensions are at maximum or minimum conditions. As a result, there will be a large allowance between the hole and pin at $B$ in the $Y$ direction. This will permit an undesirable amount of radial rotation around the axis $A$ and defeat the purpose for which pin $B$ is intended.

To achieve more accurate radial location, $B$ may be a diamond pin as shown in the inset in *Figure 4-42*. It is relieved on two sides to allow for variations in the $C$ direction, and has two cylindrical portions to locate the hole in the $Y$ direction. The minimum radial movement of the workpiece occurs when the diameter of the cylindrical portion of the pin is smaller than the diameter of the hole by the allowance necessary to slip the minimum-size hole over the pin. When positioning these locators in the tool body, the bearing surface of the diamond pin must be positioned to restrict movement of the part.

In some cases, the part may be completely located using diamond pins. In *Figure 4-43*, the part is completely located by using two diamond pins placed to restrict the rotational movement of the part. When used in this fashion, the pins must be positioned to restrict the movement permitted by the other pin.

**Floating locating pin.** One other style of locating pin that will correct slight differences between locating holes is the floating locating pin

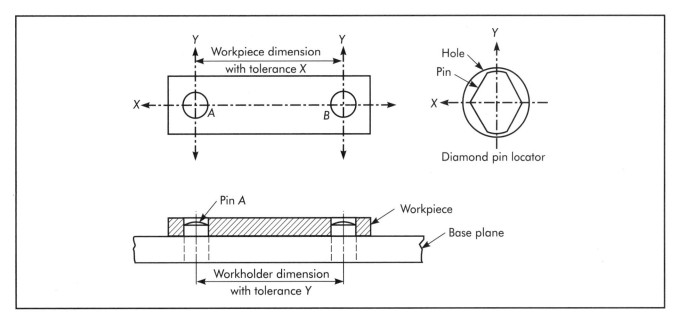

Figure 4-42. *Radial location with internal pins or plugs.*

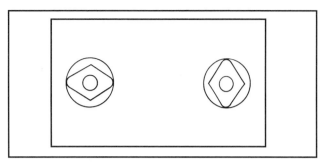

Figure 4-43. *Locating completely with diamond pins.*

(*Figure 4-44*). This pin provides precise location in one axis while allowing up to 0.125 in. (3.18 mm) movement in the perpendicular axis. The body of the locator is pressed into the tool body. It is referenced to the tool body and both the fixed and movable axes with a roll pin.

The floating locating pin generally works like a diamond pin. Due to the increased movement, however, this pin should be used for parts with somewhat looser locational tolerances of the mounting holes. The floating locating pin permits greater variation than that typically allowed by a diamond pin.

Figure 4-44. *Floating locating pin.* (Courtesy Carr Lane Manufacturing Co.)

In *Figure 4-45*, the floating locating pin is used along with a round locating pin. The part is first positioned on the round locator, then located on the floating pin locator. This pin, when positioned as shown, prevents any radial movement about the round locator. It also compensates for differences of up to 0.125 in. (3.18 mm) in mounting-hole positions.

Figure 4-45. *Floating locating pin used in combination with round locating pins.* (Courtesy Carr Lane Manufacturing Co.)

## Conical

Conical locators are centralizing locators that compensate for variations in part sizes, as well as centering a part in the workholder. The most efficient types of conical locators are spring-loaded or threaded (*Figures 4-46a* and *4-46b*). Conical locators, while normally used for internal location, also may be used for external location with a conical cup (*Figure 4-46c*).

## Self-adjusting Locators

Self-adjusting locators are used in applications such as sand casting, where there is great variation in the size of the holes to be located. These locators can be made in a wide variety of styles. *Figure 4-47* illustrates one example of a self-adjusting locator that can be used for internal location.

## Spring Locating Pins

Unless parts are properly positioned against locators, errors will result no matter how well a locating system is designed or made. One type of

Figure 4-46. *Conical locators.*

Figure 4-47. *Self-adjusting locator.*

locating device that helps reduce these locational errors is the spring locating pin (*Figure 4-48*).

Spring locating pins are designed to push the part against the fixed locators. This will ensure proper contact during the clamping operation. Although spring locating pins are not actually locating devices, they help reduce locational errors by correctly positioning the part against the locators. In addition, these pins can eliminate the need for a third hand when positioning and clamping some parts. Their small size and compact design makes them useful for smaller parts or confined space. A protective rubber seal around the contact pin helps seal out debris and coolant.

Spring locator pins may be installed directly in a hole or mounted in an eccentric liner (*Figure 4-49*). The liner permits pin adjustment to suit parts with looser tolerances (*Figure 4-50*).

When positioning a flat plate (*Figure 4-51*), the first step is to position the part over the workholder. The part then is placed against the solid locator and pushed down against the spring pin. When seated, the spring locating pins push the workpiece against the solid locator to ensure proper contact. These locating pins are well-suited for a variety of applications and part shapes.

**Spring-stop buttons.** *Figure 4-52* illustrates spring-stop buttons, another spring-loaded

Figure 4-48. *Spring locating pin.* (Courtesy Carr Lane Manufacturing Co.)

Figure 4-49. *Spring locating pin mounted in an eccentric liner.* (Courtesy Carr Lane Manufacturing Co.)

Figure 4-50. *Eccentric liner permits adjustment of the spring locating pins for loosely toleranced parts.* (Courtesy Carr Lane Manufacturing Co.)

Figure 4-51. *Positioning a flat plate.* (Courtesy Carr Lane Manufacturing Co.)

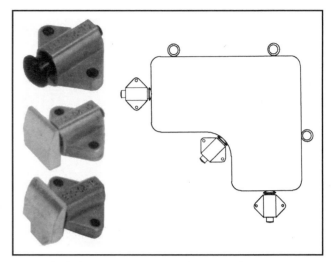

Figure 4-52. *Spring-stop buttons.* (Courtesy Carr Lane Manufacturing Co.)

workholding device. They work much like the spring pins, but are designed for larger parts or where more force is needed. Spring-stop buttons are made with three different contact faces. The first is a spherical button contact. The other two have flat contacts. Flat face contacts are made with or without a tang (*Figure 4-53*).

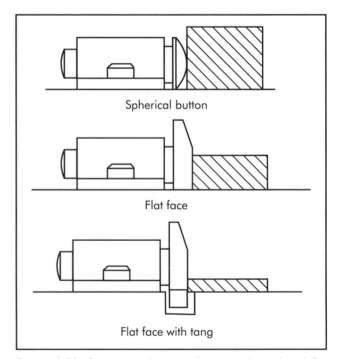

Figure 4-53. *Spring-stop buttons showing spherical and flat contacts.* (Courtesy Carr Lane Manufacturing Co.)

## CHIP AND BURR PROBLEMS

Chips, burrs, and dirt on locating surfaces cause wear and disturb proper location. Every means must be provided to keep locating surfaces and points free from foreign matter. There are three general methods of chip and dirt control:

- make locators easy to clean;
- make them self-cleaning, and
- protect them.

For easy cleaning, make locators as small as possible and consistent with adequate wearing qualities (rest buttons are popular). Jigs and fixtures should be as open as possible so supports are readily accessible and visible. Raise supports above surrounding surfaces so chips fall or can be readily swept off. Provide easy exit or passage avenues for chip ejection. Avoid pockets or obstructions where chips can collect and be difficult to clear.

Self-cleaning locators can have sharp edges or grooves that scrape dirt off of the workpiece surface as they slide across it. Relief around locating surfaces is essential as a means of escape for unwanted chips and dirt (*Figure 4-54*). For corner relief, suitable recessor grooves are provided so dirt and chips do not pack into corners and burrs do not bear against locating surfaces.

Fixed wipers may push chips along as a fixture is traversed on the table of a machine tool, or coolant may flush them away. Indiscriminate use of compressed air for blowing chips has its drawbacks because chips can be quite harmful when blown into ways and other bearing surfaces of machine tools. Shields and guards control and gather blown chips.

The drill jig shown in *Figure 4-55* automatically cleans the jig and locating points prior to loading. An air valve is actuated by contact of the jig top plate in the full-open position. The air blast is shut off as the top plate is lowered for each cycle. The chips are stopped by the shield. An air blast in front of the cutter in *Figure 4-55b* removes the chips in a milling operation. Continuous airflow of sufficient volume and pressure blows the chips into the catcher.

Suction is a means of removing light, discontinuous chips from the tool work area, particularly in dry grinding. Sometimes it is applicable when milling nonmetallic materials, such as phenolic and other compositions. A suitable mesh screen is provided at the mouth of the suction tube to keep clothes and other objects from entering.

Gravity slides can be used to control chips in a high-production setup for milling cast-iron parts. Machine vibration induces the chips to slide into the trough between table and column where hinged pushers, fastened to the table, push the chips along. The chips accumulate in barrels at both ends of the trough.

A locating surface should be entirely covered by a workpiece when the fixture is loaded. If part of the locator is exposed, dirt and chips can collect on it during cutting and may move over the locator as the workpiece is removed. Bearing surfaces, such as ways, and indexing mechanisms also should be protected from dirt and chips, which can cause excessive wear. Thus, slides, indexing pins, and buttons should be enclosed.

A burr that is raised on the work at the start of a cut is termed a minor burr, and at the end of a cut is a major burr. Jigs should be designed so that removal of the workpiece is not hindered by burrs. In *Figure 4-56*, work removal does not tend to shear the burrs since it parts the work and jig directly. In contrast, if the work must be slid across the burr, sticking complicates removal. To avoid this, suitable clearance grooves or slots should be provided (*Figure 4-57*).

## CLAMPING PRINCIPLES

For a specific operation, the selection of general clamping—simple hand-operated clamps, quick-acting hand-operated clamps, power-operated clamps, etc.—should primarily be a function of operation analysis. This selection is based on an effort to balance the cost of the clamp against the cost of the operation to obtain the lowest possible total cost for both fixture and operation.

The purpose of a clamp is to exert force and press a workpiece against the locating surfaces and hold it there in opposition to the actions of cutting or other processing forces. Clamping forces should be directed within the locating area, preferably through heavy sections of the workpiece

Figure 4-54. *Examples of proper chip clearance around locating pins and blocks.*

Figure 4-57. *Use of burr relief for the minor burr.*

Figure 4-55. *Air blast used (a) to clean drill jig and (b) in front of milling cutter to remove chips.*

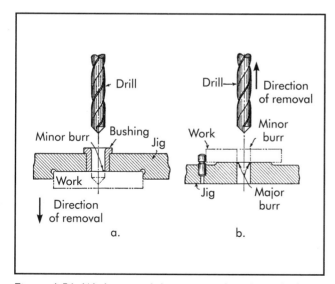

Figure 4-56. *Work removal does not tend to shear the burrs. The direction of workpiece removal eliminates sticking of minor and major burrs.*

directly upon locating spots or supports. Cutting forces should be taken by the fixed locators in a jig or fixture as much as possible, but generally some components of, or moments set up by, the cutting forces must be counteracted by clamping forces. To be effective, a clamp should be designed to exert a minimum force equal to the largest force imposed upon it in the operation. Sound judgment by the tool designer when applying these specific clamping principles to the job at hand is essential. In general, clamping arrangements should be as simple as possible.

The following design and operational factors should be considered:

- Simple clamps are preferred because complicated ones lose effectiveness as they wear. Complicated arrangements tend to lose their effectiveness as the parts become worn, necessitating excessive maintenance, which might readily offset the savings of a faster operation.
- Some clamps are more suitable for large and heavy work, others for small pieces.
- Rough workpieces call for a longer travel of the clamp in the clamping range, but clamps may be made to dig into rough surfaces to hold them firmly.
- The type of clamp required is determined by the kind of operation to which it is applied. A clamp suitable for holding a drill jig leaf may not be strong enough for a milling fixture.
- Clamps should not make loading and unloading the work difficult, nor should they interfere with the use of hoists and lifting devices for heavy work.

- Clamps that are apt to move on tightening, such as plain straps, should be avoided for production work.
- The anticipated frequency of setups may influence the clamping means. For example, the use of hydraulic clamps, even if simple and of low cost, might be inadvisable if frequent installation and removal of piping and valves are necessary.

## Tool Forces

A clear understanding of the direction and magnitude of cutting forces may eliminate the need to restrain all 12 directional movements of a workpiece. *Figure 4-58* shows how two pins and a table absorb the torque and thrust of a drilling operation. Although the workpiece is free to turn in a direction opposite to the torque, this freedom is insignificant unless a force is applied in that direction. No such force will be encountered in the planned drilling operation, and the remaining freedom need not be restrained. If, however, as part of the drilling operation, the spindle rotation is reversed for tool removal, such force will be encountered, and the freedom must be restrained.

Theoretically, there is no need to hold the workpiece down, as this is accomplished by the thrust of the drill. When the drill breaks through the thickness of the workpiece, an upward force may be created by interaction between the drill flutes and material remaining around the periphery of the hole. If there is no restraint in this upward direction, the workpiece may be lifted above the pins, creating a very dangerous condition. An upward force also may be produced when a drill or reamer gets lodged in a workpiece and the tool is to be withdrawn.

*Figure 4-59* shows how a workpiece must be restrained for tapping. Torque in both directions must be absorbed and the lifting pull of the tap must be resisted by some form of a hold-down, such as a clamp (not shown). For lead screw tapping, torque resistance is still required, but no hold-down is needed because the lead screw, having the same lead as the tap, eliminates all thrust. When two holes are tapped simultaneously on a two-spindle setup, each tap prevents workpiece rotation by the torque exerted by the other tap. Without a lead screw, a hold-down would still be needed to prevent accidental lifting of the workpiece by the spindles.

*Figure 4-60* shows another instance where the cutting force holds the workpiece, in this case

Figure 4-59. *Designing a workholder to resist torque and thrust in a tapping operation.*

Figure 4-58. *Pin-type drill fixture resisting torque and thrust.*

Figure 4-60. *Workholder for broaching operation.*

Figure 4-62. *Cutting force resisted only by friction.*

against the support plate of a horizontal broaching machine. The broach is guided in the support plate and, to some extent, the workpiece. The broach in turn also holds the workpiece. The cutting tool and cutting force both contribute to the workholding operation.

Once the designer of a workholder has identified the possible direction and magnitude of forces, he or she has two ways to restrain the workpiece to counteract these forces. Strength and rigidity of some part of the workholder is used, against which the workpiece rests or is forced by a clamp, screw, or wedge. The other way uses friction between the workpiece surface contacting a surface of the workholder under pressure.

*Figure 4-61* illustrates a workpiece held in a vise. The horizontal component of the cutting force is absorbed by the solid jaw of the vise. The vertical component is resisted by friction between the workpiece and the jaws. *Figure 4-62* shows the cutting force absorbed only by the friction between the jaws and the workpiece. Wherever possible, cutting forces should be opposed by the structure of the workholder and, preferably, by the strongest and most rigid parts of it. If necessary, a movable element may be used to absorb cutting forces, but only if properly designed for strength and rigidity.

Complete analysis of tool forces in a proposed operation will disclose which of the 12 directional movements must be restrained and to what extent. Quite often, tool forces are of such magnitude and direction that a workpiece may be dislodged or moved from its required location. If the locating elements of a fixture cannot ensure adequate restraint, it may be necessary to clamp the workpiece against them.

## Clamping Forces

Clamps hold a workpiece against a locator. Perhaps the most common application is the bench vise, where a movable jaw exerts pressure on a workpiece, thereby holding it in a precise location determined by a fixed jaw. The bench vise uses a screw to convert actuating force into holding force. *Figure 4-63* shows a number of commonly used mechanical methods for transmitting a multiplying force.

Clamping forces applied against a workpiece must counteract tool forces. Having accomplished this, further force is unnecessary and may be detrimental. The physical characteristics of the workpiece greatly influence clamping pressure. Hard vise jaws can crush a soft, fragile workpiece.

Figure 4-61. *Cutting force resisted by solid jaw of vise.*

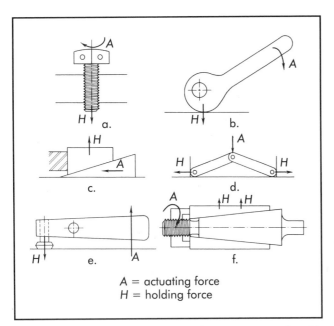

Figure 4-63. *Mechanical methods of transmitting and multiplying force: (a) screw, (b) cam, (c) wedge, (d) toggle linkage, (e) lever, (f) combined screw and wedge.*

The clamping pressure must hold, but not damage, deform, or impose too great a load on the workpiece.

The direction and magnitude of clamping pressure must be consistent with the purpose of the operation. An example is the boring of a precise round hole with the workpiece clamped in a heavy vise. Excessive clamping pressure can compress the workpiece. The bored hole may be perfect in size and roundness while the workpiece is compressed. The release of clamping pressure might permit the workpiece to return to its normal condition, and the hole might then be offsize and elongated. Another example of excessive or misdirected clamping pressure may be found in a cutoff operation. If a workpiece clamped in a vise is to be cut between the jaws, in a direction parallel to the jaws, the removal of metal by the saw blade permits the remaining metal to flow into a cut area. This metal movement can effectively wedge and stop the saw blade.

Clamping pressure should not be directed toward a cutting operation, but should, wherever possible, be parallel to it. Clamping pressure should never be great enough to change any dimension of the workpiece.

## Positioning Clamps

Clamps must be positioned to contact a workpiece at its most rigid point. When possible, they should be located over a locator or support. In cases where a part cannot be clamped over a locator, a secondary support must be installed to counteract the clamping forces and prevent damage to the part. As shown in *Figure 4-64*, the flanged part is located by its center. If the part was clamped as shown in (*Figure 4-64a*), the part would distort. Therefore, a secondary support must be added (*Figure 4-64b*). This additional support provides the required backing to prevent distorting the part. Remember, when adding a secondary support, allow enough space between the support and the part to prevent redundant locating.

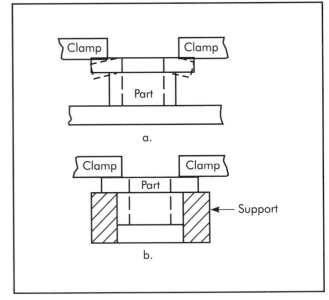

Figure 4-64. *Positioning clamps.*

When clamping large rings or similar parts, the location of the clamping devices is very important to prevent warping or distorting the part. *Figure 4-65* shows two methods of clamping this type of workpiece. If the part were clamped as shown in (*Figure 4-65a*), the clamping forces could easily distort the part. However, if the clamps are positioned as shown in (*Figure 4-65b*), the chance of part distortion is greatly minimized.

Figure 4-65. *Clamping large rings.*

## Rigid Versus Elastic Workholders

Workholders may be either rigid or elastic. Since there are no absolutely rigid bodies and materials, rigid means the holding elements are preset to a fixed position. *Figure 4-66* shows a screw holding a workpiece as an example of rigid workholding. There is some elasticity in the screw and nut, but it is not intentionally provided. *Figure 4-67* illustrates a pressure-supported piston,

Figure 4-66. *Rigid workholding.*

Figure 4-67. *Elastic workholding.*

which bears down on and holds the workpiece as an example of elastic workholding.

In *Figure 4-66*, a sideways shift will bring the screw out of contact with a workpiece that is not of uniform thickness, causing complete loosening. The workpiece, however, resists any upward force against the screw because the screw is a self-locking, mechanically irreversible element. In *Figure 4-67*, the piston clamp will continue to exert holding force in case of a sideways shift. An upward force is resisted only by hydraulic pressure exerted on the piston.

*Figure 4-68* shows examples of elastic workholding using self-contained, hydraulically operated clamping cylinders. The workholder may be clamped by hydraulic pressure and released by spring pressure (*a*), or may be held by spring pressure and released by hydraulic pressure (*b*).

Figure 4-68. *Elastic workholding with self-contained hydraulically operated clamp cylinders.*

Air pressure may be used, but usually requires considerably larger piston areas to obtain a sufficient holding force. The size of the cylinders may be reduced by the use of force-multiplying mechanical elements. Before elastic workholding devices can be used safely, forces on the work and their direction must be determined.

*Figure 4-69* compares a hydraulic mandrel for elastic workholding with a mechanical mandrel for rigid workholding. Hydraulic pressure is produced by a screw-piston arrangement. Line A-A traces a path through a solid workpiece, a solid expandable shell, and an elastic layer of a hydraulic compound. A similar path A-A on the split-collet expanding mandrel passes through only

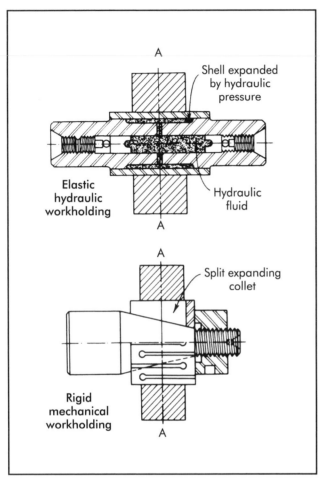

Figure 4-69. *Elastic and rigid mandrels.*

consecutive layers of rigid metal. This does not mean that the elastic hydraulic mandrel is not as positive as the rigid mandrel. In fact, the opposite may be true. The hydraulic mandrel's torsional stiffness may be greater and it may possess many other desirable features, such as higher inherent precision and less vulnerability to dirt.

## Clamp Types

There are several basic types and styles of clamps and clamping devices commonly used for jigs and fixtures. The specific type of clamp selected for a particular application normally depends on the type of tool, part shape and size, and the operation to be performed. Other considerations, such as speed of operation and permanence, are factors that must be considered for long or high-speed, high-volume production runs.

### Strap Clamps

Strap clamps are the simplest and least expensive type of clamping device used for jigs and fixtures. In *Figure 4-70*, the basic strap clamp consists of a bar, a heel pin (or block), and either a threaded rod or cam lever to apply the holding force. Additional accessories for this type of clamp include hand knobs and spherical seat nut and washer sets (*Figure 4-71*). Variations of the basic strap clamping system are illustrated in *Figure 4-72*. The force, fulcrum, and workpiece positions are altered to make differences in the level classes of the strap clamps. Other variations of this basic type of clamp include the latch strap, sliding strap, and hinged strap clamp (*Figure 4-73*).

### Screw Clamps

Screw clamps are a type of mechanical clamp that uses a screw thread to apply the holding force. The two types of screw clamps generally used for jig and fixture work are classified as direct-pressure and indirect-pressure clamps. Direct-pressure clamps use the pressure of the screw thread to

Figure 4-70. *Strap clamps.*

Figure 4-71. *Accessories for strap clamps.*

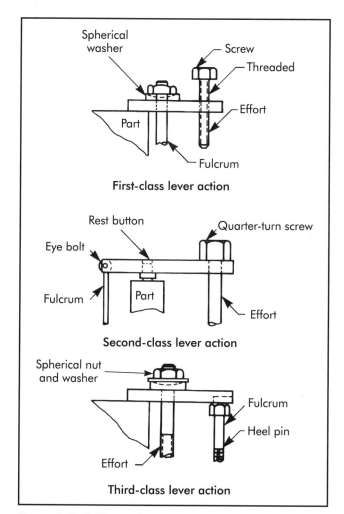

Figure 4-72. *Different classes of strap clamps.*

Figure 4-73. *Variations of the strap clamp.*

hold the workpiece. Typical examples are hook and swing clamps, and quick-acting hand knobs (*Figure 4-74*). Indirect-pressure clamps use the screw thread in combination with a secondary device to transmit the holding force of the thread. The arrangement in *Figure 4-75* is a typical application of an indirect-pressure screw clamp. One additional benefit of using indirect-pressure screw clamps is that the holding force can be

Figure 4-74. *Direct-pressure screw clamps.*

magnified by simply increasing the leverage of the holding member. This will permit the holding force to be two, three, or more times greater than the actuating force.

### Cam-action Clamps

Cam-action clamps frequently are used for fast-operating clamping devices. The three principle types of cam-action clamps used for workholders are the flat eccentric, flat spiral, and cylindrical (*Figure 4-76*). Flat eccentric cams operate on a high center principle and must be positioned exactly at the high center to hold properly. Flat spiral cams, on the other hand, have a locking range, which permits them to hold at any point within the range along the cam surface. Of these two styles, the flat spiral cam is the easiest and safest to use. All commercial cam-action clamps are made with this flat spiral cam design.

Both flat eccentric and flat spiral cams can be used for direct- or indirect-pressure applications (*Figure 4-77*). Indirect pressure is the most efficient and safest design for jig and fixture work. Direct-pressure cam clamps have a tendency to loosen during machining. Since indirect-pressure cams do not contact the workpiece, there is less chance for the vibration of the machining operation to loosen the cam.

Figure 4-75. *Indirect-pressure clamp.*

Figure 4-76. *Cam clamps.*

Figure 4-77. *Spiral cams used for direct (a) and indirect (b) pressure.*

Cylindrical cams also are used for workholding applications. With these clamps, the cam surface is generated on a cylindrical surface.

## Toggle-action Clamps

Toggle-action clamps are commercially available and work with four general clamping motions: hold-down, push-pull, squeeze, and straight-pull. The main advantages of using toggle clamps are their fast clamping and release actions, ability to move completely clear of the workpiece, and high ratio of holding force to actuation force. Several variations of toggle clamps are available to suit almost every workholding application. *Figure 4-78* shows several different styles of toggle clamps commonly used for jig and fixture work.

For all their advantages, standard toggle clamps have always caused problems because of their limited range of movement and inability to compensate for different thicknesses. Once set to a clamping height, the standard toggle clamp

Figure 4-78. *Toggle clamps.*

can only accommodate very slight changes in workpiece thicknesses. Larger variations usually require adjustment of the clamp spindle. Now, however, with the automatic toggle clamp (*Figure 4-79*), both these problems have been solved.

The standard toggle-clamp design uses a four-bar linkage arrangement with fixed pivots and levers to produce the clamping action. Though adequate for many clamping operations, this design does not permit the automatic height ad-

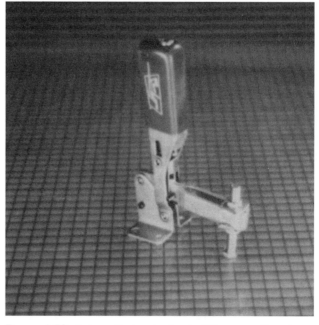

Figure 4-79. *Automatic toggle clamp.* (Courtesy Carr Lane Manufacturing Co.)

justment needed for some parts. Rather than using only fixed points, the automatic toggle clamp has a self-adjusting feature to readjust the clamp to different workpiece heights. With this clamp, the handle, normally a fixed component, has a variable length. The self-adjusting, self-locking wedge arrangement within the handle alters the pivot length to suit a variety of workpiece heights.

The clamp arm can accommodate differences in clamping heights of up to 15°. This results in a total automatic adjusting range of over 1.25 in. (31.8 mm). Added adjustment is permitted by manually moving the threaded spindle. Together, these adjustments result in a substantial amount of clamping capacity. To set up these clamps, the spindle extension first must be set to the average workpiece height (*Figure 4-80a*). Once set, the clamp automatically adjusts to a considerable range of workpiece variations (*Figures 4-80b* and *4-80c*). Adjustments to the clamping force are made by turning a screw located in the end of the handle. This adjustment is made with a standard screwdriver. The holding capacity of this clamp is 750 lbf (3,336 N). The total movement of the vertical handle is 60° and the arm moves a total of 105° to permit easy workpiece loading and unloading.

### Wedge-action Clamps

Wedge-action clamps use the basic principle of the inclined plane to securely hold and clamp a workpiece. The two principal types of wedge clamps are flat and conical. The flat-wedge clamp (*Figure 4-81*) works as a flat cam to provide hold-

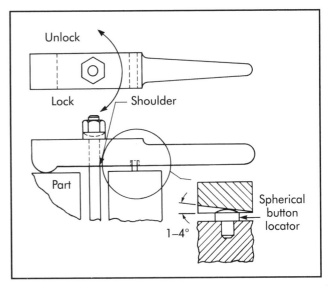

Figure 4-81. *Flat wedge clamp.*

ing force. The clamp is tightened and released by swinging the lever around the fulcrum pivot and contacting the inclined wedge against the spherical head pin. Conical wedges, or mandrels, normally are used in two styles, the solid mandrel and expansion mandrel (*Figure 4-82*). The wedging action of the conical surface directly or indirectly holds the workpiece.

### Specialty

In addition to the standard clamp variations there is a group of clamping devices not as easily classified, which use a variety of nonstandard clamping actions. Incorporating the concept of a universal clamping system for today's multipurpose machine tools offers an almost endless range

Figure 4-80. *Automatic adjustment to workpiece variations.* (Courtesy Carr Lane Manufacturing Co.)

Figure 4-82. *Conical wedges.*

of possible applications. Unlike the limited purpose clamping devices usually found on machine tools, these can perform a majority of the necessary workholding operations while substantially reducing setup and part changeover time.

The Mono-Bloc® clamp in *Figure 4-83* consists of a self-contained clamping unit that uses a worm and worm-wheel arrangement to perform the clamping action (*Figure 4-84*). This allows the clamps to have a wide clamping range and complete adjustability. These clamps are available in three different types or grades. The most common is the standard-duty model, as shown in *Figure 4-83*. Light-duty and heavy-duty models are also available.

The standard-duty clamp is available with two different arm lengths, and has an extension arm that can be installed on the basic clamp. The clamping range is increased by a series of available 3-in. (76.2-mm) riser blocks, which are installed under the base of the clamp and can be used to build it to a maximum height of 12 in. (304.8 mm). Both the basic clamp and riser blocks are assembled with a single screw. This allows the clamp assembly to be swiveled to any angle necessary to clamp the workpiece.

The Terrific 30® clamp (*Figure 4-85*) is another form of universal clamp. It is retractable and operates from a single point. By simply turning the handle 1-½ turns, both the extension and clamping action are achieved (*Figure 4-86*). Turning the handle in the opposite direction releases and re-

Figure 4-84. *Worm and worm-wheel clamp arrangement.* (Courtesy Royal Products)

Figure 4-85. *Terrific 30® clamp.* (Courtesy Royal Products)

Figure 4-86. *Extension of clamp arm to hold workpiece.* (Courtesy Royal Products)

tracts the clamp. This clamp may be mounted either directly to a machine table or on a series of riser elements. It also uses a single mounting screw, which allows the clamp to be positioned wherever necessary.

Edge-clamping small or multiple parts may be difficult. Often, standard-edge gripping clamps are too large to suit the workpiece or setup. This usually forces a design compromise. The workholder

Figure 4-83. *Mono-Bloc® clamp.* (Courtesy Royal Products)

must either use larger standard clamps, thus reducing the number of parts clamped, or expensive special-purpose clamps must be custom made to hold the workpieces. An alternative is the MITEE-BITE® clamping system (*Figure 4-87*). This clamping system has the design advantages of space-saving edge clamps with the cost-savings of standard clamps. The clamps combine the security of a screw clamp and speed of a cam-action clamp into a single clamping device. The main element is the eccentric socket-head cap screw and hexagonal brass clamping element (*Figure 4-88*). The cam-action clamping movement is achieved by turning the screw inside the clamping element. This hexagonal element is made of brass so there is little chance of damaging the clamped workpiece. For special-purpose applications, stainless-steel clamping elements also are available. The cam action of this system results in a significant mechanical advantage that can apply up to 4,000 lbf (17,793 N) of holding force directly to the workpiece.

Figure 4-88. *Eccentric socket-head cap screw inside hexagonal clamping element.* (Courtesy MITEE-BITE Products Co.)

The clamps may be mounted directly to the tool body, or in a tee nut for use on a machine table. For applications where the workpiece must be elevated off the mounting surface, a riser clamp variation (*Figure 4-89*) also is available. The riser clamp incorporates the benefits of the original basic clamp unit with the added flexibility of a small toe clamp. Rather than applying only lateral clamping force to hold the workpiece, the workpiece is held with both lateral and downward forces.

The basic clamps in the riser clamp set are mounted on a base element. This permits the

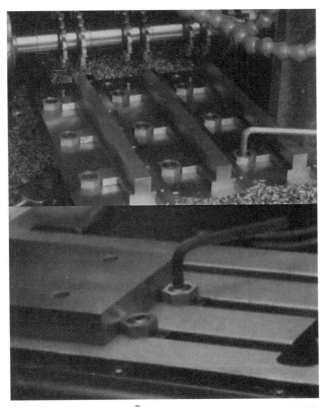

Figure 4-87. *MITEE-BITE® clamping system.* (Courtesy MITEE-BITE Products Co.)

Figure 4-89. *Riser clamp variation of MITEE-BITE clamp.* (Courtesy MITEE-BITE Products Co.)

clamps to be mounted directly in the tee slots of the machine table and provides the 10° clamping angle of the clamping element. The base element also establishes the mounting surface that holds the workpiece 0.50 in. (12.7 mm) off the machine table. The clamping element in the riser clamp is made of steel and has two clamping surfaces that can hold the workpiece. The first is a smooth surface that holds the workpiece without leaving clamp marks. The second is serrated to provide a better grip on the clamped surface. A clamp bar that works with the clamp to locate the opposite side of the workpiece during the clamping is included with the basic clamp. It also has an end-stop unit mounted at the end of the bar (*Figure 4-90*). This stop can be set to suit the desired location of the workpiece.

Figure 4-90. *Clamp bar with end-stop unit.*

## CHUCKS

Chucks are some of the most popular workholders used for jig and fixture work. These commercially made components allow a single workholder to service an infinite number of parts by simply changing or modifying the holding and locating surfaces of the tool.

## Operations

Chucking operations can be distinguished by the location at which they are performed on the workpiece. Common methods shown in *Figure 4-91* include (a) external chucking, (b) internal chucking, (c, d, k, l) endwise chucking, (e, f, g, h)

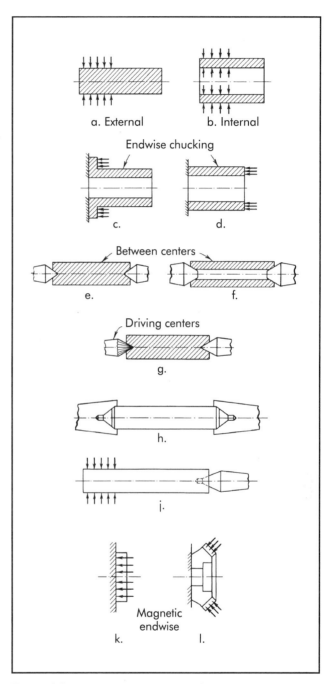

Figure 4-91. *Location and clamping of round workpieces.*

holding or driving on centers, and (j) combinations of these methods. Although different workholders may be used, the differences are primarily in name and appearance rather than function or mechanical principles employed.

Motion is imparted to the workpiece either by friction or positive means. In chucking, the jaw elements of the workholder bear on but do not positively engage the workpiece. The friction between the jaws and the workpiece rotates or drives the workpiece. Positive means for driving the workpiece can be provided by making use of its structure. Cutouts, keyways, gear teeth, or splines on a workpiece can be mated with matching elements on the workholder to give a positive drive. One distinguishing element can be found should failure occur: in a positive drive something breaks, while in a friction drive something slips.

## Nomenclature

A *chuck* is a workholder generally used for gripping the outside or end of a workpiece, and it is usually attached to a machine-tool spindle. An *arbor* is a workholder generally used for internal chucking, holding, or gripping, and is usually attached to a machine-tool spindle. A *mandrel* is a workholder that, like an arbor, is used for internal chucking, holding, or gripping. A mandrel is not generally as precise as an arbor, and is usually held between centers instead of being attached to the spindle.

*Range* is the amount of variation in workpiece diameter that can be accommodated by workholder expansion and contraction. *Interference* is the amount by which the chucking diameter in its holding condition (without the workpiece) differs from the diameter of the workpiece to be held. *Chucking area* is the area of a workholder that can be used for holding. Holding elements of workholders are called by various names depending on the type of workholder. They are the jaws on some chucks; fingers or prongs on some collets; expanding wedges or inserts on some arbors; expanding or contracting cylinders or sleeves on some special arbors and mandrels, and so on. These elements of the workholders are made to deflect by the actuating mechanism that transmits the force needed to operate the workholder. This force, in turn, may be applied manually or by power (pneumatic, hydraulic, or electric). Holding elements are backed up by the actuating elements; these are backed up by the operating mechanisms and forces.

## Lathe Chucks

*Lathe chucks* consist of a body with inserted workholding jaws that slide radially in slots and are actuated by various mechanisms such as screws, scrolls, levers, and cams, alone and in a variety of combinations (*Figure 4-92*). The number of jaws varies. Chucks in which all the jaws move together are self-centering and used primarily for round work. Two-jaw chucks operate somewhat like a vise, and may hold round and irregularly shaped workpieces by the use of suitably shaped jaws or jaw inserts. Independent jaw chucks permit each jaw to move independently for chucking irregularly shaped workpieces or to center a round workpiece. The accuracy of a chuck deteriorates with use from wear, dirt, and deformation caused by excessive tightening.

The jaws of most lathe chucks can be reversed to switch from external to internal chucking. Jaws may be adapted to fit workpiece shapes that are not round. The means of attaching a lathe chuck to different machine tools have been well standardized, so chucks made by different manufacturers can be easily interchanged.

In addition to their standard jaws, lathe chucks also may be fitted with a variety of special-purpose jaws to accommodate different types of workpiece surfaces and configurations. The principal types of chuck jaws used for these purposes are called *soft jaws* and generally are made of cast aluminum. The two standard forms of soft jaws are regular soft jaws and pie-type soft jaws (*Figure 4-93*). Regular soft jaws resemble stan-

Figure 4-92. *Lathe chucks.*

Figure 4-93. *Soft jaws.* (Courtesy Carr Lane Manufacturing Co.)

dard jaws since they are made to the same width as the jaw carrier in the chuck. Pie-type soft jaws are much larger and completely cover the face of the chuck. In use, these jaws are first attached to the chuck, and the chuck is tightened. The desired form or shape then is machined into the soft jaws, forming a type of partial or full nest. Once machined to a size slightly less than the part size, the normal operation of the chuck securely holds and locates the parts to be machined.

Another type of soft jaw sometimes employed for large chucks uses an insertable element in either the top or end of the jaw (*Figure 4-94*). This arrangement reduces the cost of replacing the entire set of jaws and the space required to stock-and-store machine jaw inserts. Since only the small insert is machined, the major body of the jaw can be used repeatedly for any number of different inserts. *Figure 4-95* shows a few examples of how these inserts may be machined to suit a variety of workpieces.

Figure 4-94. *Insertable element for soft jaws.* (Courtesy Starwood Enterprises, Inc.)

Figure 4-95. *Inserts may be machined.* (Courtesy Starwood Enterprises, Inc.)

## Solid Arbors and Mandrels

The solid, slender-taper mandrel (*Figure 4-96*) is about the simplest possible workholder for round workpieces. Its main characteristic is a slightly tapered chucking surface with a taper of

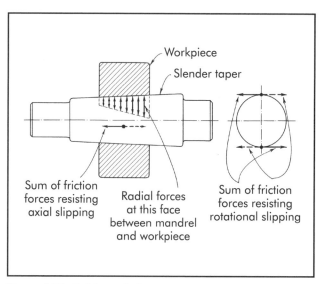

Figure 4-96. *Solid mandrel.*

0.004–0.006 in./ft (0.33–0.50 mm/m). The workpiece diameter must be smaller than the largest diameter of the mandrel. The workpiece is forcibly pushed endwise onto the mandrel. This produces a gripping force around the hole in the workpiece, decreasing axially in relation to the interference produced between the outer diameter of the mandrel and inner diameter of the workpiece. The driving torque that can be transmitted depends on the radial gripping and tangential friction forces produced. The resistance against axial slipping depends on the axial friction forces produced.

It is not always easy to obtain the same driving power or position the workpiece to a definite stop when trying to control the resulting interference between workpiece and mandrel. Pressing the workpiece on the mandrel requires an arbor press. This procedure is slow and may damage the finish of the workpiece bore and score the mandrel. Workpieces with accurate round and straight bores are held with great accuracy on the simple mandrel in *Figure 4-96*. If the bore is not round and straight, the workpiece and mandrel will mutually distort under the forces used to press on the workpiece.

### Straight Mandrel

A straight mandrel resembles the mandrel in *Figure 4-96*, but has a straight (untapered) chucking area. To produce the required press fit, the outer diameter of this mandrel is made larger than the bore of the workpiece by an amount called the interference. The amount of permissible interference is determined by the wall thickness, diameter, and material of the workpiece. It must be controlled to avoid exceeding the elastic limits of the workpiece and mandrel. Exceeding such limits could produce a change in bore diameter, especially in materials of low strength and with a low modulus of elasticity.

The obtained driving torque together with the axial resistance to slipping depends on the interference and is not easy to control. Possible mandrel wear and damage to the workpiece offset the advantages of the simple low-cost design. The range of interference-fit mandrels is limited.

### Combination Slender Taper and Straight Mandrel

The combination slender taper and straight mandrel has a short, tapered chucking length followed by a straight length. The straight area fits snugly into the workpiece without interference and helps prealign the bore. The tapered surface provides the driving area. To obtain reliable results for torque, accuracy, and axial location, hole tolerances must be carefully held. Like the taper and straight mandrels, this type also requires a press-fit condition between the workpiece and chucking surface. The larger diameter of the mandrel is forced into the smaller bore of the workpiece. This axial pressing often damages the workpiece and workholder.

### Solid Expanding Mandrels

The hydraulic mandrel and roll-lock type of arbor (*Figure 4-97*) produce a shrinkfit by expanding tubular shells that are not split length-

Figure 4-97. (a) *Elastic hydraulic workholding.* (b) *Roll-lock expanding solid mandrel.*

wise. The internal actuating mechanism expands the shell into the workpiece bore, and the amount of press-fit produced can be controlled by stops limiting the amount of expansion. The hydraulic type expands by hydraulic pressure; the roll-lock type expands by the gradual rolling and wedging action of straight rollers between the tapered inner diameter of the shell and a tapered plug, which is turned by a wrench.

Mandrels with a chucking area not weakened by axial slots generally have only small ranges. The solid mandrel will permit a resulting interference of 0.0010–0.0020 in./in. (0.001–0.002 mm/mm) of diameter. Hydraulic and roll-lock workholders expand from 0.0020–0.0030 in./in. (0.002–0.003 mm/mm) of diameter. The obtainable accuracy is approximately 10–20% of the range of the workholder. A mandrel expanding 0.002 in. (0.05 mm) may then hold a round piece within a 0.0002–0.0004 in. (0.005–0.010 mm) total indicator reading, or 0.0001–0.0002 in. (0.003–0.005 mm) eccentricity.

## Split-collet and Bushing Workholders

Solid expanding mandrels have a very small range. To increase range and hold workpieces with larger diameter variations, split collets and bushings are used. These are basically slotted shells of various shapes. The slots permit greater flexibility to increase for internal chucking and decrease for external chucking. The more flexibility provided, the greater the range. The shell splits and acts as a spring (*Figure 4-98*). *Figure 4-99* shows a popular collet in which the range is obtained by cantilever deflection. The cantilevers are produced by splitting the collet from one end only and leaving a solid ring on the other. *Figure 4-100* shows a high-range collet where great flexibility is obtained by imbedding loose individual collet jaws in a suitable rubber compound.

Many varieties of chucking elements (*Figures 4-98, 4-99,* and *4-100*) form the basis of various workholders. They are used for internal and external chucking, and are actuated in most cases by cones or tapers acting on corresponding surfaces of the chucking elements. Some workholders use small, self-locking tapers since a force must be applied to disengage the mating tapers.

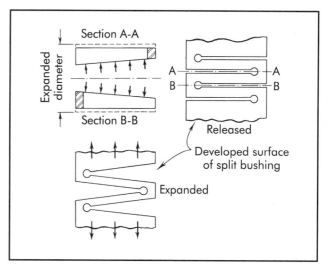

Figure 4-98. *Expansion of a split bushing.*

Figure 4-99. *Split collet.*

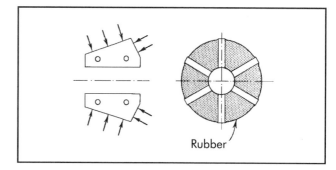

Figure 4-100. *High-range type collet.*

A taper of less than 10° is usually self-locking (*Figure 4-101*). Above 12–16°, the taper becomes self-releasing and requires force only to engage the mating tapers (*Figure 4-102*).

Split collets and bushings, made with care, give satisfactory results for most applications. However, they are hard to maintain where extreme accuracies are required, and are vulnerable to the entry of dirt and wear on the sliding surfaces.

Figure 4-101. *Collet with self-locking taper.*

Figure 4-102. *Collet with self-releasing taper.*

Standard 5C collets, step collets, and step chucks may serve as workholders for a wide variety of parts. Standard 5C collets are available in a wide range of sizes and shapes (*Figure 4-103*). In addition, unmachined, soft emergency collets, step collets, and step chucks also may be machined to suit any particular part shape or configuration (*Figure 4-104*).

Figure 4-103. *Collets are available in a wide range of sizes and standard shapes.* (Courtesy Hardinge Brothers, Inc.)

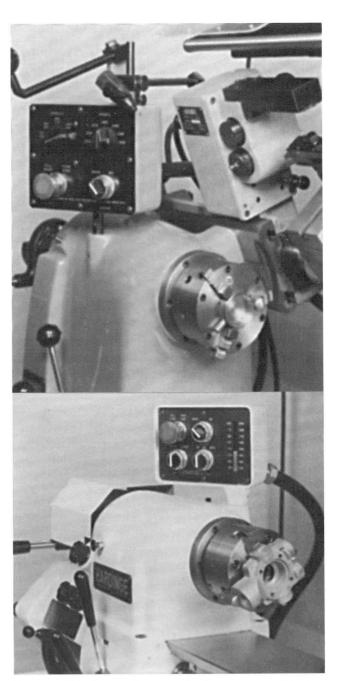

Figure 4-104. *Some collets and chucks may be machined to suit part shapes and configurations.* (Courtesy Hardinge Brothers, Inc.)

## Axial Location

Moving collets provide axial location for workpieces. *Figure 4-105a* shows a collet chuck, the diameter of which contracts upon axial motion. *Figure 4-105b* shows a collet mandrel that expands upon axial motion. At first, the clearance

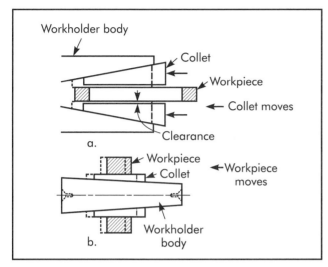

Figure 4-105. *Axial location of workpieces as affected by collets.*

between the workpiece diameter and chucking diameter is closed; upon sufficient axial motion, a shrinkfit is created between the workholder and workpiece. As soon as contact is made at the chucking surface, the workpiece has a tendency to move together with the collet relative to the workpiece holder body. This movement produces an axial shift of the workpiece and prevents precise axial location.

*Figure 4-106* shows a method of obtaining precise location by providing a stop surface. The collet moves the workpiece toward and pushes it firmly against the stop. Some slipping occurs between the collet and workpiece during the chucking operation. This slippage tends to reduce the driving power of the workholder by absorb-

ing part of the actuating force, and the mechanical efficiency of the chucking operation is thus lowered. To obtain accuracy, the stop surfaces of the workholder and faces of the workpiece must be square with the centerline. Lack of squareness may cause distortion of the arbor by one-sided loading against the stop surface (*Figure 4-107*).

*Figure 4-108* shows a collet arrangement designed to eliminate axial shift by not moving the

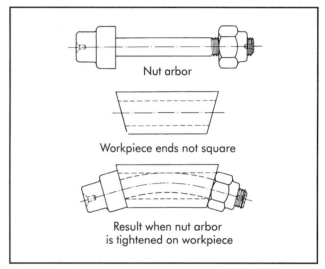

Figure 4-107. *Distortion caused by lack of squareness.*

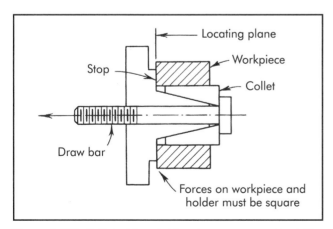

Figure 4-106. *Collet with stop plate to ensure correct axial location.*

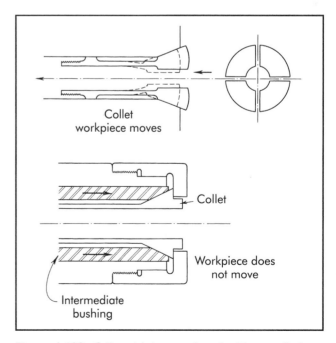

Figure 4-108. *Collet with intermediate bushing to eliminate axial shift.*

collet. An intermediate bushing interacts with the tapered surfaces of the collet. An additional element with more mating surfaces between the two cylinders is thus introduced. The need to produce additional surfaces to a high dimensional and relational accuracy will lower the resultant final accuracy of the workholder. The more mating surfaces required in any workholder, the lower the accuracy.

Some frequently used workholders have sliding elements or inserts actuated by movement on inclined planes (*Figure 4-109*). They are similar to a lathe chuck in that their jaws or driving keys are held, guided, or moved in the workholder body. The actuating mechanism consists of a bar with inclined wedge surfaces. The body, jaws, and bar are three different elements working together. Accuracy depends on excellence of workmanship and careful use. A desirable simplification is the combination of the actuating mechanism with the workholder body. These workholders usually are built as arbors and mandrels, and have good range and gripping power. They are useful tools for even the largest tubular workpieces.

Figure 4-109. *Collets for internal chucking.*

## Self-actuating Wedge Cam and Wedge-roller Workholders

Self-actuating wedge cam and wedge-roller workholders are tightened by the tangential work force of the cutting tools. They actuate themselves once the holding elements are brought even slightly into contact with the workpiece.

*Figure 4-110* illustrates an arbor using the principle of the roller clutch. One or more rollers are nested in cutouts in the body of the work-holder and retained by wire clips. Turning the workpiece in the direction shown (relative to the work-holder) wedges the rolls between the work-piece and workholder. An increase in the applied cutting force increases the wedging action almost proportionally.

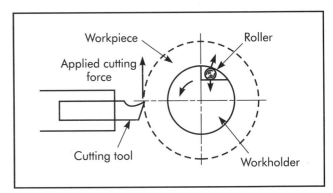

Figure 4-110. *Roller-clutch-type arbor.*

The wedge-cam chuck (*Figure 4-111*) has two inwardly spring-loaded cams. The jaws are lifted by a ring (not shown), then released until they touch the workpiece. Cam surfaces usually are serrated to obtain a better grip. Tangential cutting forces wedge the cam jaws tightly against the workpiece.

Figure 4-111. *Wedge-cam-type chuck.*

## VISES

Vises are perhaps the most widely used and best-known workholders. All vises have in common one fixed and one movable jaw that hold a workpiece between them. All other features, such as configuration of the jaws to grip particular workpieces, means of actuation and mounting, ability to position the workpiece, and sizes, contribute to an endless variety of vises commercially available. A vise is a good basic workholder element. By reworking the jaws or making special jaws, and adding details such as locating pins, bushings, and plates, vises can be easily converted into efficient, specialized workholders.

A common, heavy, plain vise is used for holding flat workpieces (*Figure 4-3*). With V-shaped grooves in the jaws, it can locate round workpieces such as bars and tubing. A vise mounted on a rotary base permits angular rotation, positioning, and locking as may be required (*Figure 4-4*). *Figure 4-112* shows a universal vise that can position a workpiece at any angle relative to the cutting tool.

*Figure 4-113* shows a workholder using a vise with a quick-acting jaw moved rapidly into contact with the workpiece. A handle-operated cam then locks the movable jaw to produce great holding force. A special plate attached to the vise pre-

Figure 4-113. *Drill jig with quick-acting vise.*

cisely positions three guide bushings in relation to the fixed (locator) jaw of the vise. Another variation of this type of vise has a solid jaw made with a drill bushing mounted in movable arms. An end stop also is incorporated into the solid jaw. This type of vise is adaptable for many different types and shapes of workpieces.

*Figure 4-114* illustrates a vise particularly well-suited for jig and fixture work. As shown, the three gripping areas of this vise are completely movable and can be locked to accommodate almost any part shape within the range of the vise.

Another form of special-purpose vise is the vertical vise (*Figure 4-115*). It is useful for applications where a standard milling-machine vise

Figure 4-112. *Universal-type vise.*

Figure 4-114. *Vise with three movable gripping areas.* (Courtesy James Morton)

Figure 4-115. *Vertical vise.* (Courtesy Mid-State Machine Products, Inc.)

cannot provide adequate support. The height of the workpiece can be regulated by the position of the stop bar (*Figure 4-116*). These bars are movable in 0.50 in. (12.7 mm) increments for the complete width of the vise jaw and made in sets to match the range of openings for the vise.

Another style of vise ideally suited for a variety of different-size workpieces is the Bi-Lok® machine vise (*Figure 4-117*). This vise is designed to hold either identical parts or two completely different-sized parts (*Figure 4-118*). The vise mechanism is totally self-compensating, and the same pressure is applied on both gripping positions regardless of the part sizes.

The same operating principle used for the Bi-Lok vise also is applied to the Multi-Lok® vise

Figure 4-116. *Regulating workpiece height using a stop bar.* (Courtesy Mid-State Machine Products, Inc.)

Figure 4-117. *Bi-Lok® machine vise.* (Courtesy Chick Machine Co.)

Figure 4-118. *Vise mechanism self-compensates to apply equal pressure regardless of part size.* (Courtesy Chick Machine Co.)

(*Figure 4-119*). A group of vises are mounted together in units of two, four, eight, or more to hold a series of identical parts, or a variety of different parts. This unit may be used for either horizontal or vertical machining operations. The individual jaws on the Multi-Lok are removable and may be replaced with soft jaws machined to suit specific part shapes.

Figure 4-119. *Multi-Lok® vise.* (Courtesy Chick Machine Co.)

Despite their many applications, most milling-machine vises cannot hold odd shapes or multiple workpieces. These parts often require expensive dedicated fixturing. The workholding system (*Figure 4-120*) corrects the problems of conventional vise designs and reduces the need for dedicated fixtures. The unit consists of two moving blocks that replace the movable jaw on any standard angle-locking-type vise. These moving blocks can hold either single or multiple parts and have the unique ability of applying holding forces in both horizontal and vertical directions. *Figure 4-121* shows several of the setup possibilities of this jaw unit. Setup (1) illustrates how this system holds a rectangular part. Two parts also can be held (2). A removable stop may be positioned on the solid jaw to act as a reference point for mounting both parts. If the part has two different clamping surfaces (3), the jaws will accommodate the variations. The jaws also may be used to hold round parts (4). Each of the jaw elements is keyed to the moving block, and the side of one jaw element may be used to establish a perpendicular mounting plane for smaller parts (5).

When the clamping elements are mounted to the pistons, a completely new set of clamping options are available. Mounting a set of swivel jaws to the piston allows this unit to hold both tapered and larger round parts (6 and 7). Odd shapes and contours may be held by simply using a set of soft swivel jaws made to conform to the part shape (8). The setup shown in (9) is useful for machining three sides of a part in one setup. A set of strap clamps are mounted in the

Figure 4-121. *Possible setups.* (Courtesy Twin Lock Tool Co.)

pistons to hold the part. Even setups involving multiple small parts (10) are easily performed using another set of swivel jaws. Clamping action is obtained by inserting an anvil plate against the sides of the parts. As the vise is tightened, swivel jaws contact the anvil plate and push it against the parts. As the pressure increases, they pivot and apply the necessary holding force to both ends of the stacked parts.

## Special Jaws

The usefulness of vise-like workholders can be enhanced by making special cast and molded jaws, which adapt to the holding and locating of complex workpieces. There are two methods for making such special jaws.

*Figure 4-122* shows the method of casting jaws for holding a nipple by pouring a low-temperature alloy around the workpiece, which is used as the pattern. Two wooden spacers locate the pattern and separate the cast jaw halves. The pattern is coated with a releasing agent for easy removal

Figure 4-120. *Twin-Lock® workholding system.* (Courtesy Twin Lock Tool Co.)

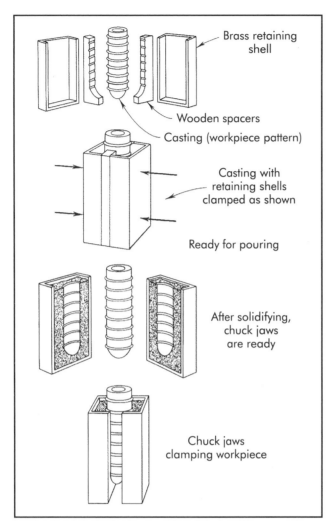

Figure 4-122. *Casting the vise jaws with low-temperature alloy.*

Figure 4-123. *Plastic vise jaws.*

## Independent Jaws

Occasionally, independent vise jaws may be used as workholders for large or odd-shaped parts. In *Figure 4-124*, these workholders are

from the cast jaws. These jaws then are attached to the vise jaws.

In the second method (*Figure 4-123*), a plastic material is used. The material has a metal filler to give it more wear resistance and strength. The plastic is of putty-like consistency, and is placed on each jaw of the vise. The workpiece, a T fitting in the example shown, again acts as the pattern. Coated with a releasing agent, it is located and pressed into the plastic material on the two sides of the vise by closing the vise jaws. The plastic hardens within two hours to form two precise half-impressions of the workpiece, which make an excellent locating and holding arrangement. The plastic material also may be used to locate and secure pins, bushings, and other details used in workholders.

Figure 4-124. *Independent vise jaws.* (Courtesy De-Sta-Co Products)

available in two general styles. The bigger of the two is useful for larger parts, while the smaller is well-suited for edge gripping on thinner parts. A third style of independent vise jaws is illustrated in *Figure 4-125*. This style of vise is an adjustable-range vise. Its jaws may be set at any convenient distance and held in place with the rack gear. This style of vise is very useful for machines with T slots in the table since the rack must be located below the working surface of the vise.

Figure 4-125. *Adjustable-range vise.* (Courtesy Universal Vise and Tool Co.)

## NONMECHANICAL CLAMPING

Sometimes it is impractical to hold a workpiece by direct clamping pressure because of possible distortion or the size of the workpiece. Magnetic, vacuum, and electrostatic workholders may be of value in such cases.

### Magnetic Chucks

Magnetic chucks are available in a variety of shapes. They can hold only ferrous workpieces unless intermediate mechanical workholders permit the holding of workpieces made of nonmagnetic material. Magnetic chucks are suitable for light machining operations such as grinding. Strongly magnetic materials and better utilization of magnetic force permit their use for heavier operations, such as light milling and turning. Magnetic chucks can be operated by permanent magnets or electromagnets powered by direct

current. The gripping power attainable depends on the strength of the magnets and amount of magnetic flux that can be directed through the workpiece. *Figures 4-126* and *4-127* illustrate various magnetic chucks.

A magnetic chuck is fast-acting and, by holding a large surface of the workpiece, causes a minimum of distortion. Magnetic chucks are available in rectangular shapes, circular shapes as rotary chucks, and as V blocks. Magnetic chucks impart some residual magnetism to

Figure 4-126. *Magnetic chucks.*

Figure 4-127. *Magnetic chuck and angle plate used as workholder.*

workpieces. This must be removed by demagnetizing if it interferes with proper functioning of the workpiece.

## Vacuum Chucking

Quite often, workpieces of nonmagnetic materials or special shapes and dimensions must be securely held flat without any mechanical clamping. An example might be a large flat plate. In such cases, vacuum chucking may be the only practical holding method.

The basic principle of vacuum chucking is simple. On each square inch (square millimeter) of a surface, a pressure of approximately 14.7 psi (101 kPa) is exerted by the atmosphere. This represents approximately $144 \times 14.7 = 2{,}116.8$ psi (146 MPa). Part of this pressure is used by creating a vacuum in a closed chamber made up of the locating surface of the workpiece and mating surface of the workholder. At first, pressures on the outside and inside of the chamber are equal. As a vacuum is produced inside the chamber, the outside pressure holds the workpiece against the locating surface (*Figure 4-128*). *Figure 4-129* illustrates a typical vacuum chuck. An O-ring seal laid in a groove around the chucking area creates the closed chamber to be evacuated. Holes

Figure 4-129. *Vacuum chuck.*

and a grid of small connecting channels in the chucking surface assist in the speedy creation of the needed vacuum.

## Electrostatic Chucks

The attraction of electrically opposite charged parts can hold flat or flat-sided workpieces that cannot be magnetized. Electrostatic chucks (*Figure 4-130*) can hold any electrically conductive material. Glass, ceramics, and plastics also may be held by flash metal plating on one flat side to provide a suitable electrical contact.

Figure 4-128. *Vacuum-chucking principle.*

Figure 4-130. *Electrostatic chuck.*

## POWER CLAMPING

Power clamping devices frequently are used for applications where speed and uniform clamping pressures are important considerations. *Figure 4-131* shows how a typical power clamping system is constructed. Power clamps normally are operated by hydraulic or pneumatic pressure, or a combination of both. *Figure 4-132* shows several typical setups for power clamping.

Figure 4-131. *Power clamping.*

Figure 4-132. *Typical setups for power clamping.* (Courtesy Jergens, Inc.)

One problem in using any power clamping system is the basic law of "no hydraulic pressure—no clamping pressure." If any clamping element leaks, or is disconnected, pressure is lost and the clamps loosen. One system of components that uses a mechanical locking principle in combination with a pressurized hydraulic system is the Stay-Lock® clamping system (*Figure 4-133*). These clamps are built using a mechanical lock activated by hydraulic pressure. Once locked, the hydraulic pressure and hoses may be removed and the clamps stay locked firmly against the part.

Figure 4-133. *Stay-Lock® clamping system.* (Courtesy Jergens, Inc.)

The locking action of the clamping system is similar to a wedge-action clamp. Working on an inclined plane principle, this locking action operates in much the same way as a tapered shank drill. The basic operation of this clamp is illustrated in *Figure 4-134*. In use, the piston is driven to one side of the clamp body by hydraulic pressure. The locking angles on the piston and locking pin engage and lock together. This provides a mechanical lock. Thus, once the clamp is activated, the hydraulic connections can be removed and the clamp will not loosen. The clamp is released by switching the hydraulic hoses to the release port. The pressure then drives the piston back off the locking pin.

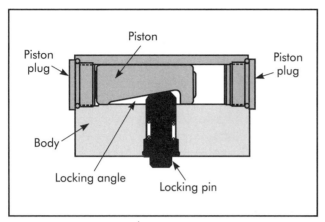

Figure 4-134. *Locking action.* (Courtesy Jergens, Inc.)

## MULTIPLE-PART CLAMPING

When designing tools for large production runs, it is often desirable to machine more than one part at a time. In these cases, jigs and fixtures should be designed to permit multiple-part clamping. The two principal points to remember when designing workholders for multiple clamping are: (1) the clamping pressure must be equal on all parts, and (2) the tool should have a minimum number of operating points. In *Figure 4-135*, a little forethought and ingenuity make it possible to clamp almost any part in a multiple-part workholder.

## BASIC CONSTRUCTION PRINCIPLES

Every jig and fixture must be constructed properly to function as intended. The degree of accuracy and durability of the workholder is directly related to the way the tool is constructed.

### Tool Bodies

Tool bodies are the major element of jigs and fixtures. They form the general size and shape of the tool, as well as provide the mounting surfaces for locators, supports, and workholding devices. The three principal types of tool bodies are cast, welded, and built-up (*Figure 4-136*).

### Cast

Cast tool bodies normally are made from cast aluminum, cast iron, or cast epoxy resin. These

Figure 4-135. *Multiple-part clamping.*

Figure 4-136. *Three types of tool bodies.*

tool bodies are well-suited for permanent, high-volume production workholders. The principal advantages in using cast tool bodies include: stability and vibration dampening, good material distribution, and reductions in machining time. But cast tool bodies normally cost more to make since a pattern must be made for each tool. Another disadvantage of cast tool bodies is the long lead time requirement between design and fabrication of the tool.

### Welded

Welded tool bodies normally are made from any weldable materials, such as steel, aluminum, magnesium, etc. This type of tool body can be easily fabricated with minimum lead time. Welded tool bodies are strong, rigid, and easily modified when required. The major disadvantage of this type of tool body is the additional machining time required to repair and remove heat-distorted areas.

### Built-up

Built-up tool bodies are made from a wide variety of different materials. Steel, aluminum, cast iron, wood, and epoxy resins are commonly used for built-up tool bodies. They are the most versatile and frequently used type for general jig and fixture work. Built-up tool bodies are inexpensive, easily modified, and require less machining after assembly than cast or welded tool bodies. This type of tool body normally uses dowel pins and socket-head cap screws to align and fasten each tool member.

## Tooling Materials

In addition to fabricating tool bodies, many different tooling materials are used for locators, supports, or other jig and fixture elements. To keep the cost of a tool to a minimum, the tool designer should be familiar with the wide variety of preformed materials available.

### Precision-ground Materials

Precision-ground materials are available in many shapes and sizes. The two primary variations are ground flat stock and drill rod. These materials are commercially available in either standard or oversized lengths of 18–36 in. (46–91 cm). Precision-ground materials normally are available in low-carbon and high-carbon steel, and oil-hardening and air-hardening tool steel. The tolerance values of these materials vary with the size and shape of the burr, but they normally are ground to within a few thousandths of an inch (μm) accuracy.

### Aluminum Tooling Plate

Aluminum tooling plate, like precision-ground materials, is available in many sizes and shapes. The major advantage to using this material is the close tolerance of the finished stock. Many times, tools assembled from these materials require no finish-machining prior to use.

### Precast Bracket Material

Precast bracket materials are commercially available in several shapes and sizes as illustrated in *Figure 4-137*. These materials generally are made from either cast iron or cast aluminum and

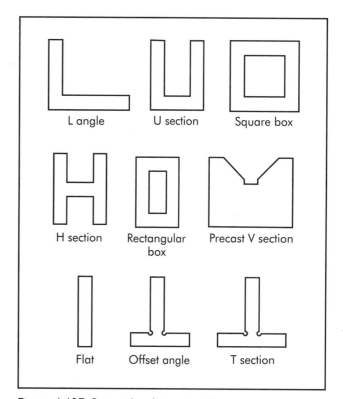

Figure 4-137. *Precast bracket materials.*

available in lengths of 18–36 in. (46–91 cm). Pre-cast bracket materials frequently are used for locators, supports, or complete tool bodies. The principal advantage to using these sections is the ease with which tools can be made using the convenient forms and shapes available.

## Structural Sections

Structural sections are another type of material sometimes used for workholder components. These sections are available in several different shapes and lengths of up to 16 ft (4.9 m) (*Figure 4-138*). This makes structural sections useful for large or oversized workholders. The most common material used for these structural sections is low-carbon steel, but those made from aluminum or magnesium also are available. Structural sections normally are made by rolling and are not for close tolerance specifications. For this reason, these sections must be machined to suit close tolerance applications. Nevertheless, these sections are versatile and will suit a wide range of tooling applications.

## Fasteners

Almost every jig and fixture uses some type of fastener to align and hold the various members and components in their proper position and relationship. The specific type of fastener used to assemble a particular workholder normally is determined by the joint and required tolerance of the assembled parts. The following are a few fasteners commonly used for jig and fixture work, and the applications where each is best-suited.

## Screws and Bolts

Screws and bolts are the most common types of fastening devices used to construct assembled workholders. The principal difference between these fasteners is the method used to hold the assembled parts together. Screws normally are driven by their heads and turned into threaded holes. Bolts, on the other hand, are installed through holes and held in place with a nut.

The most common type of screw used for jig and fixture work is the socket-head cap screw (*Figure 4-139*). In addition, variations of the socket-head cap screw (*Figure 4-140*) also are frequently used to assemble workholders. The principal type of bolt used with jigs and fixtures is the T bolt (*Figure 4-141*). These bolts primarily secure the workholder to the machine table. Other types of bolts include those with hex and square heads, and studs.

Figure 4-139. *Socket-head cap screw.*

Figure 4-140. *Variations of the socket-head cap screw.*

Figure 4-138. *Structural sections.*

Figure 4-141. *T-bolt.* (Courtesy Jergens, Inc.)

## Nuts, Washers, and Inserts

Nuts, washers, and inserts are used for some workholder designs. The most commonly used nuts are hardened hex, check, T slot, acorn, flange, and coupling. *Figure 4-142* illustrates a few of these types. The most frequently used washers are plain flat washers, self-aligning

washer sets, C washers, and swinging C washers (*Figure 4-143*). Thread inserts (*Figure 4-144*) also are used for some applications. These inserts are available in several styles, and serve basically the same purpose—to provide a replaceable thread in an area where threads will not hold up well, such as in plastic or soft metals, or places where frequent installation and removal of the threaded fastener would rapidly wear out a normal thread.

## Dowels and Pins

Dowels and pins are another form of fastener often found in workholders. These fasteners, unlike screws and bolts, are intended to align assembled components rather than secure them. The most common type of pin used for alignment is the hardened-steel dowel pin (*Figure 4-145*). Other variations include pull dowels, for use in blind holes, and roll pins. Roll pins frequently are used in place of dowel pins to eliminate the

Figure 4-142. (a) *Hardened hex nut,* (b) *T-slot nut,* (c) *acorn nut, and* (d) *spherical flange nut.* (Courtesy Jergens, Inc.)

Figure 4-143. (a) *swinging C washer,* (b) *flat washer,* (c) *C washer.* (Courtesy Jergens, Inc.)

Figure 4-144. *Thread insert.* (Courtesy Jergens, Inc.)

Figure 4-145. *Hardened-steel dowel pin.* (Courtesy Jergens, Inc.)

need to ream the mounting holes. However, these pins should only be used in areas where the assembled tolerance will permit some variation in the position of the assembled members.

### Retaining Rings

Retaining rings (*Figure 4-146*) normally are used for fastening in areas with a minimum amount of thrust against the fastener. These fasteners are faster and easier to install than standard threaded fasteners and will work equally well for both internal and external applications.

### Fixture Keys

Fixture keys, while not actually a fastening device, are an integral part of many workholders. The two principal styles of fixture keys used for jigs and fixtures are the plain, or standard, and interchangeable (*Figure 4-147*). Generally, an interchangeable style fixture key is easier to use

External retaining rings Internal retaining rings

Figure 4-146. *A variety of retaining rings.*

and permits a single set of keys to be used for several workholders. However, standard keys are well-suited for permanent installations.

### Ball Lock® Mounting System

Many different methods are used to mount workholders to machine tools during machining. The main problem always has been finding a way to reduce the needed setup time. One method of reducing this time is use of the Ball Lock mounting system. These units accurately position and securely contain the workholder on the machine table. In *Figure 4-148*, this mounting system has

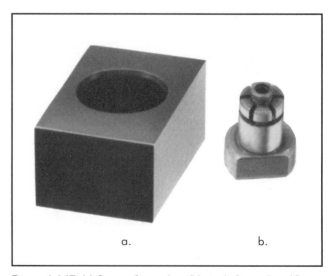

a.    b.

Figure 4-147. (a) *Square fixture key,* (b) *single fixture key.* (Courtesy Jergens, Inc.)

three major parts, a flanged shank, a liner bushing, and a Ball Lock bushing. In use, the bushings are installed in both the workholder and machine base or mounting subplate. The flanged shank then is installed through the liner bushing into the bushing and tightened down using the socket-head cap screw (*Figure 4-149*).

The operation of the Ball Lock unit is quite simple. As the socket head cap screw in the head of the flanged shank is tightened, a large ball is forced against three smaller balls. This action moves the three locking balls outward in their sockets against the conical sides of the Ball Lock bushing. When the three balls contact the conical area of the Ball Lock bushing, they both center the flanged shank and apply a downward force against the shank, locking it against the fixture base. Rather than locating on the cylindrical diameter of the shank, the actual location is achieved by the locking balls against the conical area of the Ball Lock bushing.

Figure 4-149. *Installation of bushings and flanged shank.* (Courtesy Jergens, Inc.)

## REVIEW QUESTIONS

1. What are the basic elements that make up a workholder?
2. What are the three fundamental categories of workpiece location?
3. Design a locating system to locate and support the part shown in *Figure A*.
4. What is the simplest and most basic type of location?
5. What type of materials are used for cast, nest locators?

Figure 4-148. *Ball Lock® mounting system.* (Courtesy Jergens, Inc.)

Figure A.

6. Design a locating method for the part shown in *Figure B* using a diamond pin to restrict radial movement.
7. How should clamping elements be positioned with relation to the workpiece?
8. What is the simplest and least expensive type of clamping device used for jigs and fixtures?
9. What type of nonmechanical clamping device is well-suited to hold nonferrous parts?
10. What type of tool body is most common?

Figure B.

# 5

# Jig Design

Jigs are workholders designed to hold, locate, and support a workpiece while guiding the cutting tool throughout its cutting cycle. Jigs can be divided into two general classifications: drill jigs and boring jigs. Of these, drill jigs are by far the most common. Drill jigs are generally used for drilling, tapping, and reaming, but also may be used for countersinking, counterboring, chamfering, and spotfacing. Boring jigs, on the other hand, are used mostly for boring holes to a precise, predetermined size. The basic design of both classes of jigs is essentially the same. The only major difference is that boring jigs are normally fitted with a pilot bushing or bearing to support the outer end of the boring bar during the machining operation.

There are numerous considerations that must be addressed when designing any jig. Although several of these points, such as locating, supporting, and clamping, have already been covered, they are included in this section because they apply to jig design. Since all jigs have a similar construction, the points covered for one type of jig normally apply to the other types as well. Jig design and selection begins with an analysis of the workpiece and the manufacturing operation to be performed.

## GENERAL CONSIDERATIONS

One of the first considerations in the design of any workholder is the relative balance between the cost of the tool and the expected benefits of using the tool for production. All workholders should save more in production costs than they cost to design and construct. In many instances, tool designers may have to complete detailed estimates to justify the cost of special tooling. This involves a close look at the part drawing, process specifications, and other related documents.

Typically, the complexity of the part, location and number of holes, required accuracy, and the number of parts to be made all must be considered to determine if the cost of a particular jig is warranted. Once the tool designer is satisfied that the cost of special tooling is justified, the remaining data required to produce a suitable workholder is compiled and analyzed.

## MACHINE CONSIDERATIONS

The size, type, and capabilities of the machine tool specified for a particular operation have a direct bearing on how a particular jig should be made. For example, a jig intended for small holes drilled on a sensitive drill press will not normally have the same features as a jig designed for use with a larger radial drill press. While both may share many details, the machines they are used with, as well as the specific tooling required, will generally dictate the overall construction of the workholder.

Often, the tool designer is not the individual who specifies a particular machine tool for a machining operation. In most instances, the process engineer is charged with this responsibility. However, the tool designer should check the specified machines to ensure they are the best possible choices. The following are a few machine considerations to keep in mind when designing any jig.

- The machine should be large and rigid enough to perform the desired operations.
- Production capabilities and accuracy of the machine tool should be suitable for the operation.
- The machine tool must be able to safely accommodate the workholding device.
- The machine selected should be close to subsequent machining operations. Here, the shop layout is important to minimize the distance between machining stations and prevent backtracking across the shop.
- Specified cutting tools must be compatible with the machine tool. A lathe with a #2 Morse taper in the tailstock spindle cannot hold #3 Morse taper tools without an adapter.
- Whenever possible, standard-size cutting tools should be specified to reduce costs and simplify procurement of drills, reamers, and other cutting tools.

When possible, a tool designer should have all the relevant specifications for each type of machine tool used for part production. This information should include table sizes, T-slot sizes, machine travel in all axes, and similar data. Such information is normally available in the maintenance manuals furnished with each machine, and saves the designer the time required to take measurements directly on the machine tool. A similar specification sheet should be maintained for each standard cutter normally used in the shop. This information is available from the cutter manufacturer, and should include the length of tool, length of effective cutting area, shank size, and similar data.

## PROCESS CONSIDERATIONS

Process considerations include the type of jig, number of jigs needed, and specific step-by-step processing of the workpiece. While the proposed processing of a part is normally a function of process engineering, the tool designer should always

double-check it to ensure the proposed tooling will be compatible.

First-operation jigs are normally used for workpieces without prior machining or any reliable reference surface. Here, the first holes are put into the part and act as a reference point for any subsequent machining. First-operation jigs generally use adjustable supports, adjustable locators, or sight locators to set the initial position of the part to machine the first holes.

More than one jig may be used when several different operations must be performed on the same part. The most important point to consider is the repeatability of the location in each of the jigs. In these cases, the same location point should be used for all machining operations on the part.

Other considerations that should be analyzed before a final design is determined include the actual processing methods and chip control and disposal. Several parts may, in some cases, be stacked or aligned so more than one can be drilled at a time. Likewise, when large drills must be used to produce the required size hole, smaller step drills can be used to lessen the thrust and torque required to drill the holes. Here, a smaller or lighter-duty drill press must be used if a drill press of the proper size is unavailable. Chip control and disposal are considerations that must be remembered when designing a workholder. Chips and coolant are normally removed with a brush or air flow, so adequate slots or open areas should be provided to permit easier removal (*Figure 5-1*).

Figure 5-1. *Areas to permit easy coolant and chip expulsion.*

# DRILL JIGS

The workpiece, production rates, and machine availability normally determine the size, shape, and construction details of any jig. However, all jigs must conform to certain design principles for the efficient and productive manufacture of quality workpieces. There must be a method to:

- correctly locate the workpiece with respect to the tool;
- securely clamp and rigidly support the workpiece during the operation;
- guide the tool, and
- position and/or fasten the jig on a machine.

These features will ensure interchangeability and accuracy of parts, plus provide the following advantages:

- minimize tool breakage;
- minimize the possibility of human error;
- permit the use of less skilled labor;
- reduce manufacturing time, and
- eliminate retooling for repeat orders.

Jigs are often divided into two broad categories—open and closed. Open jigs are used when machining a single surface of a workpiece, whereas closed jigs are used when machining multiple surfaces. Examples of open and closed jigs are shown in *Figure 5-2*. More often, jig types are identified by the method used to construct the jig (for example, template, plate, leaf, channel, etc.). The main types are discussed in the following sections.

Figure 5-2. *Open and closed types of drill jigs.*

## Template Drill Jigs

Template drill jigs are not actually true jigs because they do not incorporate a clamping device. However, they can be used on a wide variety of parts and are among the simplest and least expensive drill jigs to build. Template drill jigs are simply plates containing holes or bushings to guide a drill. They are placed directly on a feature of the part to permit drilling holes at a desired location. When this is impractical, they are located on the part by measurement or sight lines scribed on the template.

Two flat-plate template drill jigs are shown in *Figure 5-3*. Both are designed to drill a hole through the center of the rounded end of a lever. The jig shown in *a* consists mainly of a drill guide plate and locating V-block. The jig in *b* does the same job, but has been further simplified by using dowel pins to accomplish the same centralizing action as the V-block.

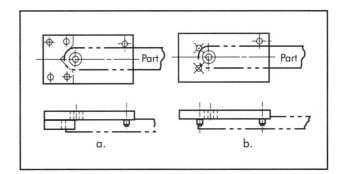

Figure 5-3. *Flat-plate template drill jigs.*

Another flat-plate template drill jig is shown in *Figure 5-4*. This jig was designed to drill a three-hole pattern into either a left- or right-hand version of a workpiece. Pins protrude from both sides of the drill plate, thereby permitting it to be flip-flopped to suit the workpiece being drilled. A common practice with template drill jigs is to place a pin into the first hole drilled to prevent excessive movement of the jig while drilling the remaining holes.

Three circular-type template drill jigs are shown in *Figure 5-5*. All are designed to locate from the maximum material condition (MMC) of the part

Figure 5-4. *Flat-plate template drill jig—left- and right-hand.*

diameter. Jig *a* was designed to locate from the outside diameter (OD) of a shaft, jig *b* from the inside diameter (ID) of a part, and jig *c* from a boss diameter. In all cases, a pin of the proper size was placed into the first hole drilled to properly position the jig to drill the second hole.

*Figure 5-6* illustrates two nesting-type template drill jigs. Jig *a* is designed to locate a small sheet-metal workpiece in a cavity to permit drilling two holes, which are located from the periphery of the workpiece. Jig *b* is designed to perform the same operation by using five dowel pins press-fitted into the jig in lieu of the cavity to locate the workpiece, reducing the cost to build the jig. A template drill jig is often used to drill holes in one portion of a large workpiece where a conventional jig large enough to hold the entire part would be impractical and costly. Template jigs usually cost much less than conventional jigs, often making the use of two or three template drill jigs more economical than using one large conventional jig.

Some of the main disadvantages of template jigs are:

• They are not as foolproof as most other types, which may result in inaccurate machining by a careless operator.

Figure 5-5. *Circular-plate template drill jigs.*

• Orientation of the hole pattern to workpiece datums may not be as accurate as with other types. However, the accuracy of the hole pattern within the template jig itself is comparable to that of any conventional jig.
• They are not practical when locating datums are dimensioned regardless of feature size (RFS).

## Plate Jigs

Plate jigs are basically template jigs equipped with a workpiece clamping system. Initial construction costs are greater for plate jigs than for

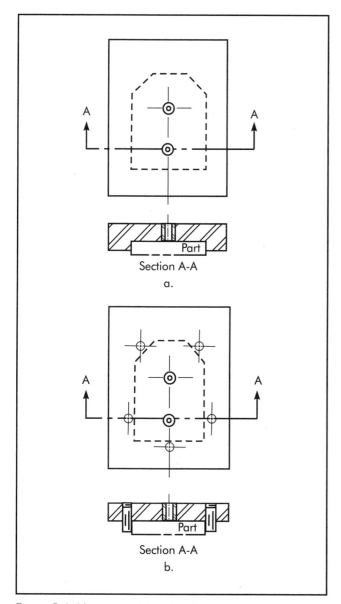

Figure 5-6. *Nesting template drill jigs.*

Three different types of plate jigs are shown in *Figure 5-7*. The open-plate jig shown in *a* is basically a template jig equipped with a means to clamp the workpiece, with work supported by the drill press table. The table-plate jig in *b* consists of a drill plate, locating stud, and clamping screw with standard screws used as jig feet. This type of jig is usually handheld on the drill press table, rather than clamped to the table, so it may be easily inverted for loading and unloading. Consequently, table-plate jigs are generally used for small parts. Note in particular that tool thrust in this type of jig is directed toward the clamps rather than the rigid portions of the jig. Therefore, it is imperative that the clamping method be strong enough to resist the thrust of the drill.

Figure 5-7. *Plate jigs.*

template jigs, but plate jigs are generally more accurate and last longer.

A plate jig incorporates a plate, usually the main structural member, which carries the drill or liner bushings. Slip bushings of various sizes can be used with liner bushings, allowing a series of drilling and related operations without the need to relocate or reclamp the workpiece. The plate jig's open construction makes it easy to load and unload a workpiece and get rid of chips.

For obvious reasons, the jig shown in *c* is called a sandwich-plate jig. With this type of jig, the workpiece is positioned between two plates: a drill plate containing the drill jig bushings, locators, and clamps, and a backup plate to provide support only. The backup plate has clearance holes for the drill and is aligned with the drill plate by two pins. The backup plate is ideal for supporting thin parts that would otherwise buckle from the thrust of the drill.

The angle-plate jigs shown in *Figure 5-8* are primarily used to drill workpieces at an angle to the part locators. The plain-angle-plate jig in *a* is designed to drill holes perpendicular to the locating surface, while the modified-angle-plate jig, *b,* is designed to drill holes at angles other than 90° to the locating surface.

Plate-type jigs are usually moved around the table by hand. Therefore, special safety precautions should be provided to prevent the jig from

whirling around the spindle whenever a cutting tool jams. The best way to prevent this is to build the jig with an extension handle long enough for the machine operator to overcome the torque of the jammed tool. When a plate jig is used with a radial drill, a provision can be made to clamp the drill jig or workpiece to the machine table.

## Universal Jigs

Universal jigs (sometimes called pump jigs) utilize a handle connected to a cam or rack-and-pinion to move either a bushing plate or a nest plate, often vertically, to clamp the workpiece. Parts held in universal jigs have surfaces adaptable to fitting against the surfaces of the bushing plate and nest.

Universal jigs are readily available in a wide variety of styles and sizes. They are well-designed, ruggedly built, and can be easily prepared to drill a specific part. Usually, all that is required is to add part locators and drill jig bushings.

Some typical applications are shown in *Figure 5-9*. Jig *a* was set up to drill a hole through the center of a small rod. It features a self-locating V-bushing liner, an adjustable end stop, and a riser block. This same operation could be accomplished with another type of universal jig called a cross-hole jig, *b,* also shown in *Figure 5-9*. In this jig, the drill is guided by a standard slip-fixed renewable bushing, which is fitted into the jig clamping plate and precisely centered above a hardened and ground V-block. Either of these jigs could be easily adapted to drill a wide variety of parts, each having a different diameter or hole size, merely by changing the renewable slip-fixed jig bushing and adjusting the end stop.

Compare the jigs in *Figure 5-9* with the plate jig shown in *Figure 5-10*, which was specifically designed and built to perform a similar operation. Needless to say, the universal-type jig could be put on the job in a fraction of the time and at a lower cost. By utilizing the wide variety of standard drill bushings, liners, and tooling components, universal jigs can be adapted to drill a wide variety of parts. Also, they are reusable for other jobs, although this may require replacement of the drill plate, which is interchangeable and available separately from the manufacturer. Because

Figure 5-8. *Angle-plate jigs.*

Figure 5-9. (a) *Universal jig ready for production,* (b) *adjustable cross-hole drill jig.* (Courtesy, (a) Acme Industrial Company, (b) Heinrich Tools, Inc.)

of this versatility, universal jigs are ideal for limited production manufacturing. One manufacturer states that tooling costs can be reduced by one-third. With few exceptions, universal jigs can be designed and adapted to do a job in a fraction of the time required to design and build a conventional jig to perform the same function.

## Leaf Jigs

A leaf jig is generally small and incorporates a hinged leaf that carries the bushings, and through which clamping pressure is applied. Although the leaf jig can be used for large and cumbersome workpieces, most designs are limited in size and weight for easy handling. A leaf jig can be box-like in shape, with four or more sides for

drilling holes perpendicular to each side. Leaf jigs with additional feet are often called tumble jigs. They permit operations from more than one side.

Off-the-shelf, tumble-type leaf jigs are available in a variety of sizes from many manufacturers. Construction consists of a drill plate (leaf) attached to the jig body with a precise-fitting hinge at one end and a positive positioning clamp at the other end. As with universal jigs, all that is required to prepare it for use is to provide a means to locate the part and add drill bushings. Lids swing up to provide easy loading from three sides. Occasionally, a side plate is attached to the lid or base to permit cross-drilling.

The leaf jig shown in *Figure 5-11* was specifically designed and built to drill two holes in a small connecting rod. The hinged drill plate contains the drill bushings and is precisely located at both ends by the slots in the body of the jig. The workpiece is located and clamped between two V-blocks, one fixed and the other movable. The V-blocks are tapered to force the workpiece down against the base of the jig body.

## Channel and Tumble-box Jigs

Channel and tumble-box jigs permit drilling into more than one surface of a workpiece without relocating the workpiece in the jig. This results in greater accuracy with less handling than using several separate jigs. These jigs can be quite complicated and more expensive to build than simpler types, but they can still be cost-effective if properly designed. An example of a channel jig is shown in *Figure 5-12*.

The channel jig in *Figure 5-12* was designed to drill holes into three surfaces. A U-shaped channel was used as the main body, along with press-fit drill bushings, locators, and clamping details. The U-shaped channels used to construct this jig can be cast, built-up, or of welded construction. However, to keep building costs to a minimum, the designer should consider using the standard U-shaped sections discussed previously.

Tumble-box jigs permit drilling and similar operations from all six sides. Tumble-box jigs are commercially available in a variety of sizes, which can be prepared to machine a particular part by adding drill bushings and a means to locate and

Figure 5-10. *Plate jig for cross-hole drilling.*

Figure 5-11. *Leaf jigs for drilling two holes in a small connecting rod.*

clamp the part. Because of this off-the-shelf availability, designers often choose a tumble-box jig over a channel jig when machining only two or three sides of a workpiece.

Figure 5-12. *Cross section of a typical channel jig (clamping details not shown).*

## Indexing Jigs

Indexing jigs are used to drill holes in a pattern, usually radial. Location for the holes is generally taken from the first hole drilled, a datum hole in the part, or registry with an indexing device incorporated in the jig.

The simple jig shown in *Figure 5-13* features a base made from a standard angle-iron section,

Figure 5-13. *Simple indexing jig with standard angle-iron base.*

into which a locating stud (2) has been placed to position a bored cylindrical workpiece (1), which is clamped on the stud with a C-washer (4) and a hex nut (3). A drill bushing (5) is press-fit into the bushing plate (8). In use, the hex nut is loosened after the first hole is drilled, the workpiece is revolved, the index pin (6), which is held in place with a flat spring (7), is pushed into the hole, and the second of four holes 90° apart, is drilled after the nut is tightened. Indexing is repeated until all four holes have been drilled.

Another indexing jig, shown in *Figure 5-14*, utilizes an angle plate of welded construction. A spindle pressed into the jig's pivot point is threaded on one end for the locking wheel. Clamping for the workpiece is not shown. The workpiece, bushing plate, and bushings are rotated for drilling holes at various angles, which is determined by the location of the index holes in the jig and angle plate.

## Boring Jigs

Boring jigs are quite often constructed similar to box jigs. The principal difference between boring jigs and box drill jigs is the bearing point on

Figure 5-14. *Indexing jig of welded construction.*

the side opposite the surface to be machined. As shown in *Figure 5-15*, the workpiece is located and clamped inside the box. The boring bar is then passed through the part and the pilot bearing is aligned in the bore on the opposite side of the jig. The part is then bored by feeding the rotating boring bar into the part. In some designs, both the outer end of the boring bar and the driven end of the bar are supported during the boring cycle. This design is normally used where extreme accuracy is of prime importance.

## Miscellaneous Drill Jigs

Drill jigs can be made from many standard commercial items, such as: premachined section forms and bases; angle irons, parallels, and V-blocks; section components; and standard structural forms. Following are some examples of drill jigs constructed of these materials:

- Premachined sections were used to construct the drill jig shown in *Figure 5-16*. The base was made from a standard commercial T-section, to which a bushing plate made

Figure 5-15. *Boring jig.*

Figure 5-16. *Drill jig made from a standard T-section.* (Courtesy Standard Parts Co.)

from a flat section was attached with screws and dowels. The addition of standard jig feet, drill bushings, and clamps completed the assembly.

- The drill jig shown in *Figure 5-17* was built from a U-shaped, premachined cast-iron section and a flat section. It was built in 9.5 hours with a savings of 30% in build time.
- The drill jig shown in *Figure 5-18* was built from standard structural form materials. It consists of two pieces of angle iron welded to a channel iron to form a V-section. A larger angle iron was welded over the V-section to serve as a bushing plate and clamping mount. The addition of a standard thumbscrew, drill bushing, and section of rod completed the assembly.

## Wooden Drill Jigs

Wooden tooling is often overlooked. An example of its use for drilling is shown in *Figure 5-19*. The jig consists of a plywood base, two wood-side spacer rails, a plywood bushing plate joined together with wood screws, and four bolts that also serve as jig feet. To reduce wear and provide additional support, a sheet-metal plate is bonded to the top of the base. The part is located against a pressed-in roll pin and two more bolts through the side rail. The hand-knob clamp, thumbscrew,

Figure 5-17. *Drill jig made from a standard U-section.*

Figure 5-18. *Drill jig made from a standard structural form material (channel and angle iron).*

threaded socket, and serrated-type bushings are all standard tooling components.

## Polymer Drill Jigs

Some issues that require special consideration when using polymer materials for drill jigs include the following:

• Conventional drill bushings should not be used. Castable drill bushings are designed for this purpose and should be used.
• Drill bushings must be positioned in the exact location before pouring liquid epoxy, thermoplastic compound, or low-melt alloy.
• When designing epoxy tooling, the tool designer should understand the various construction methods available, along with the advantages and disadvantages of each method. He or she also should be familar with the many grades of epoxies, or seek help from the epoxy formulators, especially when designing complex tooling.

General construction of a cast epoxy drill jig is shown in *Figure 5-20.* A drill jig constructed by the laminating method is shown in *Figure 5-21.*

## Modified Vises

The drill jigs shown in *Figure 5-22* along with their respective workpieces were made by adding details to a commercially available vise. The first jig, *a,* was designed to drill a locating hole in a sprocket shaft. The second jig, *b,* was designed to both drill one 0.277-in. (7.04-mm) diameter

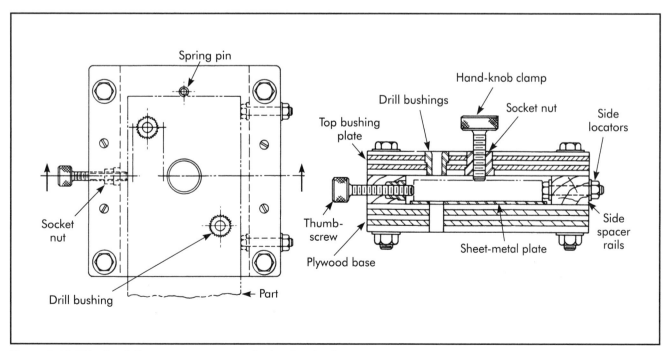

Figure 5-19. *Wooden drill jig incorporating standard tooling hardware.*

| Stock List | | | | |
|---|---|---|---|---|
| Detail No. | Number Required | Name | Stock Size | Material |
| 5 | 1 | Drill bushing (plastic type) | American type DGV 0.156 ID × 0.406 × 1 LG | Standard |
| 4 | 1 | Drill bushing (plastic type) | American type DGV 0.256 ID × 0.5 × 0.875 LG | Standard |
| 3 | 1 | Drill bushing (plastic type) | American type DGV 0.125 ID × 0.312× 0.75 LG | Standard |
| 2 | 1 | Body | Plastics, Inc. #RP-320-2 | |
| 1 | 1 | Assembly | | |
| Drill Jig | | | | |
| Drawn by: E.C. Key | | | Approved by: CD | |
| Date: 5-1-02 | | | Date: 5-6-02 | |
| | | | B—81236 | Rev. 1 |

Notes:

1. Apply plastic per instructions supplied with material

2. Mark—B.—81326

Figure 5-20. *Drawing of an epoxy drill jig.*

Figure 5-21. *Plastic drill jigs and scribe templates.* (Courtesy Ciba-Geigy Tooling Systems)

hole and also to drill, countersink, and tap a 0.3125–24 in. hole in a pivot block. The third jig, *c*, was designed to drill (1) No. 38 and (1) 0.094-in. (2.39-mm) diameter hole in a pivot pin. For positioning on the machine table, the vises are bolted to precision-ground sub-bases and located against a rail and stop pins as shown in *d*.

## Collet Fixtures

As shown in *Figure 5-23*, collet blocks and collet fixtures mounted vertically can be used to hold workpieces for drilling, tapping, and related operations. The collet fixture shown has been a.

b.

c.

d.

Figure 5-22. *Modified vises.* (Courtesy Heinrich Tools, Inc.)

mounted on a sub-base to permit long workpieces to pass through the fixture.

## Self-centering Vises

The self-centering vise shown in *Figure 5-24* was fitted with false jaws to center and clamp a cast-iron elbow for drilling and tapping. Form-fitting cast jaws perform a similar function.

## Drilling Accessories

The safety quick-acting drill vise shown in *Figure 5-25* features built-in recessed parallel jaw inserts to hold work square. A vertical V-groove in the jaw is provided to grip round workpieces. Also shown is the method of mounting the vise

Figure 5-23. *Air collet fixture.* (Courtesy Heinrich Tools, Inc.)

Figure 5-24. *Self-centering vise with false jaws.* (Courtesy Heinrich Tools, Inc.)

Figure 5-25. *Safety drill vise.* (Courtesy Heinrich Tools, Inc.)

Figure 5-26. *Quick divider.* (Courtesy Willis Machinery & Tools Co.)

Figure 5-27. *Universal drill jig.* (Courtesy Vise & Tool Co.)

to a table to prevent it from spinning or breaking loose from the table. The quick-dividing device shown mounted to an angle knee in *Figure 5-26* can speed up the drilling of equally spaced patterns of 2, 3, 4, 6, 8, 12, or 24 holes on a workpiece outside diameter. Or, when used without the knee, it can facilitate drilling equally spaced hole patterns on any bolt-circle diameter on the face of the workpiece.

The universal drill jig shown in *Figure 5-27* can be readily adapted for drilling a wide variety of workpieces. It features a solid jaw with an adjustable end stop and several movable bushing carriers, which can be oriented for different workpieces.

A mounting table for various drilling applications can be used for light drilling as well as electrical discharge machining (EDM), milling, boring, and jig grinding. The workpiece is located and held on the square ledges of the rails and a movable support block, which permits the fixture to be used for a wide range of workpiece sizes.

The application shown is a setup for a numerically controlled (NC) drill press.

A drill press clamp can be very useful for clamping a workpiece directly on the drill press table. It also can be used to guide a drill by replacing the clamp screw with a suitable drill bushing.

## DRILL JIG BUSHINGS

Drill jig bushings position and guide the tools that do the cutting. They are available in hardened steel or carbide. Bushings made of other materials, such as bronze and stainless steel, are available from bushing manufacturers by special order. Most bushings and liners can be supplied with the OD ground to industry standards, or left unground (type U) for custom grinding.

Drill jig bushings and liners may be specified by either the individual manufacturer's identification system or by the universally accepted American Society of Mechanical Engineers (ASME) designation system (ASME B94.33), which, regardless of the manufacturer, will ensure delivery of the proper bushing. Specifying the type of cutting tool and its diameter will ensure correct tool clearance.

### Headless-press-fit Bushings

Headless-press-fit bushings (types P and PU) are the most popular and least expensive. They are used in single-size cutting-tool applications where light axial loads are expected. Since they are permanently pressed into the jig plate, they are generally used where replacement is not anticipated during the expected life of the tool. They are ideal where the top of the bushing must be flush with the jig plate, or where hole spacing is too close to use headed bushings.

### Head-press-fit Bushings

Head-press-fit bushings (types H and HU) are used for permanent installations requiring greater bearing area or where heavy axial loads that could force the bushing through the jig hole are anticipated. Although designed for permanent installations, press-fit bushings are easily replaced, but at the expense of losing some mounting hole accuracy with each replacement. Since the bearing area extends beyond the jig plate, the thickness of the jig plate often can be reduced, thereby lightening the overall weight of the jig. Cases where the head must be flush with the jig plate require counterboring of the mounting hole in the jig plates.

### Slip Renewable and Slip-fixed Renewable Bushings

Slip renewable bushings (type S) and slip-fixed renewable bushings (type SF) are used with a headless liner (type L) or a head liner (type HL) where multiple operations, such as drilling and reaming or drilling and tapping, are to be performed on the same hole; or, where long production runs require occasional changing of the bushing to maintain jig integrity. They are available for use with lock screws (types LS and TW); round clamps (type CL); round end clamps (type RE); or flat clamps (type FC). Both diameters of the bushings and the ID of the liner are finish-ground to industry standards, while liners are available with the OD finish ground or unground.

When used for multiple operations, the first operation bushing is installed and the hole drilled. The bushing is then replaced with the second operation bushing and the second operation performed. This process is repeated until the hole is completely machined. The first operation bushing is then reinstalled and the procedure repeated on the next workpiece.

When slip-fixed renewable bushings are used on jigs to perform single operations on long production runs, they are usually secured in the fixed mode by a lock screw. When used to perform more than one operation on the same hole, they are secured in the slip mode for easy changing.

Type UL (UN-A-LOK®) liners are used with ASME slip-renewable bushings. These special liners eliminate the need for a bushing locking device. In use, the bushing is locked tight in the liner by the torque of the drill bit. UL liners must be installed with an arbor press.

### Oil-groove Bushings

Oil-groove bushings are designed to provide complete lubrication between the cutting tool and bushing when maximum cooling is required during a machining operation. Over two dozen separate groove patterns on the inside diameter (ID) are available.

## Bushings and Liners for Polymer, Castable, and Soft Material Tooling

Some bushing types have serrated or grooved ODs for casting in-place with epoxy resins, thermoplastic tooling compounds, or low-melt alloys. Another type is used for press-in installation in soft materials, such as aluminum, magnesium, wood, or masonite. One particular liner type serves the same function as the type UL, except that it is knurled for use in plastic tooling.

## Template Bushings

Template (Type TB) bushings are used with thin template materials ranging from 0.063–0.375-in. (1.60–9.53-mm) thick, to provide low-cost tooling. Installation is shown in *Figure 5-28*.

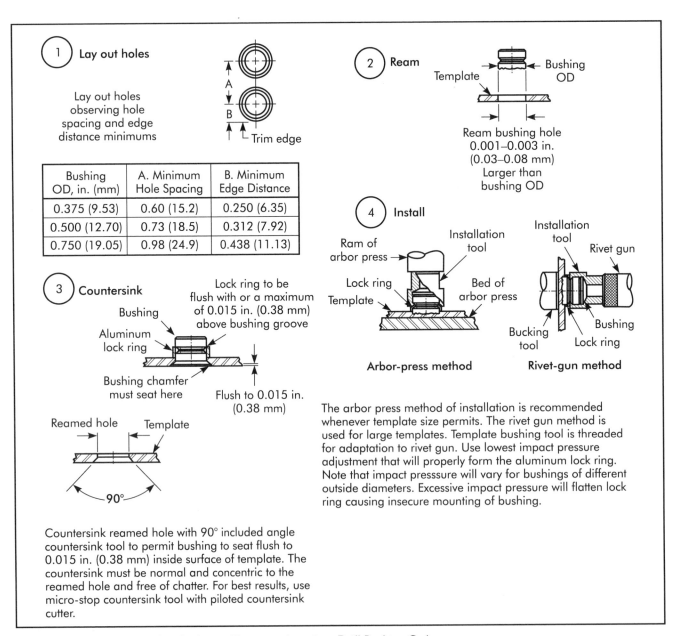

The arbor press method of installation is recommended whenever template size permits. The rivet gun method is used for large templates. Template bushing tool is threaded for adaptation to rivet gun. Use lowest impact pressure adjustment that will properly form the aluminum lock ring. Note that impact presssure will vary for bushings of different outside diameters. Excessive impact pressure will flatten lock ring causing insecure mounting of bushing.

Figure 5-28. *Installing template bushings.* (Courtesy American Drill Bushing Co.)

## Rotary Bushings

Rotary bushings feature precision-tapered roller or ball bearings capable of handling high thrust and/or radial loads encountered in some jig applications, such as supporting a piloted cutting tool for extremely close machining.

## Drill Bushing Tips and Accessories

Drill bushing tips (see *Figure 5-29*) are used with automatic self-feed drill motors. Most are constructed of two pieces consisting of a collar that contains the screw threads and alignment diameter, and lock flange. It also contains a pressed-in shank, which is a plain bushing with an ID ground to guide and support the cutting tool, and an OD ground to suit the liner bushing mounted on the jig.

## Mounting of Bushings

Use of a lock-liner bushing and lock nut is the conventional method of mounting the drilling unit to a jig or fixture. To install it, a hole is bored into the jig or fixture to suit the lock-liner bushing. It is then assembled to the jig or fixture and held in place with a lock nut.

Direct jig mounting is an alternate method, used when holes are so closely spaced that lock-liner bushings cannot be used. In this case, lock screws are mounted directly on the jig and a hardened, headless liner is pressed into the jig to accept the drill bushing tip shank.

Lock-strip mounting is another mounting method used for holes that are closely spaced. It features a lock strip along each side of a row of holes in the jig. A hardened, headless liner to accept the shank of the drill bushing tip is also pressed into the jig with this method.

## DRILL BUSHING INSTALLATION

To ensure accuracy in the workpiece, drill bushings must be properly located and installed using the following rules:

- Mounting holes should be round and properly sized to prevent bushing closure and jig-plate distortion. For this reason, it is recommended that the mounting holes be jig-bored or reamed to size. Headless-press-fit and liner bushings generally are installed with a diametral interference of 0.0005–0.0008 in. (0.013–0.020 mm), while headed-

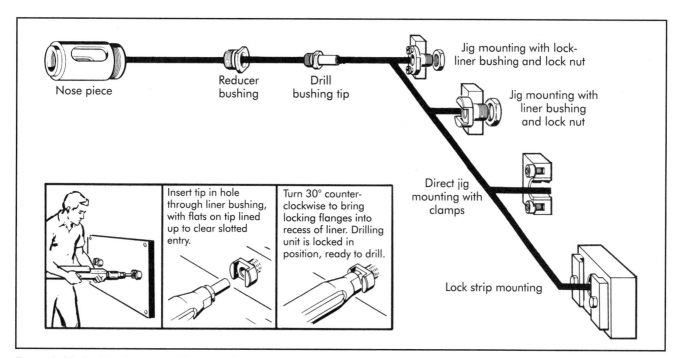

Figure 5-29. *Drill bushing tips*. (Courtesy American Drill Bushing Co.)

press-fit bushings are generally installed with a diametral interference of 0.0003–0.0005 in. (0.008–0.013 mm). Interference greater than this may reduce the ID of the bushing to the point where the tool may seize, or, in the case of liners, prevent insertion of a renewable bushing. On the other hand, too little interference will result in a loosely installed bushing that may spin or be forced out of place. Drill bushings for special applications should be installed in accordance with the individual manufacturer's recommendations.

- Sufficient chip clearance, as illustrated in *Figure 5-30*, should be provided between the bushing and workpiece to enable chip removal, except in cases where extreme accuracy is required. In this case, the bushing should be in direct contact with the workpiece. In most cases, a clearance of 1–1.5 times the

Figure 5-30. *Recommended clearance between workpiece and bushing.* (Courtesy American Drill Bushing Co.)

bushing ID should be used when machining materials such as cold-rolled steel, which produces long stringy chips. A clearance of one-half the bushing ID is recommended when machining materials such as cast iron, which produces small chips. Excessive chip clearance should be avoided because most cutting tools are slightly larger at the cutting end due to back taper, and excessive clearance will reduce the guiding effect of the bushing, resulting in less accurate drilling.

Chip control should be closely monitored after the jig is put into production. As a general rule, if the chips have a tendency to lift the bushing, more clearance is needed. If the cutting tool wanders or bends, less clearance is needed.

When performing multiple operations, such as drilling and reaming, slip-renewable bushings of different lengths should be used to provide optimum chip clearance for each operation as shown in *Figure 5-31*.

Clearance between the bushing and the workpiece also must be provided when drilling wiry metals, such as copper, which tends to produce secondary, or minor, burrs around the top of the drilled hole, as shown in *Figure 5-32*. This, in turn, causes the jig to lift from the workpiece or makes it difficult to remove the workpiece from side-loaded jigs. A burr clearance of one-half the bushing ID is recommended. The jig itself must also provide clearance for the primary, or major, burrs that form as the tool exits the workpiece.

To properly support and guide the cutting tool, the length of the drill bushing under normal circumstances should range between 1.5–2.5 times the bushing ID. The jig plate supporting the bushing should be thick enough to sustain the bushing. Usually, between 1–2 times the cutting tool thickness will be sufficient.

When drilling irregular work surfaces, the ends of the bushings should be formed to the contour of the workpieces as shown in *Figure 5-33*. This will prevent the tool from pushing off center. The distance between the workpiece and the bushing should be held to a minimum under these circumstances.

Sometimes, hole center distances are so close that there is not enough room for the bushings.

Figure 5-31. *Slip-renewable bushings provide chip clearance for multiple operations.* (Courtesy American Drill Bushing Co.)

Figure 5-32. *Burr clearance.* (Courtesy American Drill Bushing Co.)

In these cases, extra thin wall bushings or standard headed or headless bushings that have flats ground on them might work. Sometimes a single, hardened steel insert with two or more holes must be used.

## JIG DESIGN EXAMPLE

The most efficient method of designing any jig is to follow a systematic approach. *Figure 5-34* shows a part in which a systematic approach is being used. The considerations and their effects during a systematic approach to jig design are intended to illustrate, in a tabular form, the step-by-step process before determining a final jig design. The list in *Table 5-1* is only meant as an example and not as a rule for all jigs. Just as each part has its peculiarities, each tool also has its own specific characteristics that must be addressed.

Each individual element of the jig must first be thought out and sketched. The elements then are brought together in a sketch of the complete tool. The following represents one method to systematically design a jig around the proposed workpiece.

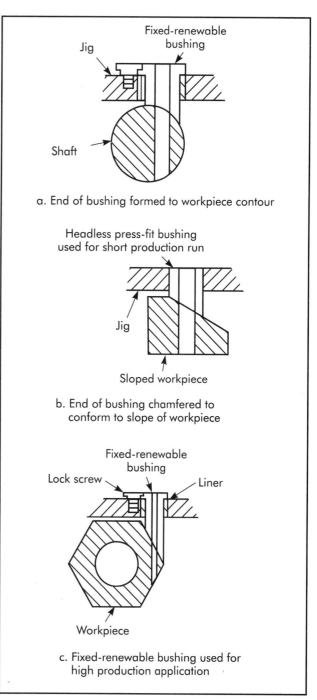

a. End of bushing formed to workpiece contour

b. End of bushing chamfered to conform to slope of workpiece

c. Fixed-renewable bushing used for high production application

Figure 5-33. *Drilling irregular work surfaces.* (Courtesy American Drill Bushing Co.)

Figure 5-34. *Preliminary jig design.*

The simplest and easiest way to begin a jig design is to sketch the part in three or more views (*Figure 5-35*). The sketch should be made to a scale size of the actual part, and the jig must be sketched to the same scale size. This sketch can be made either manually or by using a CAD sketching program. Make sure to allow sufficient room between each view to permit the remaining elements to be sketched without crowding. It is sometimes better to sketch the part in a different color, such as red, to allow it to stand out in the sketch.

Once the part is sketched, the next step is to add the locators and supports (*Figure 5-36*). Again, these details must be drawn to the proper scale and size. If commercial locating devices are specified, check a catalog to find the exact size of each area of the locator or support. Each locating element must be drawn as it would appear in each view. Many tooling suppliers provide computer-aided design (CAD) files of their products, making the design process easier.

Next, the tool body should be added to the sketch. Since this element is an integral part of the entire tool, it should be drawn to the proper proportion and size. Once the part, locators, supports, and tool body are sketched, a thorough inspection must be made to ensure there is no

interference between these elements. Look for areas where the elements are positioned close together. When constructed, these points could interfere with each other. Make sure the locator design will permit adequate space for chip and coolant removal. As a preliminary check, make sure there are no redundant locators and that the part can only be loaded correctly. Any location problem must be resolved before the clamping element is added to the jig.

Once the jig design has been checked and no conflict between the elements exists, the clamping device can be added to the design (*Figure 5-37*). When designing the clamping device, remember that this is the area of the tool that can either save or add time in production. Always specify clamping devices that are simple to use and fast-operating. It is important to make sure that the clamps do not affect the parts surface in a negative way, especially on machined surfaces. Spending an additional hour in construction time to make a better clamp could easily save thousands of dollars in production costs. Once a clamp design is determined, it should be sketched into the existing drawing of the jig. Remember to include the clamping device in each view. With the clamp sketched in place, once again begin a thorough inspection of the design to ensure there is no conflict or interference between the various elements.

The next step in the initial design of the jig should be to add the drill bushings (*Figure 5-38*). The location of these bushings is critical to the accuracy of the jig and they should be placed carefully. The tolerance values specified for the bushing placement should reflect a positional relationship of about 50% of the part tolerance. So, if a part dimension is specified as 1.000 in. ±0.010 in. (25.40 mm ±0.25 mm), the bushing location should be specified as 1.000 in. ±0.005 in. (25.40 mm ±0.13 mm). A general rule of 50% of the part tolerance applied to the tool will not only provide parts within the required accuracy, but also reduce the cost of constructing the jig since the tolerance value allows the toolmaker more space to place the bushings. In cases where extreme accuracy or high-volume production tools must be made, a tight tolerance

## Table 5-1. Considerations for jig design

| Product Analysis Considerations | Effect on Jig Design |
|---|---|
| Size, weight | Relatively small; lightweight construction. |
| Wall thickness and general shape provide rigidity | No special supporting and clamping methods are necessary to prevent part distortion. |
| Surface finish | Clamps, locators, or other details must not incorporate sharp points or edges to mar the flat surfaces of the part. |
| Machinability index | Permits metal removal with moderate machining forces. |
| Angular surface relation (95° ±15') | Clamps and locators must not change close angular tolerance. |
| Surface flatness tolerance, ±0.010 in. (0.25 mm) | Clamping forces must not alter flatness tolerance. |
| Normality of hole axes | Clamping and locating must maintain axes normal to respective flat surfaces. |
| Location (±0.005 in. [0.13 mm]) and diameter (±0.0156 in. [0.396 mm]) tolerances on 0.1405 in. (3.569 mm) holes | Standard drill bushings are satisfactory. |
| Diameter tolerance (+0.001, –0.000 in. [+0.03, –0.00 mm]) on 0.50 in. (12.7 mm) hole | Tolerances can be held with drill and ream jig, suitably bushed. |
| Location tolerance (1.000 ±0.001 in. [25.40 ±0.03 mm]) on 0.500 in. (12.70 mm) hole | Tolerance: Cannot be held with single jig. Can be held with a drill jig and a separate jig with a jig ground bushing for reaming. Can be held with a drill jig for rough-drilling all holes and a separate simple holding fixture for finish boring of the 0.500 in. (12.70 mm) hole. |
| Maximum possible mislocations, in. (mm)<br>1. From true position of liner in jig  0.0005 (0.013 mm)<br>2. Due to fit of bushing and liner  0.0004 (0.010 mm) hole.<br>3. Due to fit of reamer and bushing  <u>0.0002 (0.005 mm)</u><br>Total  0.0011 (0.028 mm) | Design decision: most accurate results will be obtained with a tumble jig for drilling all holes and a separate holding fixture for boring the 0.500 in. (12.70 mm) |

| Operation Considerations | Effect on Jig Design |
|---|---|
| Operation 1, grinding 2 × 3, 2 × 1-3/4 in. (50.8 × 76.2, 50.8 × 44.5 mm) surfaces, with horizontal disk grinder, according to process sheet | No fixture required. Grinder table provides adequate holding, positioning, and locating facilities. |
| Operation 2, drilling (10) 3.569 mm diameter holes (1) 12 mm diameter pilot hole for 12.7 mm diameter hole on number 2 two-spindle drill press according to process sheet | Torque and feed force allow handheld tumble-jig and design. Clamps must clear drills. Adequate three-point support must be provided for stability of jig on feet and normality of hole axes. |
| Operation 3, bore 12.7 mm diameter hole on machine according to process sheet | Simple holding fixture or vise, not shown; normality of hole axis is provided by shims or pins. |

| Machine Considerations | Effect on Jig Design |
|---|---|
| Option 2, number 2 two-spindle drill press according to process sheet | Tooling area, bed, size, and kind of chuck will not limit tumble-jig design. |
| Operation 3, machine according to process sheet | Simple clamping, positioning, and supporting provided to ensure 51 × 44 mm face at 90° to boring bar (this fixture not shown). |

Table 5-1. (continued)

| Operator Considerations* | Effect on Jig Design |
|---|---|
| Operation 2, loading, unloading, and fixture handling | Operator loads, unloads, and clamps with lower thumbscrew with left hand; tightens hand knob with right hand. The small tumble jig is easily turned for drilling second set of holes. |
| Operation 3, loading, unloading, and fixture handling | The simple fixture (not shown) will not be handled; unloading, loading, locating, and clamping are simple, requiring nontiring operator motions. |

| Production Considerations | Effect on Jig Design |
|---|---|
| 200 parts per month, 2,400 per year; possible future production of 8,000 per year | Air clamping, indexing, etc., and various automated designs are not justified by production rates and quantity. |

| Economic Evaluation | Effect on Jig Design |
|---|---|
| Jig cost, $360 design and make; $0.15 per part, at 2,400 annual rate<br><br>Boring fixture cost, $146.40 design and make; $0.061 per part, at 2,400 annual rate. Total fixture costs are $506.40; $0.21 per part<br><br>Operations 2 and 3 costs total $100.65; $0.5032 per part, at 2,400 annual rate | For 2,400 annual production rate, reducing the time to operate the jig and fixture by faster clamping, etc., would not be justified because of increased fixture cost. Setup and run time for both operations is 5.49 min per piece. Cost studies show that the cost of timesaving fixture details, such as air cylinder clamping, can be absorbed by the reduction possible just in labor cost to operate the fixtures, when the setup and run time are considerably longer and production is 5 to 10, or more, times the present annual production. |

* There are no particular problems of operator and machine safety or operator fatigue.

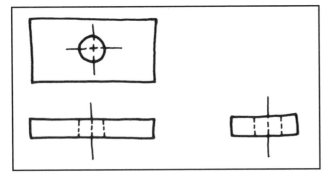

Figure 5-35. *Sketch of the part.*

Figure 5-36. *Sketch in locators and supports.*

could easily nullify the cost-effectiveness of the jig and result in a loss of money rather than an increase in profit from using the tool.

When selecting the bushings to use in the jig, always specify bushings that will permit adequate support for the cutting tool. If space is not a problem, use head-type bushings. If, however, space is limited, headless bushings should be used. In those cases where the bushings will be replaced frequently, either fixed-renewable or slip-renewable bushings should be specified. If these bushings are used, remember to include liner bushings. Once the bushings are added to the sketch, check the complete tool once again to ensure the elements will not interfere with each

Figure 5-37. *Add the clamping device to the design.*

Figure 5-38. *Add the drill bushings.*

other or the operation of the jig. At this stage, the clearance between the drill bushings must be determined. If clearance is desired, provisions should be made to ensure adequate space exists in the tool. If, however, no clearance is desired, the designer should make sure the jig plate and bushings will fit flush against the part.

The initial sketch of the basic jig is now complete; however, the relationship between the jig and the machine tool it is to be used with must now be checked. Using the size specifications for the machine tool selected for the operation, sketch the relationship between the jig and the machine table, spindle, and frame. Look for areas where any interference might occur. Sketch the proposed mounting device on the table to make sure the T-slots are the proper size, or in

cases where strap clamps are used, make sure they will not interfere with the operation of the jig. Determine if the specified cutting tool will perform the desired operation without interfering with the jig body or other elements. Check the clearance between the cutting tool and machine tool members. Always specify standard, off-the-shelf cutting tools whenever possible. Do not modify a standard tool to accommodate a flaw in the jig design. It is simpler to alter the jig design than to justify the additional costs of modifying standard tools. All tooling problems must be resolved on the drawing board, not in the shop during production.

Once the initial sketches are complete and the designer is convinced the design is sound and workable, the final tool drawings are initiated. Here again, the designer should look for any problem areas where the jig elements interfere with each other or the operation of the machine tool. In cases where the jig is to be used with numerically controlled or other automatic machinery, the designer should check the program to ensure the jig will not interfere with the operation of the machine tool. Look for points such as the height of the tool during the return or tool change cycle, or when moving from hole to hole. Make sure the actual height of the highest element of the jig is considered, not just the height of the jig plate. When manual machines are specified to use the jig, make sure the length of spindle travel is sufficient to permit easy use of the tool. Again, make sure the actual height of the tool will permit the jig to be easily moved under the spindle. Little would be saved if the table of the machine tool had to be lowered and raised to clear an obstruction or move the jig from hole to hole.

## Plate Jig Design Example

Using the methods previously outlined, the process of designing a plate jig for the part shown in *Figure 5-39* is as follows.

The first step is to inspect the part to determine the basic requirements and condition of the workpiece material. In this case, the requirement is for a jig to drill two 0.188-in. (4.78-mm) diameter holes in the locations shown in the figure. The part is 4.00 × 2.00 × 0.75 in. (101.6 × 50.8

Figure 5-39. *Part to be designed.*

Figure 5-41. *Position the locators in the sketch.*

× 19.1 mm) cold-drawn AISI 1030 barstock that weighs 1.71 lb (0.8 kg). The production plan calls for a one-time production run of 600 pieces, so the jig must be made as inexpensively as possible, yet the production rate must be rapid.

Next, sketch the part in three views as shown in *Figure 5-40*. If the part is small enough, the sketch should be made full size. This permits the designer to see the full-size tool and eliminates any errors when converting to a scale size. Remember, it is often a good idea to sketch the part in a different color to allow it to stand out and to minimize the chance of misinterpreting the meaning of the lines in the sketch.

Once the part is sketched, the locators should be positioned in the sketch (*Figure 5-41*). Here, a simple 3-2-1, or six-point locational arrange-

ment is used, since the part does not present any spacial locational problems. The three locators on the primary datum surface, in addition to establishing the location of this plane, also serve as risers to provide the necessary clearance between the drill bushings and the part. If a zero clearance is desired, the three locators on this surface can be eliminated and the part located against the flat jig plate. Secondary locators, positioned on the long side of the part, should be spaced about 3.00 in. (76.2 mm) apart to gain the maximum practical distance between these locators. The tertiary locator should be positioned approximately 1.50 in. (38.1 mm) from the secondary locators to find the end position of the part. The length of the secondary and tertiary locators is specified as 1.375 in. (34.93 mm). This will permit the accurate location of the workpiece and prevent the ends of the locators from interfering with the machine table while drilling the part. With the locators in place, the remainder of the tool body, or in this case, the jig plate, is sketched in (*Figure 5-42*). The thickness specified for this jig plate is 0.50 in. (12.7 mm). This thickness will accommodate the headless-press-fit bushings that are specified, and provide adequate support for the drill. If head-press-fit bushings are specified, a thinner jig plate could be used.

The next step in the design of the jig is to add the clamping device. This type of plate jig can use several different clamping devices; however,

Figure 5-40. *Preliminary sketch.*

Figure 5-42. *Sketch in the jig plate.*

the simplest and fastest-acting clamp would be an L-shaped clamp, as shown in *Figure 5-43*. The clamp is mounted as shown and is activated by the screw in the long leg of the L. As the screw is tightened, the part is pressed against the secondary locators. At the same time, the short leg of the L presses against the part, forcing it against the tertiary locator. This type of locator may take a little longer to make than a simple screw clamp positioned at the side and end, but the faster operation of this clamp will more than offset the additional costs of its manufacture.

To add drill bushings to the design, the first step is to locate the bushings over the holes

sketched in the part (*Figure 5-44*). Next, the dimensions and tolerances of the bushing location must be established using the 50% rule.

The initial sketch of the jig is now complete. Since the holes are only 0.188 in. (4.78 mm) in diameter, the jig can be held by hand and does not require an additional holding device. Likewise, since the jig is so small, there is little chance of its interfering with the operation of the machine tool, so the machine elements may or may not be sketched, depending on the desire of the designer. The final step in the design of this jig is to complete the final tool drawings. Remember, look for any interferences that may have been overlooked in the sketch.

## Reaming

Jig design for reaming is basically the same as for drilling, which has been discussed throughout the chapter. The main difference is the need to hold closer tolerances on the jigs and bushings, and provide additional support to guide the reamer. For long holes, it is essential to guide the reamer at both ends, as shown in *Figure 5-45a*, using special piloted reamers designed for this purpose. Jigs should be designed so the pilot enters the bushing before the reamer enters the workpiece, and it remains piloted until the reaming operation is completed. For short holes, the reamer is usually guided at one end only, as illustrated in *Figure 5-45b*, with the bushing sized to fit the OD of the reamer. Additionally, bushings

Figure 5-43. *The simplest and fastest clamp for this part is the L-shaped clamp.*

Figure 5-44. *Add drill bushings.*

Figure 5-45. *Fixtures for guiding reamers.*

for reaming are generally longer than for drilling, usually three or four times the reamer diameter. Chip clearance is generally less for reaming than for drilling, varying from one-fourth to one-half the tool diameter down to a maximum of 0.125–0.240 in. (3.18–6.10 mm), regardless of the reamer diameter.

Bushing bores must be closely controlled. Bushings that are too small can cause tool sei-

zure and breakage. Bushings that are too large will result in bell-mouthed or out-of-round holes. Data in another text can be used as a guide (Boyes 1989).

Carbide bushings should be considered for long production runs or when abrasive conditions are present. Roller or ball bearing, rotary-type bushings also provide maximum wear while maintaining close tolerances under high loads.

## REFERENCE

Boyes, W.E. 1989. *Handbook of Jig and Fixture Design*, 2nd ed. Dearborn, MI: Society of Manufacturing Engineers.

## REVIEW QUESTIONS

1. What are the two general classifications of jigs?
2. What are the two categories of jigs?
3. How does a table-plate jig differ from a plate jig?
4. What is the function of a drill bushing?
5. What is the solution when hole center-distances are so close that there is not enough room for drill bushings?
6. What is the normal clearance between the bushing and workpiece for materials that produce small chips? Long chips?
7. In cases where a drill bushing is not used, how is the jig treated in order to guide the drill?
8. What is the simplest and easiest first step in designing a jig for the workpiece shown in *Figure A*?
9. Develop a preliminary design for a simple jig for the workpiece shown in *Figure B* for a nonrepeating run of 300 parts.
10. What type of bushing should be used in conjunction with a renewable bushing?
11. What are the main disadvantages of template jigs?

Figure A.

Figure B.

# ◄6►

# Fixture Design

Fixtures are workholders designed to hold, locate, and support a workpiece during the machining cycle. Unlike jigs, fixtures do not guide the cutting tool, but they provide a means to reference and align the cutting tool to the workpiece. Fixtures normally are classified by the machine with which they are designed to be used. Sometimes a subclassification is added to further specify the fixture classification. This subclassification identifies the specific type of machining operation the fixture is intended to perform. For example, a fixture used with a milling machine is called a milling fixture; however, if the operation it is to perform is gang milling, it may also be called a gang-milling fixture. Likewise, a bandsawing fixture designed for slotting operations may be referred to as a bandsaw-slotting fixture.

The similarity between jigs and fixtures normally ends with the design of the tool body. For the most part, fixtures are designed to withstand much greater stresses and tool forces than jigs, and should always be securely clamped to the machine. For these reasons, the designer must be aware of proper locating, supporting, and clamping methods when fixturing any part.

When designing any fixture, several considerations in addition to the part must be addressed to complete a successful design. Cost, production capabilities, production processing, and tool longevity are some of the points that must share attention with the workpiece when a fixture is designed. This chapter provides an explanation of the basic considerations that must be addressed before a design is completed and a fixture built. As with other special tools, these should be modified to suit the particular workpiece or production situation.

## GENERAL CONSIDERATIONS

As with all tooling, the first consideration in fixture design is the cost versus the benefit. The production quantity, rate, or accuracy must warrant the added expense of special tooling. In addition, the fixture must pay for itself with savings derived from its use in as short a time as possible.

Once the decision is made that a fixture is required, the part print, process specifications, and other production documents are studied to determine the best and least expensive type of tool. Details such as locating and supporting the part, clamping, and tool referencing must all be considered in the initial design.

The tool designer should know enough of economics to determine the relative cost/benefit relationship of any tool design. Whether temporary or permanent tooling should be used or just how much money should be requested for a workholder are decisions the designer must make to present management with a realistic picture of production costs. Not only is a tool designer expected to prepare accurate cost estimates, but these estimates must be completely defensible.

### Machine

The machine tool specified for a machining operation should be examined. Every detail that might affect the mounting and operation of the fixture must be addressed. Table sizes, table travel on all axes, spindle size and movement, spindle swing, distance between centers, and workholder mounting methods are typical examples of points the designer must consider before beginning a fixture design.

A major consideration is the condition of the workpiece to be held by the proposed workholder for the next operation to be performed. This includes the physical characteristics of the workpiece (that is, whether it is round, irregular, large, heavy, of weak or strong sections, and the like). The operation to be performed and the physical characteristics will dictate whether the workpiece must remain stationary or move along a definite path relative to the cutting tool. Different machine tools have the means to provide the needed motions for the workpiece and the cutting tools. Most of these motions are straight-line or rotary, or combinations of both. Some operations appear to be producing and requiring an irregular path of the cutting tool on the workpiece. This seemingly irregular path is often the result of the combination of several straight-line and rotary motions. Except for speed consideration and the

Figure 6-1. *Machining a valve body by either (a) workpiece rotation or (b) tool rotation.*

degree of obtainable simplicity, only the relative motion of the cutting tool to the workpiece is of importance. For instance, the turning of flanges on a valve body or a flanged-T pipe fitting could be performed with either the workpiece or the tool revolving (*Figure 6-1*).

Whether the workpiece or cutting tool moves in a straight line, revolves, or moves in some combination of both, the design requires careful coordination of the workholder to the workpiece and the workholder to the machine tool. Operations in which the workpiece revolves require great care in the attachment of the workholder to the machine tool and the means of actuation of the workholder. Unbalanced masses in the workholder and the workpiece must be minimized by proper balancing. This is particularly true in high-speed applications, such as turning with tungsten carbide, cermet, ceramic, cubic-boron nitride (CBN), and diamond cutting tools.

The weight and size of a workpiece influence what type and size of machine tool can or should be used for a particular operation. The combined weight of the workholder and part must be carefully matched to the capacity of the ways, bed, table, and spindles of the machine tools. Excessive weight may cause distortion in the machine tool, safety concerns, and produce inaccurate work.

In cases where the specified machine is not the best choice, or where the machine cannot perform the required operation, the designer should consult the process engineer to determine if another machine tool can be used. While it is good practice to fully utilize a machine's capabilities, overloading a small machine tool is not efficient. Often, money can be saved by using a larger machine tool. Running a small machine to its maximum limits or capabilities over a long production run will not only shorten the life expectancy of the machine tool, but may also fail to give the desired degree of precision. If there is any question as to the estimated power required for a job and the available power in the machine tool, the designer or process engineer should select the machine with more power available than that required for the operation. Before any fixture design is finalized, the designer must make sure the machine tool will accommodate the fixture in every respect.

When designing a fixture, the cutting tool is another factor that must be carefully evaluated. As with jigs, cutters specified for fixtures should, whenever possible, be standard, off-the-shelf sizes. Designing a special cutter when a standard cutter would work is very costly. Even if the standard cutter requires a slight modification of the fixture, the benefits and savings derived over the life of the fixture would more than pay for the additional design time.

## Process

The tool designer and process engineer should work together to determine the best method to produce the part. The optimum situation in fixturing is where the part can be clamped one time and not removed from the fixture until the part is completely machined. This reduces costly part handling and minimizes the chance of errors when switching the part from one fixture to another. In cases where the part must be removed from the fixture, the tool designer and process engineer should find methods to process the part that require the least amount of fixture changes. When more than one machine is required, consider moving the fixture and part from machine to machine.

To achieve the lowest cost per part in production, the fixture must be fast-operating, easy to load and unload, and have a positive, foolproof method of locating the part. In addition, when the volume of production permits more money to be spent, semiautomatic or automatic tooling should be considered. As the volume of production increases, the opportunity to save more per part increases dramatically. Multi-part machining and power clamping are other areas that could be explored to reduce the cost per part in production.

On all operations, the magnitude and direction of the forces produced by the material-removing operation determine the necessary holding forces. Cutting forces must be held within limits so the part cannot be distorted to an amount that would affect the required accuracy. Rigidity and strength of the workpiece limit the applicable holding forces and the speed and amount of metal removal per unit of time. A thin-walled part may not be able to sustain heavy cutting and holding forces without distortion or damage. The mounting or attachment of the workholder or the workpiece to the machine tool should be arranged so the forces produced in the material-removing operation are absorbed by the strongest and most rigid parts of the machine tool. Cutting forces should tend to hold the workholder against the bed of the machine rather than lift it away. Projection between the point at which the cutting force is applied and the nearest support should be minimized. Cutting forces that act parallel to the bed, table, and face plate should be applied as close to the bed as possible. Cutting forces should never be permitted the advantage of a large lever arm that could increase the tendency to loosen or pry away the workpiece and the workholder from their attachment. This is in contrast to holding forces, where the effect of a large lever arm or mechanical advantage is always desirable.

*Figure 6-2* illustrates a lathe chuck and a short workpiece to be turned. The cut is relatively close to the supporting spindle bearing and no difficulty is expected. Increasing the workpiece length, however, will give the cutting tool greater leverage; higher side thrust will be produced and may cause difficulties in the spindle bearing. The workpiece will deflect much more if the same size

Figure 6-2. *Minimizing cutting force by applying holding force as near as possible to the point of tool application.*

chip is removed as on the short piece. The results may be inefficient material removal, tool chatter, low accuracy, and poor tool life.

## Product

In conjunction with general, machine, and process considerations, the part also must be a major factor in the design of a fixture. All relevant details, such as location and clamping surfaces, clamping methods, areas requiring machining, the amount of machining required, and the degree of surface finish desired all must be considered in the initial phase of fixture design. While each of these design considerations has been treated separately in practice, all are usually considered simultaneously. So, when designing a fixture, constantly consider all of the effects a design decision may have on the complete, overall function and operation of the tool. Remember, changes in a fixture design are much easier and less expensive to make while the tool is still on paper. Changes or corrections in a fixture design are very expensive when they must be made during production or after the tool is built.

## TYPES OF FIXTURES

Fixtures are classified either by the machine they are used on, or by the process they perform on a particular machine tool. However, fixtures also may be identified by their basic construction features. For example, a lathe fixture made to turn radii is classified as a lathe-radius turning fixture. But if this same fixture were a simple plate with a variety of locators and clamps mounted on a faceplate, it is also a plate fixture. Like jigs, fixtures are made in a variety of forms. While many fixtures use a combination of different features, almost all can be divided into five distinct groups. These include plate fixtures, angle-plate fixtures, vise-jaw fixtures, indexing fixtures, and multipart or multistation fixtures.

### Plate Fixtures

Plate fixtures, as their name implies, are constructed from a plate with a variety of locators, supports, and clamps (*Figure 6-3*). They are the

Figure 6-3. *Plate fixtures.*

most common type of fixture. Their versatility makes them adaptable for a wide range of different machine tools.

Plate fixtures may be made from any number of different materials, depending on the application of the fixture. For example, if a very large fixture is needed, and it is only required to make a few parts, aluminum or magnesium plate may be selected to keep the weight of the fixture to a minimum. If, however, weight is not a factor and a very large number of parts must be made, another material such as tool steel may be selected. Likewise, if the part or process requires a material resistant to corrosion, a nickel-based alloy might be selected. But, as is usually the case, a combination of different materials may be used to construct the plate fixture. The part being machined and the process being performed are the only guides the tool designer has when selecting material.

### Angle-plate Fixtures

The angle-plate fixture (*Figure 6-4*) is a modified form of plate fixture. Here, rather than having a reference surface parallel to the mounting surface, the angle-plate fixture has a reference surface perpendicular to its mounting surface. This construction is useful for machining operations performed perpendicular to the primary reference surface of the fixture.

Figure 6-4. *Angle-plate fixture.*

Another variation of the basic angle-plate fixture is the modified angle-plate fixture (*Figure 6-5*). This design differs from the basic angle plate in that where the angle plate fixture is designed to be at a 90° angle to its mounting surface, the modified angle plate is made to accommodate angles other than 90°.

## Vise-jaw Fixtures

Vise-jaw fixtures are basically modified vise-jaw inserts machined to suit a particular workpiece. In use, modified vise jaws are installed in place of the standard, hardened jaws normally furnished with milling-machine vises. Vise-jaw fixtures are the least expensive type of fixture to produce, and since there are so few parts involved, they are also the simplest to modify. *Figure 6-6* shows several examples of parts easily fixtured with this type of workholder.

The principal advantage of using a vise-jaw fixture is that only the locating elements need to be constructed to suit each part. The milling-machine vise contains the clamping elements and the means to attach the fixture to the machine

Figure 6-6. *Vise-jaw fixture.*

table. If simple or compound angles need to be machined, a vise with these capabilities may be used. The only limitations to using this type of fixture are the size of the part and the capacities of the available vises.

## Indexing Fixtures

Indexing fixtures (*Figure 6-7*), like indexing jigs, are used to reference workpieces that must have machine details located at prescribed spacings. The typical indexing fixture will normally divide a part into any number of equal spacings, such as those used for geometric shapes or gears. Some may be used to locate and reference a

Figure 6-5. *Modified angle-plate fixture.*

Figure 6-7. *Indexing fixtures.*

workpiece for unequal spacings. Regardless of the configuration of the workpiece, indexing fixtures must have a positive means to accurately locate and maintain the indexed position. The most common device used for location and indexing is a simple indexing pin, as shown in *Figure 6-7*.

## Multipart Fixtures

Multipart or multistation fixtures (*Figure 6-8*) are normally used for one of two purposes: to machine multiple parts in a single setting or

Figure 6-8. *Multipart or multistation fixture.*

machine individual parts in sequence, performing different operations at each station.

## FIXTURE CLASSIFICATIONS

As previously mentioned, fixtures are normally classified by the machine tool with which they are designed to be used. The following is a brief description of the major fixture classifications and their basic design characteristics.

### Milling Fixtures

Milling fixtures are the most common type of fixture in use. The simplest type of milling fixture is a milling vise mounted on the machine table. However, as the workpiece size, shape, or complexity becomes more sophisticated so does the fixture. The following are a few points to be considered when designing fixtures for milling operations.

- The design should permit as many surfaces of the part to be machined as possible, without removing the part.
- Whenever possible, the tool should be changed to suit the part. Moving the part to accommodate one cutter for several operations is not as accurate or efficient as changing cutters.
- Locators must be designed to resist all tool forces and thrusts. Clamps should not be used to resist tool forces.
- Clearance space or sufficient room must be allotted to provide adequate space to change cutters or load and unload the part.
- Milling fixtures should be designed and built with a low profile to prevent unnecessary twisting or springing while in operation.
- The entire workpiece must be located within the fixture's area. In cases where it is either impossible or impractical, additional supports or jacks must be provided.
- Chip removal and coolant drainage must be considered when designing the fixture. Sufficient space should be permitted to easily remove chips with a brush.
- Set blocks or cutter-setting gages must be provided in the fixture design to aid the op-

erator in properly setting up the tool in production.

## Lathe Fixtures

The same basic principles that apply to the design of milling fixtures also apply when designing lathe (turning) fixtures. The only major difference between the two is the relationship between the workpiece and the cutting tool. In milling, the workpiece is stationary and the cutting tool revolves. However, with turning operations, the workpiece revolves and the cutting tool is stationary. This situation creates another condition the tool designer must deal with—centrifugal force. The complete fixture must be designed and constructed to resist the effects of the rotational, or centrifugal, forces present in turning. The following are a few basic design principles that apply to lathe fixtures.

- Since lathe fixtures are designed to rotate, they should be as lightweight as possible.
- Lathe fixtures must be balanced. While perfect balance is not normally required for slow-speed turning operations, high rotational speeds require the fixture to be well-balanced. In most fixtures, balance is achieved by using counterweights positioned opposite the heaviest part (or area) of the workpiece.
- Projections and sharp corners should be avoided, since these areas will become almost invisible as the tool rotates and they could cause serious injury.
- Parts to be fixtured should, whenever possible, be gripped by their largest diameter or cross section.
- The part should be positioned in the fixture so that most of the machine operation can be performed in the first fixturing.
- Clamps should be positioned on rigid surfaces or areas before and after machining. Clamping over an area to be bored to a very thin-wall thickness could cause the part to warp or deform, thus causing the hole to be bored incorrectly.
- As with other fixtures, some means of cutter setting should be incorporated into the design. However, since the workholder will

be rotating, this setting device should later be removed.
- Whenever possible, standard lathe accessories should be adapted in the design of turning fixtures. Lathe faceplates are an ideal method to mount large fixtures. Likewise, a standard lathe chuck, or collets, can and should be modified for many fixturing applications.

## Grinding Fixtures

Grinding fixtures are a family of fixtures rather than a single classification. The major types of fixtures used for grinding are surface grinding fixtures and cylindrical grinding fixtures.

Surface grinding fixtures have the following design characteristics.

- Surface grinding fixtures are similar in design to milling fixtures, but made to much closer tolerances.
- Whenever practical, use magnetic chucks to hold the workpiece. In these cases, the fixture is simply a device to contain the workpiece and prevent any lateral or transverse movement of the part.
- Provide adequate room or slots to permit the escape of coolant and easy removal of built-up grinding sludge.
- Provide coolant containment devices or splash guards to keep the fixture from spilling coolant on the floor around the machine.
- Fixture elements in contact with the magnetic chuck should be made from ferrous materials if they are to be held on the chuck. If they are not to be held to the chuck, then a nonferrous metal should be specified.
- Include provisions for rapid wheel dressing and truing in the design of the fixture, if not built into the machine.
- All locators must be accurately and positively positioned.

Cylindrical grinding fixtures contain the following design characteristics.

- Cylindrical grinding fixtures often are very similar to lathe fixtures.

- Since cylindrical grinding is normally a secondary operation, performed after turning, it is often desirable to use the same center holes for grinding as turning the part.
- Coolant buildup is seldom a problem with cylindrical grinding; however, sludge removal must always be considered.
- Cylindrical grinding fixtures should always be perfectly balanced to achieve the desired results.
- When possible, use standard accessories and attachments. These include grinding collets, chucks, and drive plates with special right-angled holders called dogs.
- Incorporate provisions for wheel dressing and truing into the design.

## Boring Fixtures

Boring fixtures are designed to hold the workpiece while the part is bored. These fixtures differ from boring jigs in that they do not have any provision for guiding or supporting the boring bar. Boring fixtures are normally used for large parts with large holes where the boring bar is rigid enough to provide additional support. A pilot bushing is not needed.

Boring fixtures, like milling fixtures, should have some provision for setting the position of the cutting tool relative to the part. In cases where a boring fixture is to be used on a very large machine, such as a boring mill or vertical turret lathe, it is also good practice to include alignment areas on the fixture to ensure proper alignment with the machine.

## Broaching Fixtures

Broaching fixtures are designed to simply hold and locate a part relative to either an internal or external broach. Since there is a great deal of cutting force exerted during broaching, the complete fixture must be built more substantially than those for other processes.

Internal broaching fixtures need only locate and hold the part in proper position relative to the hole in the broaching machine. Most broaching performed today is of the pull type and tends to keep the part firmly seated on the fixture.

However, clamping devices are necessary to establish the proper relationship and maintain the position of the part until the broaching pressure pulls the part against the table.

External, or surface broaching, requires a different approach to fixturing. Since this type of broaching is performed on the outside of a part, the fixture must be designed to resist both pulling and perpendicular thrust that tends to try to push the part away from the broach. In either case, the principal purpose for a broaching fixture is to maintain the proper relationship between the part and the cutting tool, and to prevent the part from moving.

## Sawing Fixtures

Two primary machines commonly used for sawing operations are the vertical bandsaw and the horizontal bandsaw. With both types of machines, the main intent is to accurately position and hold the workpiece so it can be either sawed into pieces or slotted with the saw blade. The following are a few design characteristics peculiar to these saws and the sawing process in general.

- When possible, standard saw accessories and attachments should be used in conjunction with fixturing elements.
- Clamps, locators, supports, or similar fixture details must be kept clear of the blade path. Since the area occupied by the saw blade on both types of bandsaws extends above and below the actual working area, any overhanging fixture elements could interfere with normal operation of the saw.
- Provisions for coolant drainage and chip disposal must be planned into the fixture design. While most bandsaws have an internal chip disposal system, a significant amount of chips also will collect in the fixture unless some means for their elimination is planned.
- When practical, table slots should be used to reference the fixture to the saw blade.
- Use power feed whenever possible. This may require designing a means to secure the power-feed chain to the fixture.

## STANDARD FIXTURE MOUNTING

Quite often, only standard accessories such as clamps, straps, T-slot bolts, T-slot nuts, and jacks are needed to hold a workpiece. This is especially true where only a few pieces have to be produced and economic considerations do not justify more elaborate workholders. Most machine tools can accommodate such equipment. The beds and tables of machine tools, such as drill presses, boring mills, and jig bores have T-slots milled into their worktables. Various types of workholders, chucks, arbors, and collets may be attached to the spindles of machine tools, such as lathes and grinders either directly or with adapters. There are standards for the sizes and spacing of T-slots of tables, beds, and other equipment to which workholders are to be attached. Standards for spindles are established as well. Adherence to standards can result in economic interchangeability and multiple sources of tooling. Tool costs will be lower because the supplier can produce standard components in larger quantities at a lower cost.

Always be sure the fixture can be solidly located on the machine. A fixture that rocks or cannot be kept from twisting under cutting forces will not provide acceptable accuracy.

## RELATIONSHIP BETWEEN FIXTURE AND CUTTING TOOL

The direction and magnitude of forces created at the cutting tool-workpiece interface must be known. Use this knowledge to minimize the size of force moments (force × distance of force application) by reducing the moment arm. Avoid excessive projection, and provide additional supports where needed.

The relative motions between the workpiece and the cutting tool may change the tool geometry during the cutting cycle. Rake and clearance angles may change from a selected optimum condition to a bad condition.

The final objective of every material-removing operation is the removal of a certain amount of material per unit of time, to a certain depth, thickness, diameter, contour, and other related

specifications. These quantities can be obtained by controlling machine motions with stops, gages, or computer controls. Workholders may be equipped with stop surfaces such as the top of a drill bushing against which a drill stop abuts. Setting gages may be placed between a milling cutter and a reference plane on a workholder to obtain the right thickness of a workpiece after milling (*Figure 6-9*). Swing stops on lathe fixtures and arbors may be used to locate workpieces for removing stock evenly from both sides (*Figure 6-10*).

It is necessary to check for and control the interference between any cutting edge of the cutting tool and any part of the workholder during

Figure 6-9. *Use of a gage block in setting up a milling operation.*

Figure 6-10. *Lathe fixture with swing stop.*

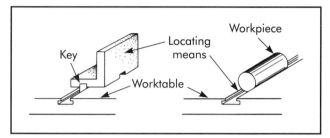

Figure 6-11. *Positioning a workpiece relative to locating means.*

possible contact of workholder and cutting tool. It is advisable to simultaneously check for, minimize, and control excessive nonproductive approaches of cutting tools to the workpiece. Planning should be done to avoid "cutting air."

Space must be provided to remove and load workpieces easily, without danger to the operator or damage to the workpiece and equipment. In addition, space is required to insert and remove cutting tools, especially with automatic tool changers. This includes space for the application of any wrenches, keys, or other tools used to change cutting tools. Such change should be possible without removing the workpiece.

## TOOL POSITIONING

Tool positioning refers specifically to locating the tool with respect to the work, or vice versa. Prior to setting up a workpiece, the blueprint and workpiece are studied to determine primary and secondary locating points or surfaces. Once these are determined, it is then necessary to visualize how these points or surfaces may be accurately located in relation to the locating means.

### Relationship to Locators

The locating means are the alignment or gaging surfaces of any angle, plate, bar, V-block, vise, or the like secured to or part of the worktable or fixture. Locators are used to properly position the work relative to the tool (*Figure 6-11*). A T-slot

may be considered a locating means, but generally such slots are only accurate enough for rough machining.

Keys are used under the base of an angle plate, vise, or workholding fixture. They provide an easy and accurate method of aligning a workholding fixture to a T-slot to the same degree of accuracy as the T-slot itself. Before using the T-slots as a basic locating reference, their accuracy should be established in relation to table or cutter movement. Removable keys are used extensively, especially in vises with slots at right angles to each other for lengthwise or crosswise mounting. Both the key and T-slots should be periodically inspected for wear to ensure proper dimensions and accuracy between the key and locating means.

The most practical procedure for establishing the relationship of the tool to the work will be governed by the type and size of the machine, type and size of work, production rate, and specified dimensions and tolerances of the cut. Correct tool positioning results in proper depth and location of the finished cut. Regardless of the type of machine involved, there are several different techniques for locating the work in relation to the tool (the exact technique is determined by the specific job requirement). Mass production may dictate a great expenditure of money to minimize the time required for locating parts, compared with less expenditure and more time for locating a single part.

### Cutter-setting Devices

Gage or setup blocks are a common means of reference for cutter setting (*Figure 6-12*). In many cases, the reference may be a designated surface on a locator (locating means). In its cor-

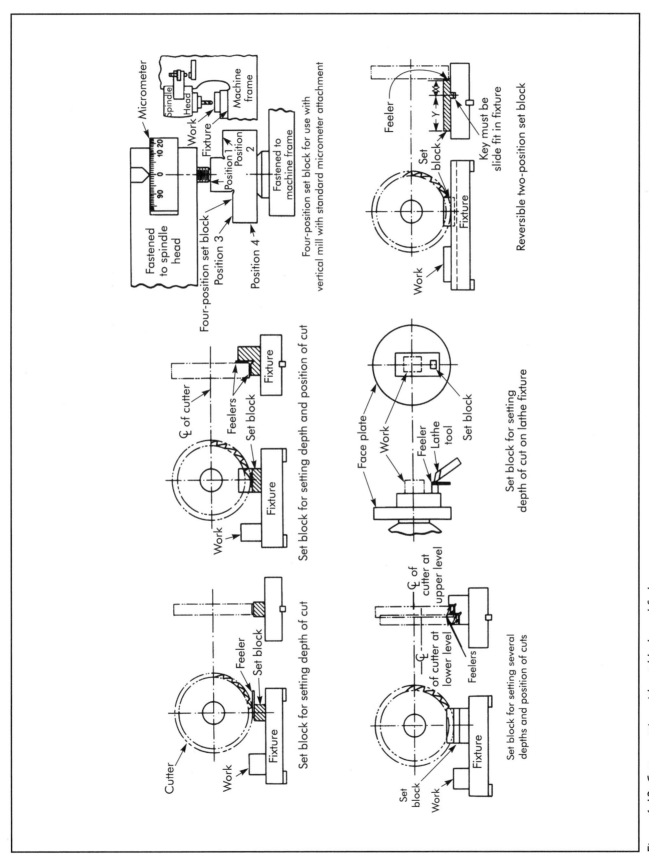

Figure 6-12. Cutter setting with set blocks and feelers.

rect position, the cutter should clear the setting surface by at least 0.03 in. (0.8 mm). Usual shop practice employs only one feeler thickness to avoid use of the wrong size on any particular operation. The thickness of the feeler should be stamped on the fixture base near the setup block.

Optimal methods are used extensively for gaging the accuracy of tool position in relation to the work and locating means. The many optical tooling equipment designs make broad applications in tool-to-work locations possible. One typical application is the establishment of an exact drill center location. An optical instrument is inserted in the chuck. Through high magnification, the eyepiece and cross-hairs are used to determine the exact center. The work is clamped securely to the locating means and rechecked for proper tool-center alignment. The optical instrument is then removed from the chuck and replaced with the drill. The same principle may be applied for gaging the indexing accuracy of a rotary fixture or index plate onto which work is positioned for machining.

## DESIGN FUNDAMENTALS

The following step-by-step approach will prove valuable for the design of all fixtures.

1. View the workpiece as a whole. A study of the complete workpiece, including its intended function, will disclose the relationship between the various workpiece features. With this insight, the fixture designer can either establish or better understand a proposed sequence of operations. Such knowledge may enable the designer to combine operations and minimize fixturing.

2. Gather all necessary data. All information that may affect jig and fixture design should be readily available. The designer must know the physical characteristics of the workpiece, such as composition of the material, condition (hardness), and rough and finished weight. If the workpiece is to be made directly from raw material (mill extrusions or sheet stock), the designer must know the shape, size, and tolerances of the mill stock. All production data, including total pieces, rate of production, tooling budget, and proposed production sequence must be available.

3. Consider standard workholders. Many operations can be performed by using available commercial workholders, such as machine vises, T-slots and bolts, jacks, and clamps. The design of a special workholding jig or fixture must be economically justifiable. If the planned operations are similar to present operations, the rework of present fixtures may be considered.

4. Determine what special workholders will be required. Every proposed operation should be carefully examined to see whether it can be performed economically with commercial workholders (machine vises, etc.), or whether available fixtures can be used. The designer should also consider whether available fixtures with minor alterations can be used. After assigning as many operations as possible to commercial or available workholders, a small number of operations will remain, for which special workholders must be provided. The number of special workholders may be further reduced by combining operations within a single fixture.

5. Study fixtures for similar operations. Every operation for which a special workholder is required should be considered individually. The designer should seek out—within his or her own plant, in other plants, in technical journals, etc.—similar operations for which fixtures are provided. By examining a number of existing fixtures, the designer can combine the best features of each.

6. Review the fixturing plan. The designer should consider, in turn, every operation of the production sequence and review fixturing decisions. There should be confirmation at every step that the proposed workholder will be structurally adequate to withstand cutting forces and as precise as required (location aspects). The preliminary design of the required special workholders should be completed.

7. Execute the fixturing plan. The designer should remain available during the execu-

tion of the fixturing plan. No plan can be regarded as final because of possible changes in workpiece dimensions and the typical cut-and-try aspects of fixture planning. After the line has been used for production, further process improvement may suggest fixturing changes.

A well-designed fixture may be used on several different machines. *Figure 6-13* shows a workpiece and its fixture as it might be processed in several different ways. The workpiece is a magnesium casting; the base and primary reference plane are established at the first milling operation. For the second milling operation, the workpiece is clamped to the fixture and repeatedly indexed to present four surfaces to the mill-

ing cutters. The rotating plate of the fixture has hardened bushings at 90° intervals, into which an index-pin plunger can nest.

## FIXTURE DESIGN EXAMPLE

Fixtures, like jigs, should be designed using a systematic approach. Using a similar sketching technique as outlined in Chapter 5 (Jig Design), the following method may be used to design a simple plate fixture.

The part to be machined is the stop block in *Figure 6-14*. The first step is to study the part and determine the basic requirements and relevant information about the part. In this case, the requirement is for a simple fixture to hold the part for gang-milling a stepped shoulder. The part

Figure 6-13. *Fixtured workpiece processed by several different methods.*

Figure 6-14. *Rough sketch of a stop block.*

is 3 × 2 × 1 in. (76.2 × 50.8 × 25.4 mm) 6061-T6 aluminum barstock. The production plan calls for the part to be cut from 1 × 2 in. (25.4 × 50.8 mm) bar to a tolerance of 3 in. ±0.005 in. (76.2 ±0.13 mm). The production run is specified as 750 parts per month. The fixture should be designed as a permanent tool, yet the cost must be kept to a minimum. Since gang milling is specified, a horizontal milling machine will be used to machine the parts.

Next, the part is sketched in three views (*Figure 6-15*). If possible, the part should be sketched full size. However, if the part is too large, half or quarter scale can be used. Make the sketch as accurate and true to scale as possible. Sketching the part in a different color will help reduce con-

fusion as the design develops and more lines are added to the initial design sketch.

Once the part is sketched, the base plate and part locators should be added to the sketch. As shown in *Figure 6-16*, a simple six-point locating method is used since the part presents no special locational problems. In *Figure 6-16*, the base plate acts as the primary reference surface. Since the primary reference surface on the part is flat, no special locators, other than the base plate itself, are installed. The secondary reference surface is located with two dowel pins positioned toward the rear of the plate. The final locator is positioned on the short side of the part, toward the

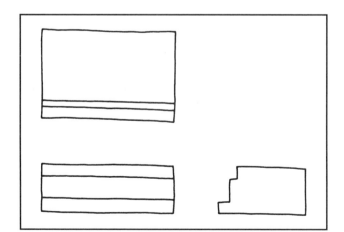

Figure 6-15. *The part is sketched in three views.*

Figure 6-16. *A six-point locating method is used.*

end where the tool thrust will be directed. This locational method will be sufficient to accurately locate the part and resist the cutting forces that can be expected with this part. If, however, a different material, such as steel, were to be machined, a more substantial locator might be better suited. In such a case, a block, screwed and doweled to the base plate, might be installed to resist the additional tool thrust.

The base plate of 1-in. (25.4-mm) steel plate is selected for this fixture due to its low cost and durability. The top and bottom surfaces should be machined to provide accurate and parallel locating surfaces for both the part and the fixture. The slots in either end provide a means to secure the fixture to the machine table. Drill and ream fixture keys are used to maintain the proper position of the fixture in the table T-slot (*Figure 6-17*). These were selected because of the relatively short time required for installation. Other standard fixture keys could be used (*Figure 6-18*).

The set block is now sketched into the design. The block selected is made 0.0625 in. (1.588 mm) smaller than the actual part on both the referencing surfaces to accommodate the feeler gage. A feeler gage made from 0.0625 in. (1.588 mm) stock will be used to position the cutters in the initial setup of the fixture. The location of the set block is determined by the dimensions on the part print. As shown, the first step is dimensioned as 1.625 in. (41.28 mm) from the secondary refer-

Figure 6-18. *Standard fixture keys.*

ence surface and 0.75 in. (19.1 mm) from the primary reference surface. The set block then is positioned to suit these dimensions (*Figure 6-19*).

The clamping device selected for this fixture is a toggle-action clamp. While any of several different clamps could be used, a toggle-action clamp offers the advantages of being fast-acting, capable of being moved completely clear of the part, and easily modified to suit the part shape. The cost

Figure 6-17. *Fixture keys.*

Figure 6-19. *The set block is positioned.*

Figure 6-20. *A clamp is added to hold the part against the base plate.*

of this clamp is also a consideration. Since this is a commercial component, the cost is much less than making a similar clamp in-house. The end of the toggle clamp is modified to suit the length of the part. This will spread the clamping force over a greater area and ensure the part is properly held against the base plate (*Figure 6-20*).

The next step is selecting the cutters to perform the gang milling. The logical choice for this operation is interlocking, staggered-tooth, side-milling cutters. The diameters of the cutters are normally determined by the size of the milling machine specified for the machining. However, for this example, the cutters specified have diameters of 5.00–6.00 in. (127.0–152.4 mm). Both cutter widths are 0.25 in. (6.4 mm). The cutters are now put into the initial design sketch (*Figure 6-21*). The entire design now must be checked to make sure there is no interference between members or between the fixture and machine tool. Once this inspection is complete, the final tool drawings are prepared.

## REVIEW QUESTIONS

1. What are fixture keys? What purpose do they serve?
2. How are fixtures classified?
3. When designing a lathe fixture, what must be done to equalize unbalanced masses in the workpiece or fixture?
4. What type of fixture is the most economical for small parts?
5. What type of fixture could be used to machine a hexagonal shape on a shaft?
6. Why should fixtures have a low profile?
7. Why should standard components and cutters be used wherever possible?
8. Based on its construction, what type of fixture is the most common?
9. Design a milling fixture for the part shown in *Figure A*. This is a temporary tool to machine slots in 250 pieces.
10. Design a fixture to perform the slotting operation in the part shown in *Figure B*. This is a high-volume part, but the cost of the fixture must be minimal.

Figure 6-21. *Cutters sketched into the initial design.*

Material: 6061-T6 aluminum
Operation: mill two slots

Figure A.

Material: SAE 1020 steel
Operation: mill slot

Figure B.

# 7

# Power Presses

A *power press* is a machine that supplies force to a die used to blank, form, or shape metal or nonmetallic material. Thus, a press is a component of a manufacturing system that combines the press, die, material, and feeding method to produce a part. A press is composed of a frame, bed, or bolster plate and a reciprocating member called a ram or slide, which exerts force upon work material through special tools mounted on the ram and bed. Energy stored in the rotating flywheel of a mechanical press (or supplied by a hydraulic system in a hydraulic press) is transferred to the ram to provide linear movement. The system designer must provide proper point-of-operation guards to safeguard pressroom personnel.

## GAP-FRAME PRESSES

*Gap-frame* or *C-frame* presses derive their name from the C-shaped throat opening. Gap- or C-frame presses have many useful features, including excellent accessibility from the front and sides for die setting and operation. The machines also cost substantially less than straight-side presses. The open back is available for feeding stock and ejection of parts and scrap.

Another advantage of a gap-frame press is that the machine is easier to set up than a straight-side press. The die setter has much greater access to locate and bolt the die in place. Gap-frame presses generally have less height than a straight-side press of comparable tonnage. This is a valuable consideration when overhead clearance is limited.

Gap-frame presses with force capacities up to 250 tons (2,224 kN) and larger are less costly than straight-side presses having the same force capacity. In the 35–60-ton (311–534-kN) force range, a gap-frame press costs approximately half that of a straight-side press.

Gap-frame presses have more deflection than a straight-side press for a given load. The deflection has both a vertical and angular component. The angular deflection or misalignment that occurs is due to the spreading of the throat opening as tonnage is developed. In many applications, this angular misalignment under load may not be objectionable.

Popular for short-run work, these presses are specified where high accuracy of die alignment or close part tolerances are not necessarily controlling factors. For low-tonnage, high-speed work, precision gap presses are widely used. Here, the work is done before the bottom of the stroke. Light loading helps avoid angular deflection problems. However, straight-side presses are generally recommended for any application in which angular machine deflection would cause unacceptable part quality and accelerated die wear. The lower cost of gap-frame construction machines may be poor economy if accelerated tooling wear and quality problems result.

### Open-back Inclinable Type

*Figure 7-1* illustrates three types of gap-frame presses. Two of these (*a* and *b*) are a style of machine known as the *open-back inclinable* (OBI) press. The press frame is secured in a cradle, which permits the machine to be inclined backward. This is done to facilitate gravity loading, as well as part and scrap discharge out of the open back of the press.

The frame of most older OBI presses is made of cast construction. The most commonly used materials are gray cast iron or steel. *Figure 7-1a* illustrates two pretensioned tie-rods across the open front of the machine. Lugs are cast into the frame of the machine to accept the tie-rods, which are installed as an option to reduce angular deflection.

The development of timed air blow-off devices and a variety of small conveyors have lessened the demand for the OBI-style press. Today, many OBI presses are operated in the vertical position. While the OBI style is not obsolete, many press builders supply OBI presses only on special order.

### Open-back Stationary Type

The open-back stationary (OBS) gap-frame press shown in *Figure 7-1c* is more compact and often a more robust machine than the older OBI style that it has largely replaced. The OBS press has a heavy, box-like structure. The OBS press in *Figure 7-1c* features a guided plunger connection and eccentric drive.

## STRAIGHT-SIDE PRESSES

Straight-side presses derive their name from the vertical columns or uprights on either side of the machine. The columns together with the bed and crown form a strong housing for the crankshaft, slide, and other mechanical components.

The housing or frame of most straight-side presses is held together in compression by prestressed tie rods. Some straight-side presses have solid frames. Generally, a solid-frame straight-side press is less expensive than one with tie rods. However, tie-rod presses are easier to ship disassembled and can better withstand overloads.

A major advantage of the straight-side press compared to the gap-frame machine is freedom from angular misalignment under load. Maintaining true vertical motion throughout the press stroke minimizes tool wear and achieves accurate part tolerances.

(a)                    (b)                    (c)

Figure 7-1. (a) *Shown are an older style, unguarded open-back inclinable (OBI) gap-frame press with tie-rods added to reduce angular deflection; (b) a modern gap-frame OBI single-pitman mechanical press; (c) a modern open-back stationary (OBS) press featuring eccentric drive and a guided plunger connection.* (Courtesy Minster Machine Company)

## Components

*Figure 7-2* illustrates some of the principle mechanical components of a straight-side press with double-end drive gears and two connections. The bed is the base of the machine. The columns support the crown. It also has gibs or gibbing attached to guide the slide. The crankshaft end bearings may be contained in the columns or crown.

The crown serves many functions, depending upon machine design. Typically, the clutch, brake, motor, and flywheel mount on the crown of the press. The gears shown in *Figure 7-2* may be open, having only a safety guard designed to contain the gear in case it should fall off due to a failure, such as a broken crankshaft. In modern designs, the gears are fully enclosed and run in a bath of lubricant. Housing them in separate enclosures permits the use of a heavier viscosity lubricant than that used for other machine parts, such as the bearings. The latter are often supplied from a recirculating lubricant system. The separate gear housing and lubricant bath system serves to lessen noise and ensure long gear life.

*Figure 7-3* illustrates the placement of pneumatic piping tanks and controls for a typical straight-side press. The system is typical of a good pneumatic arrangement for a press equipped with an air-actuated friction clutch and die cushions.

## Production of Precision Stampings

Many high-volume, close-tolerance stampings are made in straight-side presses. These include electrical connectors, snap-top beverage cans, spin-on oil filter cartridge bases and refrigeration compressor housings. Tiny computer connectors are stamped at press speeds of 1,800 strokes per minute (SPM) or more. Often, two to eight or more parts are completed per hit. Precision stampings also are produced at low speeds. For example, large refrigeration compressor housings may be stamped at press speeds of approximately 12 SPM. The housing consists of two mating halves, which must fit together precisely to properly align the internal parts.

## MECHANICAL VERSUS HYDRAULIC PRESSES

Mechanical presses are built with force capacities to 6,000 tons (53.4 MN) or more. Force capacities of 50,000 tons (445 MN) or more are available in hydraulic presses. Large hydraulic machines are used in hot- and cold-forging applications, as well as various rubber-pad and fluid-cell-forming processes. Both single- and double-action hydraulic presses are used for forming large parts for the automotive and appliance industries.

### Features

Deep drawing and forming applications often require large forces very high in the press stroke, all the way to the bottom of the stroke. Some mechanical presses do not develop enough force high enough in the downward stroke to permit severe drawing and forming applications.

In most hydraulic presses, full force is available throughout the stroke—a very important characteristic. *Figure 7-4* illustrates why the rated force capacity of a mechanical press is available only near the bottom of the stroke.

Another advantage is that the stroke may be adjusted to match the job requirements. Only enough stroke length to provide part clearance

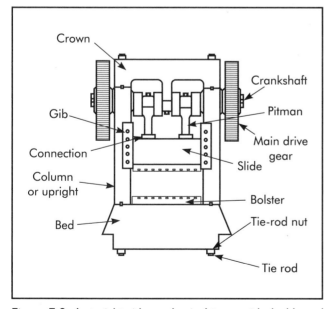

Figure 7-2. *A straight-side mechanical press with double-end drive gears and two connections* (Smith 2001).

Figure 7-3. *Pneumatic piping, tanks, and controls installed on a straight-side press.* (Courtesy Verson Corporation)

Figure 7-4. *The rated capacity of a mechanical press is available only at the bottom of the stroke. The full force of a simple hydraulic press can be delivered at any point in the stroke (Smith 1994).*

is required. Limiting the actual stroke will permit faster cycling rates and reduce energy consumption. Ram speed also can be adjusted to a rate that is best for the material requirements.

## Overload Protection

The force that a hydraulic press can exert is limited to the pressure applied to the total piston area. The applied pressure is limited by one or more relief valves. A mechanical press usually can exert several times the rated maximum force in case of an accidental overload. This extreme overload often results in severe press and die damage. Mechanical presses can become stuck on bottom due to large overloads caused by part ejection failures or die-setting errors.

Hydraulic presses may incorporate tooling safety features. The full force can be set to occur only at die closure. Should a foreign object be encountered high in the stroke, the ram can be programmed to retract quickly to avoid tooling damage.

## Force

When choosing between a mechanical and hydraulic press for an application, a number of items should be considered. The force required to do the same job is equal for each type of press. The same engineering formulas are used.

There is always a possibility that an existing job operated in a mechanical press requires 20–30% more force than the rated machine capacity. The overloading problem may go unnoticed, although excessive machine wear and possible damage may result. If the job is placed in a hydraulic press of the same rated capacity, there will not be enough force to do the job. Always make an accurate determination of true operating forces to avoid this problem.

Mechanical presses have the full rated force available only near the bottom of the stroke. The distance from bottom of stroke versus available force is determined by a force or tonnage curve. The force curves for six different mechanical presses are shown in *Figure 7-5*.

In some cases, a sharper coined impression may be obtained at a rapid forming rate. Jewelry and medallion work makes use of both high-force hydraulic presses and drop hammers. Each process has its own advantages. Often, die pressures in excess of 250,000 psi (1,724 MPa) occur in hydraulic press medallion work.

A mechanical press with high force capacity often is larger than a hydraulic press of similar force capacity. Few mechanical presses have been built with force capacities of 6,000 tons (53.4 MN) or more. Higher tonnage and/or compact construction are practical in hydraulic presses. Hydraulic presses for cold forging are built up to 50,000 tons (445 MN) or greater force capacity. Some hydraulic fluid cell presses have force capacities over 150,000 tons (1,335 MN).

The pressure at which the press delivers full tonnage is important. The most common range is from 1,000–3,000 psi (6,894–20,684 kPa). Some machines operate at substantially higher pressures. There is no set rule on the best peak operating pressure for a press design. Obviously, higher pressures permit the use of more compact cylinders and smaller volumes of fluid. However, the pump, valves, seals, and piping are more

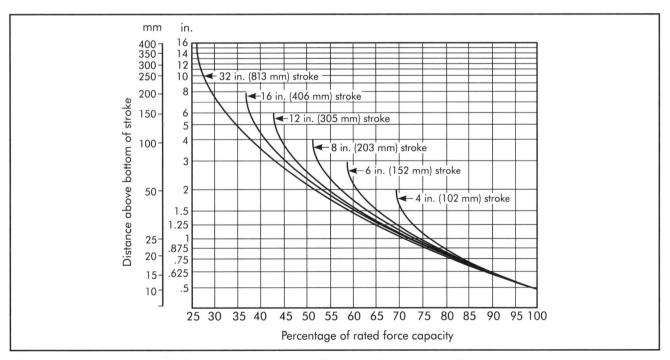

Figure 7-5. *Force curves for six different mechanical presses.* (Courtesy Danly Machine Corporation)

costly because they must be designed to operate at higher pressure.

The force of a hydraulic press can be programmed in the same way that the movements of the press are preset. In simple presses, the relief-valve system that provides overload protection may also serve to set the pressure adjustment. This allows the press to be set to exert a maximum force less than press capacity. Usually, there is a practical lower limit, typically about 20% of press capacity. At extremely low percentages of force capacity, a stick-slip phenomenon known as *stiction* in the cylinder rod and piston packing can cause jerky, erratic action.

## Press Speeds

Most press users are accustomed to describing press speeds in terms of strokes per minute. Speed is easily determined with a mechanical press. It is part of the machine specifications or can be measured with a stopwatch.

The number of strokes per minute made by a hydraulic press is determined by calculating a separate time for each phase of the ram stroke. First, the rapid advance time is calculated. Next, the pressing time or work stroke is determined. If a dwell is used, that time is added. The return stroke time is added. Finally, the hydraulic-valve reaction-delay time is added to determine the total cycle time. These factors are calculated to determine theoretical production rates when evaluating a new process. In the case of jobs already in operation, measuring the cycle rate with a stopwatch is sufficient.

The forming speed and impact at the bottom of the stroke may produce different results in mechanical presses versus their hydraulic counterparts. Each material and operation to form it has an optimal forming rate. For example, drop hammers and some mechanical presses seem to do a better job on soft jewelry pieces and some jobs where coining is required.

In deep drawing, controllable hydraulic press velocity and full force throughout the stroke may produce better results. This is necessary to prevent shearing and tearing during the drawing process. Often, parts that cannot be formed on a mechanical press with existing tooling can be formed in a hydraulic press. Hydraulic presses can be provided with controllable force throughout the press stroke and variable blank-holder pressure distribution.

Straight-side presses, such as the one illustrated in *Figure 7-6a*, are much better able to withstand off-center loading and snap-through energy release than the type shown in *Figure 7-6b*. Quality features in a press designed for severe work when ram tipping is to be minimized are a single-piston design together with a large ram or slide with long guiding and eight-point gibbing. However, loading should be carefully balanced and cutting dies timed to minimize snap-through shock to the best extent possible in any press-working operation.

## UPGRADING EXISTING PRESSES

Older presses are candidates to be upgraded for smoother, more reliable operation and reduced tooling wear. In some cases, rebuilding a damaged press can pay rapid dividends in reduced tooling repair costs and better part quality.

Electrical controls, which may no longer meet safety requirements, can be replaced. Usually, the most satisfactory way to retrofit the press is to install a new control package specially designed for the application. Such systems are available from several suppliers.

## DIE CUSHIONS

When a single-action press is used for drawing operations, the manner in which the blank-holder pressure is applied to control the flow of the metal blank is important. The application of pressure to a blank holder is one of the features of a double-action press. Single-action presses lack this feature and therefore require supplementary blank-holding equipment.

Dies are sometimes built with a blank holder using compression springs, air cylinders, or high-pressure nitrogen cylinders to supply the holding pressure. This greatly increases the cost of the die. A press cushion can serve every die with this requirement, lowering the cost of tooling.

Figure 7-6. (a) *A straight-side hydraulic press is designed for applications requiring close alignment and high forces. This design will withstand moderate lateral loads with very little side deflection. (b) A two-cylinder, four-post hydraulic press is suited for light- to medium-duty work that does not involve lateral (side) loads. (Courtesy Verson Corporation)*

## Pneumatic Die Cushions

A pneumatic die cushion is supplied with shop air pressure. Normally, the design uses either one or two pistons and cylinders. The recommended capacity of a die cushion is about 15–20% of the rated press tonnage. The size of the press-bed opening limits the size, type, and capacity of the cushion.

*Figure 7-7* shows a pneumatic die cushion. This illustration shows an inverted-type cushion in which the downward movement of the blank holder, through pressure pins, forces the cylinder against a cushion of air inside the cylinder, and moves the air back into the surge tank (not shown). The external components, such as the surge tank, regulator, and pressure gage, are essentially identical in function to a press counterbalance system. On the upstroke, the air in the surge tank returns to the cylinder. Other designs function without surge tanks.

It is very important to load the cushion evenly to avoid premature wear and cushion failure. The die designer should incorporate equalizing pins in the lower shoe if required to accomplish loading equalization. Often, these can be actuated by the upper heel blocks. In some cases, special pin drivers made of structural tubing may be attached to the slide to actuate equalizing pins.

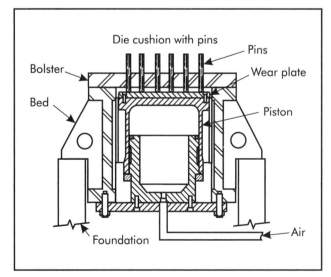

Figure 7-7. *Sectional view through a press bed and bolster illustrates a pneumatic die cushion.*

Operator safety must be considered when performing this operation because additional pinch points are created.

To avoid press and die problems, it is of extreme importance that the correct length of pins be used. In drawing and other critical operations, a pin that is 0.060 in. (1.52 mm) longer than the others can easily cause a wrinkled or fractured part. When problems are encountered, and the pressure setting is found correct, the pins should be carefully measured with a micrometer or vernier caliper to determine if they are the same length. The contact plate on the cushion and surfaces contacted by the tip ends of the pins in the die should be examined for wear.

### Hydraulic Die Cushions

Hydraulic die cushions have the advantage of taking up less space than air cushions. Also, they can be equipped with fixed or servo-actuated relief valves. Programmable controllers and servo valves are used to control resistance throughout their travel. Hydraulic die cushions can be retrofitted to existing presses.

## HYDRAULIC FORMING MACHINES AND DIES

One advantage of forming processes, in which hydraulic pressure acts on one side of the workpiece, is that only one half the die is needed. Simple dies may use rubber pads or cast shapes alone as a forming medium to transmit pressure.

The Guerin process uses a thick rubber pad contained within the ram of the press. The die is placed on the lower press platen or bed. This process has long been used to form short runs of parts from thin, soft materials, such as aluminum.

Several processes apply hydraulic pressure to the workpiece through a flexible rubber bladder or membrane. These systems combine many of the advantages of direct fluid application without the mess. The Verson Corporation Wheelon forming process uses a method of applying direct hydraulic pressure to the rubber-forming pad. The blanks are placed over simple male dies, similar to those used in the Guerin process. *Figure 7-8* illustrates a Wheelon hydraulic forming machine.

Figure 7-8. *The Verson Wheelon hydraulic forming machine.* (Courtesy Verson Corporation)

The banks and dies are moved into the press frame on a carrier, and forming pressure is applied by hydraulically inflating a rubber bladder mounted in the immobile roof of the press. *Figure 7-9* shows a cross section of the press frame with a die and

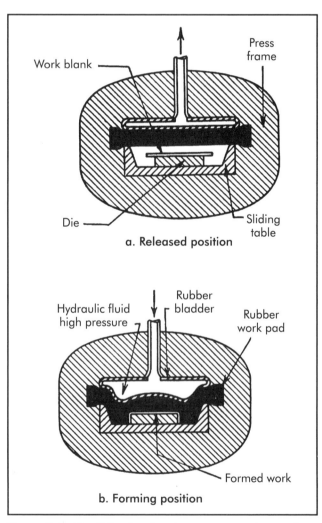

Figure 7-9. *The Wheelon forming sequence of operations.* (Courtesy Verson Corporation)

blank in place. The rubber bladder is shown in the released (*a*) and forming (*b*) positions. This method is limited in depths of draw to about the same as the Guerin process. With pressures of 6,000–10,000 psi (41.4–68.9 MPa) available, practically all wrinkling is eliminated.

## TUBULAR HYDROFORMING

Specialized press dies use oil or water under pressure to act directly on tubular workpieces. Tubular hydroforming is an old process that has found new uses, especially for automotive and truck-frame production. The tooling costs are low considering that complicated shapes can be produced in one or just a few steps. To avoid housekeeping problems, water with a rust inhibitor is used instead of oil. *Figure 7-10* illustrates operators removing a formed frame member from a hydroforming die.

### Hydroforming Press Design

Hydroforming presses nearly always use hydraulically actuated clamping. Typical clamping forces are from 700–5,000 tons (6,228–44,482 kN). Some presses employ a latching mechanism to hold the press in the closed position while forming pressure is applied to the inside of the tubes.

Pressure applied inside the tubes to form them to the inside of die cavities is typically in the range of 5,000–45,000 psi (34.5–310.2 MPa). When the part is filled with fluid, a pressure intensifier may be used to economically achieve the high forming pressures required.

### Multiple-stage Hydroforming

*Figure 7-11* illustrates a multiple-stage hydroforming operation. By adding cams and other means of actuating forming and punching details, it is feasible to form complicated parts, including piercing holes. There is little fluid spillage, due to entering the punches while the part is pressurized and relieving the pressure before the punches are withdrawn.

### Multiple-slide Forming Machines

Slide forming machines originally were built to produce wireforms. Wire was fed into the machine, severed to length, and formed into the desired configuration. From the initial dedicated machines evolved universal machines capable of producing a variety of wireforms simply by changing tooling.

Figure 7-10. *Operators removing a formed frame member from a hydroforming die. The axial plungers on either end of the die move into the tube upon die closure and form a watertight seal. (Courtesy Schuler Hydroforming)*

Figure 7-11. *A multiple-stage hydroforming operation used to make a "U"- shaped tubular automotive suspension part. (Courtesy Schuler Hydroforming)*

Strip or ribbon stock is fed into the slide forming machine and processed the same as wire. To feed, cut off, and form, a conventional press was added to the machine, as an integral member, and ribbon material could then be formed in a progressive die before the cut-off and forming sequence of the slide machine.

The four slide machine, as it is commonly known, (more correctly called a multislide machine), was introduced to the metal stamping industry for processing parts that required more than could be achieved from conventional power presses and the progressive dies that complemented them.

The four-slide machine has more than four sliding members if the cut-off slide, stripper slide, feed slide, and press slide are included. The machine has four main driving shafts arranged in a rectangle where synchronous power is transmitted from one to the next by bevel gears at the corners. The shafts carry cams of various configurations for various functions including four prominent cams, one on each shaft for forming; hence, the machine is labeled a "four slide." Later machine versions have included a provision for a sliding kingpost or mandrel and a twin front slide primarily utilizing one for blank holding and the other for forming.

A bracket known as the *kingpost holder* is fastened to the top platen of the machine and houses the kingpost plus the stripper shaft. The kingpost or forming mandrel, suspended from the kingpost holder is the center of all forming activity. Single or multiple stages of forming are feasible with the stripper slide being used for transferring the blank from one stage to the next or ejecting the finished part through a large hole in the machine platen located directly under the kingpost. *Figure 7-12* is a plan view of a typical four-slide machine.

Further enhancements to four-slide machine versatility included multiple press attachments and the availability to reverse individual presses 180° to accommodate burrside requirements or for forming or extruding in the press section toward the desired direction. Mechanical blankholders also are a standard feature to facilitate tooling.

A vertical four-slide machine is essentially a horizontal four-slide machine tilted on edge. The press sections can no longer be reversed, but all slugs from gutting and piercing have gravity to assist their ejection. The kingpost or forming mandrel is relocated to the rear of the machine along with the stripper slide. The formed parts are now transferred and ejected forward toward the front of the machine rather than through the base platen. The major advantage of the vertical four-slide machine is convenience of tool setup as well as service and maintenance of the tooling in the machine. The need to bend over a machine to service tooling is eliminated and visibility improved due to vertical proximity and the relocation of the kingpost holder to the face or rear of the platen.

The vertical shaft-driven slide machine enables the addition of auxiliary slides for angular orientation toward the forming center of the machine. The press section is located farther to the outside to permit added positioning and flexibility for the forming slides. The kingpost becomes the center mandrel and the number of motions available for transfer or ejection from the rear of the machine into the working area is increased.

## TRANSFER PRESS AND DIE OPERATIONS

Transfer presses have several distinguishing features that suit them for many types of medium- to high-volume work. Most operations use precut blanks, although there are combined operations. For example, the first station is a coil-fed-blanking die.

Many different sizes of transfer presses are used. The first type of transfer press, the eyelet machine, originally was designed to make small metal eyelets for shoes. Today, some small transfer presses are still called eyeleting machines, although great varieties of parts are produced on them.

Large transfer presses have force capacities of 3,500 tons (35 MN) or more. Transfer press operations have several common factors. These include:

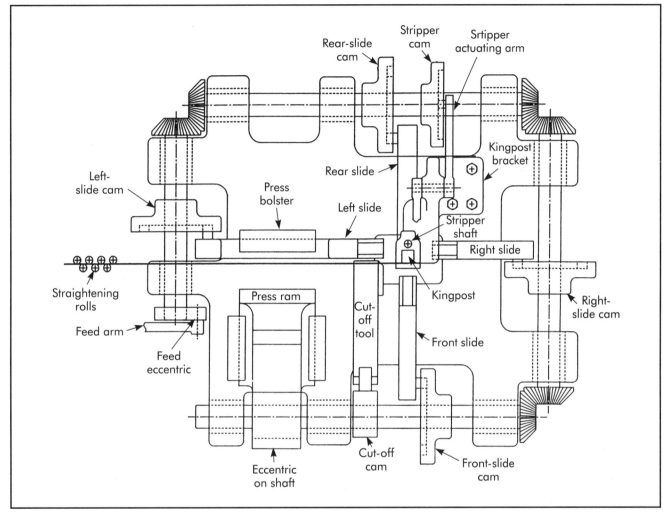

Figure 7-12. *Plan view of a typical four-slide machine looking down (Smith 1990). (Courtesy Design Standards Corporation)*

- individual dies used for each operation;
- reciprocating transfer-feed bars on each side of the press fitted with fingers that move the parts between the dies, and
- feed bars synchronized with the press ram motion.

*Figure 7-13* illustrates a very large specialized transfer press. The automotive and appliance industries are the principal users of this type of machine. All stamping operations required to complete large parts, such as automotive hoods and roof panels, are done in such presses. Flat blanks are destacked and automatically fed into the right end of the press. The transfer-feed bar fingers move the parts from die to die. Completed stampings emerge from the left end, where they are placed in storage racks or conveyed to the assembly operation.

## Multiple-slide Straight-side Presses

The straight-side press shown in *Figure 7-13* is an example of a highly specialized custom-built machine known as a Verson Transmat. The three slides and four columns are customized for the type of work to be performed. Large, multiple-slide presses have a number of advantages, including:

- each slide can be designed to provide the required force, and

Figure 7-13. *A four-column three-slide transfer press with a force capacity of 46,000 tons (409 MN). (Courtesy Verson Corporation)*

- by using multiple slides, ram tipping can be minimized.

The center slide in *Figure 7-13* is designed for a heavy, single-station forming operation. Placing a die having a high-force requirement, such as a stretch form or reforming operation, on one end of a long press slide can result in severe ram tipping, die damage, accelerated wear, and quality problems. This is often an unexpected problem in applications where large single- and double-slide transfer presses are specified.

The load on the slide should be balanced throughout the press stroke. This is an important consideration in the design shown in *Figure 7-13*. Some builders and users may have been unaware of this problem. When troubleshooting transfer-press problems, ram tipping sometimes is found to be the cause of poor quality work and excessive press and die wear. A solution may be to add a compensating load on the lightly loaded end of the slide. Nitrogen cylinders with a force capacity and stroke equal to a the draw-ring load on the opposite end of the slide springs may be required to provide the counterforce needed to balance the load.

## Transfer Feed Motion

Two types of transfer motion are used. The simplest system uses dual-axis motion. Only in-and-out motion is used to grasp the part. The second axis of motion transfers the part from die to die.

*Figure 7-14* illustrates the motion of tri-axis transfer feeder bars for indexing parts between dies in the transfer press. In older designs, the transfer feeder bars are mechanically driven synchronously with the slide motion. The fingers inserted into the transfer feeder bars hold the parts during indexing. In some cases, the fingers use pneumatic jaw clamps to grasp the parts. Wherever possible, simple scoops that rely upon gravity are used.

The application of dual-axis transfer-feeder bar motion is limited to relatively flat parts with only shallow formed features. The lack of up-and-down motion results in the parts being dragged across the top of the lower die surfaces when transferred. The system works very well within these limitations. The advantages, compared to a tri-axis system, are lower initial cost, less maintenance expense, and faster cycle times.

Tri-axis transfer is needed for parts having deeply formed features. Here, the part must be

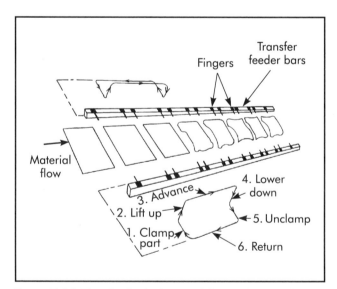

Figure 7-14. *Transfer feeder bars and fingers showing the motion sequence: (1) clamp (catch part); (2) lift up; (3) advance; (4) lower; (5) unclamp; (6) return. (Courtesy Auto Alliance International)*

lifted out of the die cavity before being transferred to the next die.

## Transfer Drive Methods

There are several basic systems for actuating transfer-feed motion. The original method, which is still in use, is to drive the transfer directly from the press crankshaft. Gears and cams transform rotary motion to the reciprocating action needed for part transfer.

Another mechanical drive system used for some new systems and retrofit applications uses plate cams attached to the press ram. The mechanism is driven by cam follower rollers.

Both pneumatic and hydraulic cylinder-driven systems are built into multistation dies. Many clever designs have been fabricated in press shops. Hydraulic systems are more costly but provide better control. The motions of pneumatic systems tend to require frequent adjustment because they are not as stable.

Most new transfer designs are powered by electrical servomotors—the best technology for most new designs. The technology includes a combination of a high-power servomotor, microprocessor-based electronic control, and solid-state power supplies. These are all mature technologies in widespread use in other industrial equipment, such as machine tools and robotics.

## Equipment for Decoiling and Straightening Coil Stock

Press auxiliary equipment includes stock decoilers, straighteners, feeders, part handling and scrap removal devices, as well as robots and dedicated die change carts. Like any pressworking equipment subject to movement, appropriate safeguarding measures are required to prevent injury to personnel.

A variety of commercially available coil handling, decoiling, straightening, and feeding equipment is used in coil-fed die operations. The equipment, available from many manufacturers, can be used interchangeably in a variety of configurations.

In some cases, the entire system is delivered as a turnkey package by the press builder or equipment supplier. However, it is common to find a mixture of used equipment working as an integrated system. Cost-conscious stampers often retrofit older equipment with modern drive systems and controls at a fraction of the cost of new machinery.

## Quick Coil Change

A rapid means to band and remove a partial coil of stock left over from the job being removed is an important feature. Time is saved if the new coil is pre-staged at the decoiler, as shown in *Figure 7-15*. It is important to have the next coil ready. During many production runs, quick coil change saves more time than quick die change.

Cradle-type decoilers may be mounted on a movable track to center different widths of stock on the press centerline. In this case, markings of the correct settings should be provided to avoid trial-and-error adjustment.

## Stock Feeders

There are many types and styles of stock feeders. Choosing between them often is dictated by what is available and affordable in the press shop. The modern feeder of choice usually is the servomotor-driven roll feeder, although a crankshaft-driven roll feeder provides good service if correctly adjusted and maintained. Air-driven

Figure 7-15. An operator loads a coil into a powered decoiling cradle. An overhead crane is used to handle the coil safely. Note that a spare coil is ready at the decoiler. (Courtesy W. C. McCurdy Company)

hitch feeders are popular based on cost and ease of use. An old but useful design for small work is the die- or press-mounted hitch or grip feed actuated by either the press ram or upper die shoe. The design is illustrated in *Figure 7-16*. The operating cam fastens to the upper shoe. The remainder of the device attaches to the lower die shoe. Provided the shut height of the die is set exactly the same each time, the feed pitch will always be correct. On the upstroke, the rollers at the left grip and move the stock a feed length determined by the adjustment nuts. Movement of the rollers, carried on a slide, is controlled by the spring shown in *Figure 7-16*. On the downstroke, stationary rollers at the right prevent backward movement of the stock. Both sets of rollers have one-way clutches so they can only turn in one direction.

## CNC LASER AND TURRET PUNCHING MACHINES

The need for building low-cost, short-run dies has been greatly lessened by the advent of the modern CNC punching and laser-cutting work cells. When these machines are used with nesting software, it is possible to produce several different parts from the same blank sheet of metal. Often, all of the stamping needed to make a product can be made quickly this way, greatly reducing inventory costs. A view of a combination laser and punching machine is illustrated in *Figure 7- 17*.

Turret punching machines started to appear in sheet-metal shops doing aircraft work decades

ago. These machines were simple hand-actuated punches with the upper and lower turrets rotated in alignment by a simple gear or roller chain and backshaft arrangement. These evolved to power-actuated machines capable of both cutting standard holes and nibbling out special shapes. Some of the turret punch holders bend and form features such as louvers and bent tabs. *Figure 7-18* illustrates a machine with that capability. *Figure 7-19* shows a part being discharged from a machine by a tilting worktable. When cutting steel, oxygen is the usual assist gas. Stainless steels also can be cut. Here, argon and/or nitrogen are used to obtain the smoothest cut edges.

## A Punch-setting Breakthrough

At one time, low-cost dies used punches set in low-melting-temperature alloys. Today, it is common to set punches that run with zero clearance. They are used for cutting materials, such as rubber and mica in filled epoxy cement, and aided by a retaining screw. This simple construction method can provide dies of high precision at a much lower cost since a precision jig ground punch holder is not required. Ways of reducing tooling costs are limited only by human imagination.

### Economic Payback

Laser CNC cutting and punching represents a large capital investment for many shops. Calculating a reasonable payback period is important. Such a machine can eliminate many short-run dies and several presses, freeing up valuable floor space. If the cost of ownership and rate of machine utilization represents a good return on investment, with machine time to spare, this time can be marketed to take on new product lines or produce short-run parts for other clients.

## FORGING

While forging may seem like an obsolete process to the public not involved in an industrial manufacturing base, it remains the best way to produce a strong part that can sustain the high stresses of many critical applications, such as an

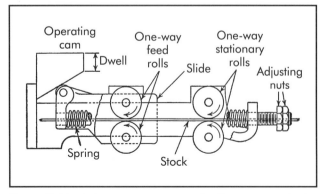

Figure 7-16. A die- or press-mounted roller-type grip feeder. (Courtesy H. E. Dickerman Manufacturing Co.)

Figure 7-17. A machine with both punching and laser-cutting capability, which is useful for fast production of small lots of parts. (Courtesy Finn Power International, Inc.)

Figure 7-18. An example of a combined CNC laser and turret punching machine equipped for forming louvers and other features. (Courtesy Finn Power International, Inc.)

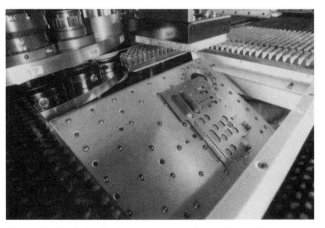

Figure 7-19. A finished part being discharged. A wide variety of different parts can be made from a single sheet of metal using the combined cutting capability of a CNC laser and turret punching. (Courtesy Finn Power International, Inc.)

aircraft engine, power turbine for electrical power generation, high-quality engine crankshafts, and connecting rods and axle shafts, which are upset on one end to produce a hub to which the driving wheels are fastened. Forging achieved the first net-shape methods of manufacturing metal parts, and it remains the most common method without wasted material.

## Powdered Metal Compaction

Many parts are made by compacting powdered metal into a near-net-shape form with slight porosity. To achieve a stronger part, that metal compaction can be heated to a welding temperature in an oven with a carefully controlled atmosphere to permit the fusion of metal particles in a process called *sintering*. Next, the hot powdered metal compactions are struck in a forging die to close all voids. This method is popular for producing extremely strong connecting rods and automotive engine uniformity.

## Nonferrous Forgings

The largest nonferrous forgings made in great quantities are aluminum semi-truck and tractor wheels. These wheels are cold-forged using a 50,000-ton (445-MN) hydraulic press. Although the wheels are costly to make, the weight savings compared with steel wheels will pay back the cost difference in additional payload capacity.

Solid brass ornamental hardware often is forged to produce a net-shape or near-net-shape part. Such forgings reduce cost and make a better part than casting by achieving a very low reject rate and simplifying the polishing process.

The aerospace industry has many essential applications for aluminum and titanium alloy forgings, such as landing gear and many other critical structural applications.

## Forging Hammers

The earliest forgings were the work of a blacksmith's hammer. This was mechanized several centuries ago as the gravity drop hammer was invented. The drop hammer blow is controllable by the height to which the hammer is lifted. A well-equipped gravity drop-hammer shop has hammers ranging from 1,000–5,000 lb (454–2,268 kg). One advantage of drop-hammer presses is that excellent repeatability is achieved by the height to which the hammer is lifted and number of blows delivered. For critical applications, the operator's skill is augmented by equipping the drop hammer with CNC controls to ensure repeatability from part to part.

## Screw-type Forging Presses

*Figure 7-20* illustrates a 900-ton (8-MN) screw press. These presses evolved from the manual screw presses that are still found in toolrooms for testing dies and light pressing applications. The screw is actuated by a flywheel brought up to operating speed by a motor. When the flywheel engages the screw, all of the kinetic energy of the flywheel is delivered to the workpiece. Thus, a limited amount of energy is available to do the work without repeating the cycle. One disadvanage to a screw press in hot forging work is that a finite amount of time is required to reverse the flywheel. This delay enables the heat in the work to transfer to the die, which both cools the work and may overheat the die in repetitive work.

## Crankshaft Forging Presses

Forging presses operated with crankshafts, clutches, flywheels, and brakes differ little, if any, from conventional mechanical stamping presses. Typically, they are more compact and robust than their conventional metal-stamping counterparts.

One danger to be aware of is that, unlike drop hammers and screw presses, which have a finite fixed amount of kinetic energy to deliver to the workpiece in the die, a crankshaft press can deliver many times more energy if overloaded by a double hit or misalignment accident. Such an accident can fracture the die, sending shrapnel at high velocity into the pressroom. The machine may require heavy bulletproof plastic or metal guarding to address this hazard.

## Hydraulic Forging Presses

Hydraulic presses are used for open-die and closed-die forging, such as the example mentioned previously of the 50,000-ton (445 MN) hydraulic press used to force an aluminum blank into a die to produce semi-truck wheels. While very few hydraulic forging presses are built over the 50,000-ton (445-MN) capacity, they are used almost universally in force capacities of 2,000 tons (18 MN) and higher.

## Forging Dies

Forging dies are made of especially tough grades of hot-work tool steel that is machined in the hardened state. Because of the high forces

Figure 7-20. A 900-ton (8-MN) screw press. (Courtesy National Machinery Company)

involved, CAD die-design programs that illustrate stress distribution under forging conditions are especially useful for designing robust dies without making them larger than necessary.

Dies subjected to extremely high forces are often reinforced with one or more shrink rings that form an interference fit around the die. These are shrunk into place by heating, hence the term "shrink ring." High forces are achieved in forging jewelry items with high detail, as well as the back extrusion of steel artillery ammunition steel casings. The subject of forging-die design is a fascinating area of work but beyond the scope of a fundamentals book.

## REFERENCES

Smith, D. 2001. *Die Maintenance Handbook*. Dearborn, MI: Society of Manufacturing Engineers.

———. 1994. *Fundamentals of Pressworking*. Dearborn, MI: Society of Manufacturing Engineers.

Smith, D., ed.1990. *Die Design Handbook,* Third Edition. Dearborn, MI: Society of Manufacturing Engineers.

## REVIEW QUESTIONS

1. What force systems are used in power presses?
2. What are the types of gap-frame presses?
3. What are the advantages and disadvantages of gap-frame presses compared to a straight-side press?
4. What are the principal contrasts between mechanical versus hydraulic presses?
5. What are three major considerations made during power press selection?
6. What is the most satisfactory way to upgrade (retrofit) an older press?
7. What are the most useful types of die cushions?
8. Which forming process is especially suitable for short runs of parts from thin, soft materials such as aluminum?
9. What are the two principal advantages of the hydroforming process?
10. What are the distinguishing features of transfer press/die operations?
11. What technological advancement has reduced the need for building low-cost, short-run dies?
12. What process remains the best way to produce a strong part that can sustain high stresses found in aircraft engines, engine crankshafts, etc.?

# Metal Cutting, Forming, and Drawing

Making a good die begins with the die designer. If the die is designed correctly, it will work properly and require infrequent, simple repairs. This chapter starts with the basics that all diemakers and designers most likely understand. It serves as a source of information to solve problems for the apprentice and trainee, as well as those with decades in the trade.

Cutting, which includes shearing, is the most common pressworking operation. A single formed stamping, such as a sieve or automobile inner door, has many holes, all of which are produced by cutting operations.

## SYSTEMS OF LENGTH, AREA, AND FORCE MEASUREMENT

In North America, many shops still carry out engineering calculations for stampings using measurements based on the inch for length and thickness. Shear or yield strengths are based on pounds per square inch (psi). Usually, the press force in short tons is based on 2,000 lb.

The metric system is in standard use throughout most of the world. Metric pressworking linear and area measurements are in terms of the meter, centimeter, and millimeter. Pressworking forces in metric tons based on 1,000 kilograms are common in Asia. Most of the metric world uses the kilo-Newton (kN) or mega-Newton (MN). The preferred metric unit for material strength is the kilo-Pascal (kPa) or mega-Pascal (MPa).

## SIMPLE DIE PUNCHING

*Figure 8-1* illustrates a sectional view of a simple die for punching a round hole in a part. Such dies may have several punches. In addition, a cut-off shearing operation may be included if the die is fed with strip or coil stock. On contact with the stock, the punch compresses slightly, as

Figure 8-1. (a) *Sectional view of a cutting die for producing round holes; (b) the punch is compressed after making initial contact with the stock (Smith 2001).*

shown in *Figure 8-1b*. The punch is fastened to the upper die shoe by means of a retainer with a hardened backing plate. A slug can be seen falling through the lower die shoe. On the upstroke of the press, the stock is stripped from the punch by a simple fixed or tunnel stripper.

## DIE-CUTTING OPERATIONS

Cutting operations essentially are a controlled process of *plastic deformation* or yielding of the material, leading to fracture. Both tensile and compressive strains are involved. Bending or distortion of the scrap metal trimmed away and metal cut out by the punch may occur. Stretching beyond the elastic limit occurs, then plastic deformation, reduction in area, and finally, fracturing starts through cleavage planes in the reduced area and becomes complete. *Figure 8-2* illustrates a sectional view of a successful hole-cutting operation.

Figure 8-2. *Section view of a successful hole-cutting operation involving normal clearances. Note that both the cut edge and slug have one third of metal thickness as a sheared edge shear and two thirds fracture (Smith 1994).*

The cutting of metal between die components is a shearing process in which metal is stressed in shear between two cutting edges to the point of fracture, or beyond its ultimate strength.

### Clearance

There are no absolute rules governing the amount of clearance between the punch and die. Clearance normally is expressed as a percentage of stock thickness per side. For mild steel, the clearance per side varies between 5–12% of stock thickness. Optimizing the amount of die clearance to best suit the material being cut may require some experimentation to minimize an uneven fractured edge condition. By increasing clearance, within reasonable limits, cutting pressure is lowered, extending tool life. A limiting factor is the amount of taper permitted in the hole and allowable burr height.

### Insufficient Clearance

In general, tight clearances will result in holes having a high ratio of shear or burnish to fracture, and less taper at the expense of accelerated tooling wear. The fracture, which starts from each side, may not meet evenly, as shown in *Figures 8-3a, b* and *8-4a, b*. One or more sharp projections may result. Double breakage sometimes occurs when too little clearance is used.

There are no absolute rules governing correct die clearances. Generally, tight clearances in the 3–5% per side range result in:

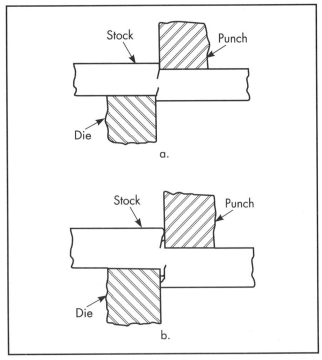

Figure 8-3. *Cutting* (a) *with insufficient clearance. As fracture continues* (b), *rough tearing occurs and the fracture paths will not meet.*

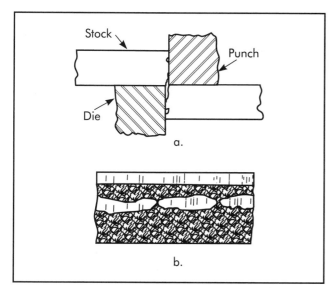

Figure 8-4. *Insufficient clearance (a) results in a double fracture or breakage condition. Appearance of a cut edge with a double-breakage condition (b).*

- parts with less taper on the cut edge;
- fewer tendencies for the slug to be pulled from the die opening;
- higher cutting forces, and
- a tendency to have double breakage problems, especially in thick materials.

As larger clearances in the 7–25% range are used, the result is often:

- longer punch and die life between resharpening;
- a need to use a means to ensure against slug pulling;
- lower cutting forces;
- avoidance of double breakage, and
- greater edge taper and more burr height.

Essential indicators for the diemaker are secondary shiny areas on the inside of the hole and/or slug and a rough, torn fracture. The fracture, which travels from both sides of the stock, may not be meeting evenly and will leave a ragged edge. The die may burnish the torn peaks of the fracture.

## Double-breakage Solutions

The solution to double breakage problems usually requires increasing the punch-to-die clearance.

Making the punch smaller is frequently the solution. This will maintain both the size of the hole and part. There will be an increase in the taper of the fracture. The cutting action is shown in *Figure 8-5.*

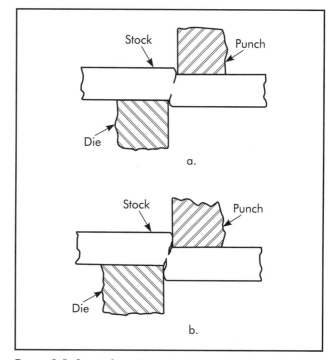

Figure 8-5. *Start of punch penetration and fracture (a) with a relatively large punch-to-die clearance. The fracture paths (b) will meet evenly.*

Within reasonable limits, increasing the amount of clearance between the punch and die will decrease the required cutting force. Lowering the cutting force will usually increase the number of parts produced before the tool requires sharpening.

Clearances of 12–15% per side may be required to eliminate double-breakage problems in soft steels. In some thick blanks, 25% per side is occasionally necessary. The cut edges may have a pronounced taper in the fractured portion of the cut edge. Die roll and burr height also may be more pronounced. These edge conditions, shown in *Figure 8-5b,* may be acceptable for many applications.

If the clearance is very large, higher than normal forces may be required. The cut-edge holes may have very pronounced die-roll, taper, and burr height (see *Figure 8-6*). Very large clearances also result in high lateral forces on the punch and die that can shorten tool life.

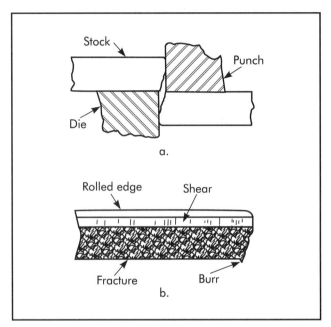

Figure 8-6. *Completion of punch penetration and fracture (a) with a generous punch-to-die clearance. View of a cut edge (b) with large punch-to-die clearances.*

## Applying Clearances to Irregularly Shaped Blanks

*Figure 8-7* illustrates how to apply clearances to obtain the correct sizes in holes or blanks. When the metal punched out is the functional part, and the metal around the opening is scrap, the die is made to the desired part size and the clearance is subtracted from (applied to) the punch size as shown in *Figure 8-7a*. When the slug is discarded and the punched opening is functional, the required clearance is applied (added) to the die opening as shown in *Figure 8-7b*.

## Center of Pressure

If the contour to be blanked is irregularly shaped, the summation of shearing forces on one side of the center of the ram may greatly exceed the forces on the other side. Such irregularity results in a bending moment in the press ram and undesirable deflections and misalignment.

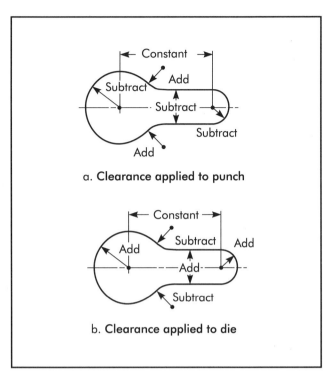

Figure 8-7. *How to apply clearances to irregularly shaped holes (Paquin 1962).*

For critical work, such as fine blanking, it is necessary to find a point about which the summation of shearing forces will be symmetrical. This point is called the *center of pressure*, and is the center of gravity of the line that is the perimeter of the blank. It is not the center of gravity of the area. The press tool will be designed so that the center of pressure will be on the central axis of the press ram when the tool is mounted in the press. If this is not possible, it may be feasible to offset the tool in the press to achieve the same goal.

## Mathematical Calculation

The center of pressure may be precisely determined by the following procedure:

1. Draw an outline of the actual cutting edges, as indicated in *Figure 8-8*.
2. Draw axes X-X and Y-Y at right angles in a convenient position. If the figure is symmetrical about a line, let this line be one of the axes. The center of pressure will, in this case, be somewhere on the latter axis.

3. Divide the cutting edges into line elements, straight lines, arcs, etc., and number each as 1, 2, 3, etc.
4. Find the length, $L_1$, $L_2$, $L_3$, etc., of these elements.
5. Find the center of gravity of these elements. Do not confuse the center of gravity of the lines with the center of gravity of the area enclosed by the lines.
6. Find the distance, $x_1$, of the center of gravity of the first element from the axis Y-Y, $x_2$ of the second, etc.
7. Find the distance, $x_1$, of the center of gravity of the first element from the axis X-X, $y_2$ of the second, etc.
8. Calculate the distance, $X$, of the center of pressure $C$ from the axis Y-Y by:

$$X = \frac{L_1 x_1 + L_2 x_2 + L_3 x_3 + L_4 x_4 + L_5 x_5 + L_6 x_6}{L_1 + L_2 + L_3 + L_4 + L_5 + L_6} \quad (8\text{-}1)$$

9. Calculate the distance, $Y$, of the center of pressure from the axis X-X by:

$$Y = \frac{L_1 y_1 + L_2 y_2 + L_3 y_3 + L_4 y_4 + L_5 y_5 + L_6 y_6}{L_1 + L_2 + L_3 + L_4 + L_5 + L_6} \quad (8\text{-}2)$$

In *Figure 8-8*, the elements are shown and numbered 1, 2, 3, etc. The length of 1 is obtained directly from the dimensions. It has a value of 4. The center of gravity is evidently at the geometrical center of the line. Therefore:

$$x_1 = 0$$

and

$$y_1 = 4.25 + 4/2$$

For the second element, $x_2$ is 1.5. The value of $y_2$ is found from:

$$C_G = 2r/\pi \quad (8\text{-}3)$$

where:

$C_G$ = center of gravity
$r$ = radius of the element, in. (mm)

To find the requirements for line 3 in *Figure 8-8*, it is necessary to solve the hypotenuse of the right triangle. The requirements of the other elements are found in a similar manner. All values are entered in *Table 8-1*. These values then are substituted in *Equations 8-1* and *8-2*.

$$X = \frac{35.92}{18.98} = 1.89 \text{ in. (48 mm)};$$

$$Y = \frac{115.98}{18.98} = 6.11 \text{ in. (155 mm)}$$

The center of pressure $C$ is therefore located as indicated in *Figure 8-8*.

## Wire Method of Location

The center of pressure of a blank contour may be located mathematically, but it is a tedious computation. Locating the center of pressure within

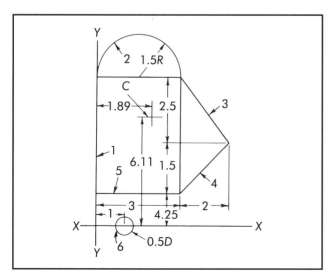

Figure 8-8. *Drawing used to determine center of pressure.*

Table 8-1. Values for elements shown in Figure 8-8

| Element | L | x | y | Lx | Ly |
|---|---|---|---|---|---|
| 1 | 4.00 | 0.00 | 6.25 | 0.00 | 25.00 |
| 2 | 4.71 | 1.50 | 9.20 | 7.05 | 43.33 |
| 3 | 3.20 | 4.00 | 7.00 | 12.80 | 22.40 |
| 4 | 2.50 | 4.00 | 5.00 | 10.00 | 12.50 |
| 5 | 3.00 | 1.50 | 4.25 | 4.50 | 12.75 |
| 6 | 1.57 | 1.00 | 0.00 | 1.57 | 00.00 |
| Total | 18.98 | | | 35.92 | 115.98 |

0.50 in. (12.7 mm) of true mathematical location normally is sufficient. One simple procedure accurate within such limits is to bend a soft wire to the blank contour. By balancing this frame across a pencil, in two coordinates, the intersection of the two axes of balance will locate the desired point. The calculated offset should be made in the die design for critical work. The use of computer-aided design (CAD) provides many methods for locating the center of an irregular area. If CAD is used when designing the part, the center of pressure (not center of area) can be located by the designer.

## Forces

The pressure required to cut through stock increases with the ultimate tensile strength of the material. Die cutting requires less energy than parting metal by tensile failure. There is no absolute relationship between tensile strength and shear strength. Typically, shear strength is 60–80% of ultimate tensile strength.

Generally, the ultimate tensile strength of the material used to calculate the cutting forces based on the area of material cut provides a substantial safety factor. For example, AISI-SAE 1010 cold-rolled steel has an approximate ultimate tensile strength of 56,000 psi (386 MPa) and shear strength of 42,000 psi (290 MPa).

The shear strength of the material increases due to the fast strain rates encountered in high-speed pressworking. The ultimate strength provides a safety factor in such cases.

### Determining Length of Cut

To calculate the force required for cutting or shearing materials, the actual measurement of the total cut length is required. The dimensions printed on the part provide a starting point. For progressive die work, all pilot holes and work done to cut the carrier strip must be included.

The die designer or engineer must calculate force requirements to determine the size and type of press required. An assumption that the pressroom can somehow fit a new job into an existing press can be a foolish blunder.

The length of cut, material thickness, and shear strength are calculation entry items. Some CAD programs can automatically calculate tonnage if all of the correct data is entered.

### Determining Theoretical Peak Cutting Force

To determine the theoretical peak cutting force, multiply the total length of cut by the stock thickness to obtain the area of material cut. Then, multiply the total area of cut by the shear strength of the material. *Equation 8-4* is useful for calculating cutting forces. Making no allowance for shear angles or timing of entry provides a safety factor. The same system of units should be used.

$$F_s = L \times t \times S_s \qquad (8\text{-}4)$$

where:

$F_s$ = force required to shear in the same system of units as $L$, $t$, and $S$, lbf (kN)
$L$ = length of cut, in. (mm)
$t$ = thickness of material, in. (mm)
$S_s$ = shear strength of the material as defined by ASTM tests

### Stripping Forces

A properly designed tool needs to have a method for holding the work while the punch is pulled back through the material. This stripping process can be by either a fixed-bridge or spring-loaded stripper. Thinner material requires a spring-loaded stripper because the material lacks the strength to prevent deformation when the punch is withdrawn from a hole.

The force required to strip the stock from a punch is difficult to determine with accuracy since it is influenced by the type of metal being cut, area of metal in contact with the punch, punch-to-die clearance, punch sharpness of the stripper spring position with respect to the punch, and numerous other factors. The following rough empirical equation is used for approximations:

$$F = L \times T \times 1.5 \qquad (8\text{-}5)$$

where:

F = stripping force, ton
L = length of cut, in.
T = material thickness, in.

When using metric units:

$$F \times L \times T \times 20{,}600 \qquad (8\text{-}6)$$

where:

F = stripping force, kN
L = length of cut, mm
T = material thickness, mm

## Press Tonnage

Press tonnage is determined by the sum of all the forces required to cut and form the part. Practical experience with presses equipped with tonnage monitors shows close agreement between calculated versus measured values. In many cases, the stripping force must be added to the cutting force if a spring-loaded stripper is used because the springs must be compressed while cutting the material. Any spring pressure for forming, draw pads, and the like will have to be added. Using fixed, or tunnel, strippers will keep the press load to a minimum, but these devices will not control the stock as well as spring-loaded strippers.

## Side-thrust or Lateral Forces

*Figure 8-9a* illustrates the shear, tensile, and compressive forces that occur during the cutting process. The amount of lateral force varies with the cutting clearance and material.

For round and symmetrical holes, the lateral forces balance out. However, the die must be sufficiently strong to withstand high spreading forces. For notching, shearing, and other unbalanced operations, the alignment system of the die must not allow excessive deflections to occur.

Side-thrust or lateral force can result in excessive deflections of die components, such as punches, heel blocks, and guide pins. As lateral deflection occurs, clearances increase.

The lateral pressure can exceed the press force by a factor of three or more, due to a wedge-like,

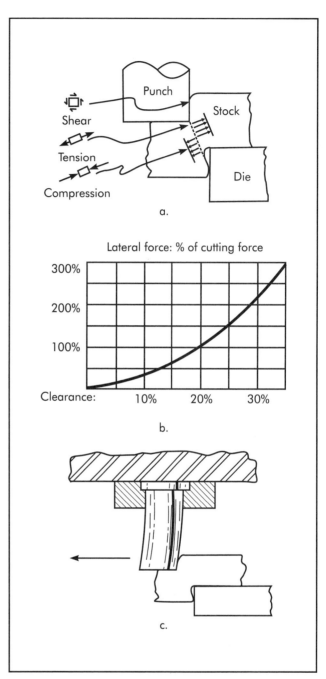

Figure 8-9. (a) *Shear, tensile, and compressive forces occur during the cutting process; (b) as punch-to-die clearance increases, the lateral force, or the side thrust, rapidly increases; (c) as side thrust increases, the cutting clearance may increase, leading to greater side thrust or lateral forces, and potential tool failure.*

mechanical advantage. If not carefully controlled, the resulting misalignment can damage the tooling and produce scrap parts.

## Effect of Die Clearance on Lateral Forces

*Equation 8-7* gives an approximation of the side thrust or lateral force generated when cutting or shearing. When applying this equation, adjustments for the type of material and die conditions must be made. *Figure 8-9a* shows that cutting operations involve some bending. The graph in *Figure 8-9b* is based on *Equation 8-7*.

$$\frac{C}{T-P} = \frac{F_H}{F_V} \qquad (8\text{-}7)$$

where:

$C$ = clearance, in. (mm)
$F_H$ = side thrust, lbf (kN)
$T$ = material thickness, in. (mm)
$P$ = penetration, typically $0.33 \times T$
$F_V$ = cutting force, lbf (kN)

Excessive clearances and dull die steels can result in very large side thrusts. In extreme cases, the side-thrust or lateral force can be so great that the die may shatter due to extreme pressures and interference of die components.

The function of die alignment system components, such as guide pins and heel blocks, is to limit deflections due to side thrusts to acceptable values. *Figures 8-9b* and *c* illustrate how side thrusts can cause punch deflection. Punch deflection increases the die clearance, thus leading to greater lateral forces.

## Reducing Cutting Forces

By their nature, cutting operations are characterized by very high forces exerted for very short periods. Sometimes it is desirable to reduce the magnitude of the forces and spread it over a longer portion of the press stroke. Punch contours of large perimeter or many smaller punches frequently will result in tonnage requirements beyond the capacity of an available press. In addition, whenever abnormally high tonnage requirements are concentrated in a small area, the likelihood of design difficulties and outright tool failure increases.

Two methods generally reduce cutting forces and smooth the shock impact of heavy loads.

(Keep in mind that during a piercing operation with proper clearance, complete fracture occurs when the punch has penetrated one-third the material thickness.)

1. By adjusting the height of the punches so they differ in length by one-third the material thickness, they can cut in sequence rather than all at once. Using three punches of the same diameter and stepped properly, one-third of the total tonnage required is used to do all three simultaneously.
2. Adding shear to the die or punch equal to one-third the material thickness reduces the tonnage required by 50% for that area being cut with shear applied. Note that shear should be applied to the die member (punch or die) that contacts the scrap. As a result, the deformation due to the shear angles will have little effect on the part *(Figure 8-10)*.

In piercing, the tool is designed so the direction of shear angles generally is such that the cut proceeds from the outer extremities of the contour toward the center. This avoids stretching the material before it is cut free. Limiting factors in how much timing can be used to reduce tonnage requirements include:

- the allowable tonnage as a function of the distance above the bottom of the press stroke, and
- the allowable loss of flywheel energy per stroke.

Timing can be important to reduce the snap-through that can damage the press.

## Analysis of Snap-through Forces

The loud boom that characterizes a snap-through problem when cutting metal directly results from the sudden release of potential energy stored in press and die members as strains or deflection occur. The deflection is a normal result of the pressure required to cut through the material. In extreme cases, the energy released can damage the press. The press connection (the attachment point of the pitman to the slide) is easily damaged by the reverse load generated by snap-through. As a rule, presses are

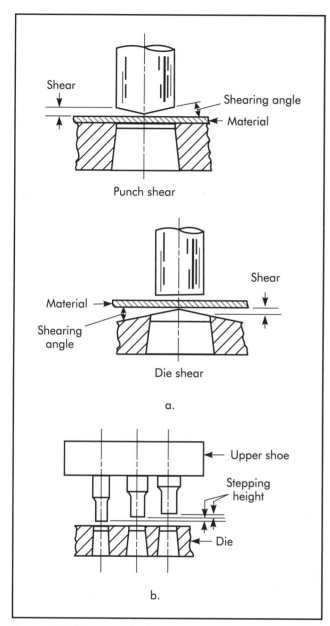

Figure 8-10. *Three methods to reduce cutting forces.*

not designed to withstand reverse loading of more than 10%, and the shock may result in die components working loose.

In timing punch entry or die shear, care must be taken to provide a gradual release of developed tonnage. The shock normally is not generated by the impact of the punches on the stock. In fact, when the punches first contact the stock, the initial work is done by the kinetic energy of the slide. To complete the cut, energy must be supplied by

the flywheel. As this occurs, the press members deflect. An analysis of the quantity of energy involved will show why a gradual reduction in cutting pressure before snap-through is important.

The magnitude of actual energy released increases as the square of the actual tonnage developed at the instant of final breakthrough. The actual energy is given by:

$$E = \frac{F \times D}{2} \qquad (8\text{-}8)$$

where:

$E$ × 166.7 = energy, ft-lbf (1 ft-lbf = 1.356 J)
$\quad F$ = pressure at moment of breakthrough, tons
$\quad D$ = amount of total deflection, in.

or

$E$ × 9.807 = energy, J (1 J = 1.356 ft-lbf)
$\quad F$ = pressure at moment of breakthrough, metric tons (kgf × 1,000)
$\quad D$ = amount of total deflection, mm

For example, if 400 tons (362.8 t), which resulted in 0.080 in. (2.03 mm) total deflection, were required to cut through a thick steel blank, the energy released at snap-through would be 2,667 ft-lbf (3,616 J).

If careful timing of the cutting sequence results in a gradual reduction of tonnage at the instant of snap-through so that only 200 tons (181.4 t) were released, the reduction in shock and noise would be dramatic. Half the tonnage would produce half as much deflection, or 0.04 in. (1.0 mm). The resultant snap-through energy would be only 667 ft-lbf (904 J), or one-fourth the former value.

**Example: snap-through reduction by die timing.** *Figure 8-11* illustrates a waveform resulting from an operation to punch two 1.625-in. (41.28-mm) holes and parting a chain side bar from fine-grained AISI-SAE 1039 steel. The steel was 0.50 in. (12.7 mm) thick by 3.00 in. (76.2 mm) wide. A 300-ton (2.7 N) straightside press was used for this operation. The allowable reverse load is 30 tons (267 kN). Point *A* on *Figure 8-11* illustrates a peak load of 191 tons (1.7 N), which

Figure 8-11. *Actual waveform signature of a combined piercing and cut-off operation having excessive snap-through, or reverse load.* (Courtesy Webster Industries)

was also displayed on the tonnage meter. This is well within press capacity. In this case, the reverse load *B* was 87 tons (774 kN), which is nearly three times the allowable amount. The die was immediately taken to the repair bench and one punch was shortened 0.312 in. (7.92 mm). Balanced angular shear was ground on the punches. Balanced shear was also ground on the parting punch.

*Figure 8-12* illustrates the improvement achieved by modifying the tool. The peak tonnage was reduced to 82.8 tons (737 kN), which is less than half the initial value. The reverse load was reduced to 22 tons (196 kN), or about one-fourth the former value. This is keeping with the square-law formula for the energy suddenly released.

## BALL-LOCK PUNCHES

The ball-lock concept of retention and quick punch replacement has been in use for many years. Style changes on a stamping can be made quickly, without removing the die from the press, simply by pulling or adding punches. In the event that a fragile punch needs replacement during

the run, the ball-lock system of punch retention makes this operation quick and easy. *Figure 8-13* illustrates the ball-lock retention system used to positively lock both a punch and die button (matrix) in a die. The ball-lock retention is an adaptation of the wedge principle. Once correctly seated, the ball locks the punch in position both vertically and radially, while permitting rapid replacement of the punches in the die assembly.

To install a ball-lock punch, it is inserted into the retainer and twisted until the spring-loaded ball drops into position, locking the punch. The punch is removed by depressing the ball with a ball-release tool inserted in a hole in the retainer provided for this purpose. This action frees the punch so it can be pulled out of the retainer. Some retainers use a ball-release screw, as illustrated in *Figure 8-13*. To ensure correct location and certain retention, it is important that the ball-locking condition be correct. There can be a problem if the ball is either too low or high in the teardrop-shaped ball seat. This problem can be avoided by always using the correct retainer specified for a given punch and using the proper size replacement ball, as shown in *Figure 8-14*.

Figure 8-12. *Waveform signature after adding timing and balanced shear.* (Courtesy Webster Industries)

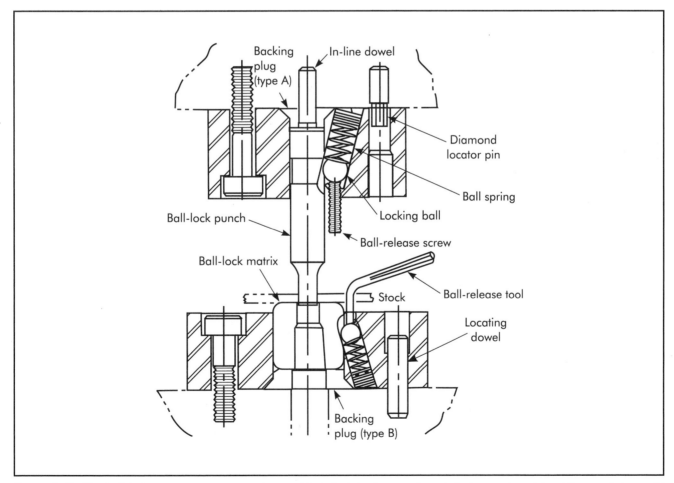

Figure 8-13. *The ball-lock retention system used to positively lock both a punch and die button (matrix), while permitting rapid replacement of these components in the die assembly.* (Courtesy Dayton Progress Corp.)

Figure 8-14. *The ball must be properly seated in the teardrop-shaped pocket in the punch. (Courtesy Dayton Progress Corp.)*

## BENDING

Bending frequently is used to increase the rigidity of shaped parts in pressworking operations. The simplest bending operation is air bending, so called because the die does not touch the outside of the bend radius. The part to be bent is supported on each side of the bend and force is applied to a forming punch in the center.

*Figure 8-15* illustrates a metal beam supported at two points, with a load applied at the midpoint. The load produces compressive stresses in the material on the inside of the bend as it is forced into compression. Tensile stress or stretching occurs on the outside of the bend.

To produce a bend in a finished part, the yield point of the material must be exceeded. If the

Figure 8-15. *A metal beam supported at two points, with a load applied at the midpoint, resulting in bending or deflection.*

bending force applied does not exceed the yield strength of the material, the beam will return to its original shape on removal of the load, as shown in *Figure 8-16*. However, if the stress exceeds the material-yield strength, the beam will retain a permanent set or bend when the load is removed, as illustrated in *Figure 8-17*.

Figure 8-16. *If the applied force does not exceed the material-yield strength, the beam returns to its undeflected shape.*

Figure 8-17. *Simple beam deflection occurs in air bending. If the applied force exceeds the material-yield strength, the beam retains a permanent set or bend when the load is removed.*

The goal of the process is to bend the material the correct amount. Springback, or elastic recovery, will occur until residual stresses in the bend are equal to the stiffness of the material. This concept is illustrated in *Figure 8-18*.

Figure 8-18. *Springback occurs until the residual stress forces are balanced by the stiffness of the material.*

Not all of the material in the bend zone is stressed equally. The material in the inner and outer surfaces is stressed the most, and the stress gradually diminishes toward a neutral axis between the two surfaces. At that point, the stress is zero and there is no length change.

## Bend Allowances

The exact length of a bend is determined by trial and error. The assumed neutral axis varies depending on the bending method used, location in the bend, and type of stock to be bent.

The direction of grain in a steel strip relative to the bend has a slight effect on the length of metal required to make a bend. Bending with the grain allows the metal to stretch more easily than bending against the grain. However, this results in a weaker stamping.

Bend allowances depend on the physical properties of the metal, such as its tensile and yield strength and ductility.

The exact bend allowance is the arc length of the true-neutral axis of the bend (metal is stretched above the neutral axis; below it, metal is compressed). The neutral axis only can be approximated. Many manufacturers assume the neutral axis is 1/3-stock thickness from the inside radius of the bend for inside radii of less than twice the stock thickness. For an inside radii of two times the stock thickness or greater, the neutral axis is assumed to be 1/2-stock thickness from the inside radii. One reason for the requirement of relatively less metal to make a tight bend is that the sharp radius tends to be drawn or stretched slightly.

Many experts believe that the location of the true-neutral axis from the inside radius varies from 0.2–0.5 times the stock thickness. An important factor that determines the neutral axis is how the bend is accomplished. Less metal is required for a bend made by a tightly wiped flange than for an air bend on a press brake. Wiping the flange tends to stretch the metal.

## Equations

Before the advent of the inexpensive handheld calculator, books of extensive data were available to determine the inside of the bend radii, from 0.015–1.250 in. (0.38–31.75 mm), including metric equivalents for various materials and types of bends. The equation used to develop these tables is based on extensive experimental data. One concern regarding the accuracy of the tables is the uncertainty of the true location of the neutral axis.

For 90° bends, the coefficient of 1.57 (the number of radians in 90°) determines the assumed amount of metal necessary to make the bend, which is multiplied by the assumed neutral axis. For bends that are not exactly 90°, it is necessary to multiply the number of degrees of bend × 0.0175 (the number of radians in 1°), and to substitute the result for the coefficient 1.57. The actual neutral axis will vary from the stock centerline to 20% from the inside of the bend for sharp bends.

## Springback or Elastic Recovery

Stiffness is a function of the material's modulus of elasticity. This explains why some materials, such as mild steel with a high modulus of elasticity (as compared to tensile strength), spring back less than materials with a lower modulus. However, some materials do have comparable tensile strength, such as hard aluminum alloys.

Some springback occurs whenever metals are formed. Springback is caused by the residual stress that is a result of cold-working metals. For example, in a simple bend, residual compressive stress remains on the inside of the bend, while residual tensile stress is present on the outside radius of the bend. When bending pressure is released, metal springs back until residual-stress forces are balanced by the material's stiffness, which resists further strain. The most common method used to compensate for springback is overbending, in which material is bent a sufficient amount beyond the desired angle and allowed to spring back to the desired angle after elastic recovery occurs. Because of the uncertainty of the exact location of the neutral axis, it is best to use trial-and-error methods when developing close-tolerance stampings.

Many complex factors determine the amount of springback that will occur in a given operation. Because the exact amount of springback is difficult to predict, data for a specific material and forming method is often developed under actual production conditions to aid process control and future product development. If the die designer and builder fail to include correct springback compensation in the die, correction will need to be done by the repair facility of the press shop that uses the tool.

### Factors Affecting Springback

Some factors that increase springback are:

- higher material strength;
- thinner material;
- lower modulus of elasticity;
- larger die radius;
- greater wipe-steel clearance;
- less irregularity in part outline, and
- flatter part-surface contour.

If a flanged part is irregular compared with either the outline or surface contour, the springback will be slight. The springback for large wipe-steel clearances can be several degrees or more.

### Air Bending

Press-brake tooling for air bending (*Figure 8-19*) is quite simple. Air bending is one of the most common press-brake operations. It requires minimum tonnage for the work performed. Exact repeatability of ram travel is required to maintain close repeatability of the bend angle. The amount of over-travel is determined experimentally to compensate for springback.

### Causes of Bend-angle Variation

There are several causes of bend-angle variation when bending materials in pressworking operations. These include:

- changes in stock yield strength;
- variation in stock thickness;
- machine variations due to temperate changes, and
- machine deflection, especially in long press-brake-bending operations.

Compensation for any change of the conditions that affect the bend angle may require adjustment of the ram travel. With press-brake-bed deflection, shimming also may be required. In addition, shims may be required to correct for additional machine deflection. Some press-brake designs have automatic deflection-compensating devices, such as hydraulic cylinders built into the bed. If high force is required in press-brake bending because of an increase in stock thickness or hardness, a simple ram adjustment may not be enough to correct the problem.

### Coining

Coining, or bottom bending, has the advantage of producing sharp, accurate bends with less sensitivity to material conditions than air bending. The disadvantages of coining compared to air bending are high-force requirements and accelerated die wear.

*Figure 8-20* illustrates a press-brake die designed to coin the bend for a precise angle. This coining action eliminates the root causes of spring-

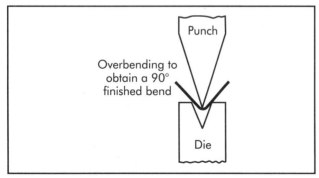

Figure 8-19. *Simple tooling of the type used to air-bend sheet-metal parts in press brakes. The upper die is lowered a little and a hit is made until the desired bend angle is obtained.*

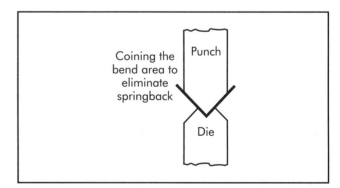

Figure 8-20. *Coining the bend requires high tonnage to obtain a sharp, accurate bend.*

back, including the tensile and compressive residual stresses on opposite sides of the bend. Coining action is accomplished with pressure that is sufficient to subject the metal to the yield point in the bend area.

The tonnage required for coining might be five to 10 times that required for simple air bending. Higher forces increase machine deflection. Air-bending jobs that produce acceptable bend angles throughout the entire length of the bend may need shimming if coining is required. The amount of machine deflection increases approximately in proportion to the developed tonnage.

## Wipe Bending

It is often not feasible to use V-bending tooling for bending. V bending is popular for press-brake work. The tooling is simple and can accomplish a variety of work. Usually, only a single bend occurs per stroke. Accurate work requires that each previous step be accurate. Skilled and experienced operators are required. A limiting factor is the cost of press-brake work, because of low throughput and the high skill required for accurate work.

### Wipe Flanging and Springback Control

*Figure 8-21* illustrates a sectional view through a wipe-flanging die. In this design, the flange steel

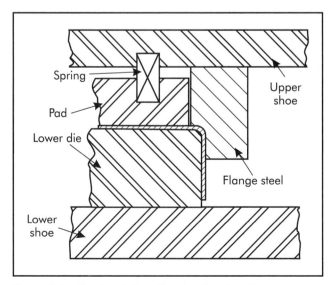

Figure 8-21. *Sectional view of a wipe-flanging die.*

attached to the upper die wipes the metal around the lower die. A popular method for controlling springback is to coin the top of the bend with the flange steel. A disadvantage is the limited springback compensation that can result in a distorted bend-angle condition. A close-up view of this is shown in *Figure 8-22*. The top thickness of the bend can be squeezed beyond the material-yield point by careful adjustment of the die-shut height. Only the top portion of the bend is coined. This can result in a score mark that might weaken the stamping and extrusion of the metal being coined.

Figure 8-22. *Close-up view of the point of flange steel contact on the bend radius in a wipe-flanging die.*

If excessive coining pressures are applied, the metal at the top of the bend will extrude and result in a weak and distorted bend condition. An improved flanging method relieves the radius in flange steel so it does not contact the top of the bend radius. One way to do this is to relieve the flange steel at an angle approximately 20° tangent to the radius. Another method is to machine the flange steel to a radius that is larger than the outside of the bend. The flange steel is positioned so the tightest point is 45–60° beyond the top of the radius. The side of the form steel is relieved 5° or more to permit the material to be overbent (*Figure 8-23*). This method is more effective than coining the top of the bend. In addition, the improved bending process is not as sensitive to variation from press adjustments and material conditions.

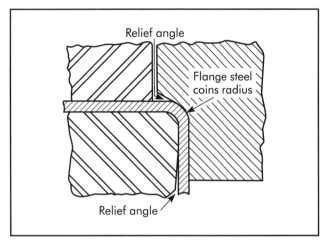

Figure 8-23. *The side of the radius is coined and a relief angle is provided in the lower-die steel in this improved springback control method.*

## Rotary-action Die Bending

A patented rotary-action bender known as the Ready™ bender combines low-tonnage requirements of air bending with the accuracy and multiple-bend capabilities of wipe-flange tooling. *Figure 8-24* illustrates a Ready bender making initial contact with the stock. As the upper die travels down, the stock is clamped and bent by the rotating bender. As the die closes, the rotary-

bending action progresses (*Figure 8-25*). An optional relief angle in the lower die permits the rotary member to overbend the stock at the bottom of the stroke to provide springback control (*Figure 8-26*).

Rotary-action benders can bend angles up to 120°. The rotary member usually is made of heat-

Figure 8-25. *A Ready™ bender bends the stock through rotary action of the circular member.* (Courtesy Ready Tools, Inc.)

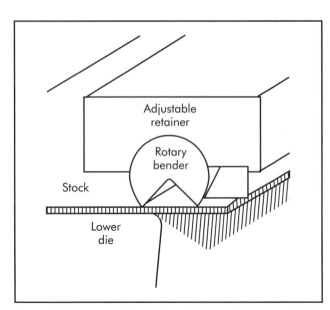

Figure 8-24. *A Ready™ bender makes initial contact with the stock. As the die closes, the bender clamps and bends the stock.* (Courtesy Ready Tools, Inc.)

Figure 8-26. *A Ready™ bender overbends the stock at the bottom of the stroke to provide for springback.* (Courtesy Ready Tools, Inc.)

treated tool steel for long wear. The bending pressure is typically 50–80% less than that required for conventional wipe bending. The lower pressure permits many types of prepainted materials to be fabricated without damaging the finish. Rotary benders also can be constructed of nonmetallic materials, such as hard thermoplastics, for work with prefinished materials.

### Fine Adjustment of Bend Angle

On conventional wipe-flange tooling, the bend-angle adjustment usually is made by adjusting the flange steel up or down with shimming. In the case of rotary-action benders, the bend angle is adjusted by moving the assembly containing the bender along the horizontal plane relative to the lower-die member or anvil. Attempts to obtain a tighter bend by excessive lowering of the press shut-height can result in tooling damage.

## FORMING

A large percentage of stampings used in the manufacturing of products require some forming operations. Some are simple forms that require tools of low cost and conventional design. Others may have complicated forms, which require dies that produce multiple forms in one stroke of the press. Some stampings may require several dies to produce the shapes and forms required.

Forming dies, often considered in the same class with bending, are tools that form or bend the blank along a curved instead of straight axis. There is little stretching or compressing of the material. The internal movement or plastic flow of the material is localized and has little or no effect on the total area or thickness of material.

A first consideration in analyzing a stamping is to select the class of die to perform the work. Next, determine the number of stampings required, which will help identify the forming process to be used.

A forming die may be designed in many ways and produce the same results. The tool that is cheapest and has the simplest design may not always be best because it may not produce the stamping to the drawing specifications. Where limited production is required, press-brake tooling often can be used.

### Solid Form Dies

The solid form die is of simple construction and design. A forming die of this kind need not be mounted on a die set as shown in *Figure 8-27*. A die set can be considered because the repeatability of the setup reduces the chances for misalignment. A great deal of side pressure is exerted on the female die block and must be considered in the design.

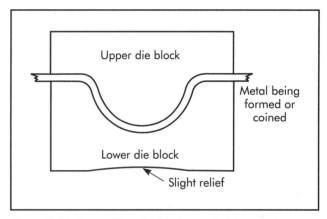

Figure 8-27. *A set of forming blocks used without a die set.*

One option is to make the female block out of two pieces of tool steel held together by prestressed tie rods. This is used as a safety measure in case the operator should feed a double blank or if metal of a thicker gage is used. The use of a computer-aided stress analysis program is advised to determine the areas of high tensile stress to proportion the die blocks accordingly.

In some cases, a relief under the center of the female die will transfer much of the tensile strain to the more robust bottom of the female die—this is the same mechanical principle as a bridge with truss work under the roadway. Operating dies of this type in a hydraulic press is advised to limit the pressure applied to the die, especially if coining pressures are achieved.

### Coining Dies

Coining dies are useful for achieving precise forms by coining—a process for bringing the entire thickness of the metal up to the yield strength of the material. Applications for coining may include sizing bearing shell halves. Any curved stamping requiring accurate form dimensions is a candidate for coining.

Flat coining often is done to produce parts with high flatness requirements. Another application for flat coining blocks is the production of medallions of materials varying from stainless steels to sterling silver and gold alloys. Surprisingly high forces are required to bring up all of the detail in some medallion work. Forces of 325 ksi (2,240 MPa) or higher are achieved.

## Form Die Design

*Figure 8-28* is an example of a two-piece forming die mounted in a die set and equipped with a positive knockout. A close-fitting cutout is machined into the lower die shoe to limit shifting of the lower form detail. The form blocks are made wider than the workpiece and their height by an approximate ratio of 1:1.5. The radius being formed should be at least twice the metal thickness to avoid excessive stress concentration. A good polish on the form blocks will ensure against marring the stamping.

The punch is made of a good grade of tool steel heat-treated for toughness. Grade S5 or S7 are good choices for severe work.

Consideration is given to stripping the formed part from the punch. This is accomplished by means of a knockout, stripper hooks, or stripper-pin (spring) construction. It is important to consider and plan for removing the formed part. *Figure 8-28b* shows these details and illustrates the gages necessary for locating the stamping.

The shoe is thick enough to withstand the pressure required for the forming operation. In selecting its thickness, the size, shape of the bolster, and plate of the press is considered. The *A* and *B* dimensions of the shoe, when placed over the hole of the bolster plate, should be long and wide enough to allow ample space for clamping it to the bolster plate. A die shoe made of steel will give good service and costs only a little more than cast iron. Select the diameter of the guide pins according to the working area of the shoe, or by consulting a die set manufacturer's catalog. The length of the pins should be at least 0.25 in. (6.4 mm) shorter than the shut height of the die, as listed on the drawing. The guide bushings may be of the shoulder type. Ball-bearing pins and bushings can be used if better accuracy and longer life than plain bushings are desired.

The upper die shoe is made to the same outer dimensions as the lower shoe. If a shank is to be used, Occupational Safety and Health Administration (OSHA) rules require toe clamps or some

1 = lower die shoe
2 = upper die shoe
3 = guide pin
4 = guide pin bushing
5 = lower form blocks
6 = gage plate
7 = sides of slide
8 = forming punch
9 = knockout rod

Figure 8-28. *Pressure-pad-type form dies.*

other means other than the shank alone to retain the die in the press.

Locate the screw and dowel holes so they will not mar the stamping. Dowel pins should be large enough in diameter to ensure correct alignment of the sections. The screws should be large enough to clamp the die-steel details to the shoes by friction. The screw- and dowel-hole locations also should locate the gage plate. Designing the gage-plate shape when designing the die blocks, then placing the screw and dowel holes accordingly, accomplishes a dual purpose. It helps eliminate some of the holes required in the die blocks, as well as the work of drilling and tapping the holes and the cost of the extra screws. The blank should be fed the long way to ensure easy loading and accurate location. When blanks are fed the short way, they have a chance to twist, and the operator loses control over them. This causes a loss of time in locating the blank in the die, and can cause mislocation of the blank.

Make the gage plate so the blank will slide into the proper location for forming. When the blanks are wide and flat, provide ribs or wires to help slide the blank into the die and reduce some of the friction caused by the oil on the blank surfaces. Design the slide long enough to prevent the operator from placing his or her fingers under the punch.

## Pressure Pads

When the forming of stampings requires accuracy, dies employing pressure pads often are designed. A pressure pad helps hold the stock securely during forming and eliminates shifting of the blank. The pressure can be applied to the pad by springs or the use of an air cushion. When springs are used, they can be located directly under the pad and confined in the die shoe. They also may be located in or under the press bolster plate. By the use of pressure pins, which are located under the pad, and through the die shoe, pressure is applied to the pressure plate.

Pressure pins also are used with an air cushion. The construction of the pressure plate and pressure pins would be the same, except that an air cushion is substituted for the springs. When an air cushion is used, the proper amount of pressure on the pressure pad is ensured as long as the air supply is set properly. It is important that the die-setter have the correct pressure in the die setup procedure, as well as proper pin placement. A consistent pressure ensures repeatable stamping quality.

When springs are used to apply pressure to a pressure pad, spring pressure increases with the pad travel. Each fraction of an inch or millimeter of travel increases the pressure on the pad. This may cause some trouble when stamping light-gage material because too much pressure may cause the metal to stretch. Care must be used to not over-compress the springs used, which can lead to loss of pressure and spring breakage. Increasingly, nitrogen manifold systems or self-contained nitrogen cylinders are used as pad pressure sources.

The pressure pad is controlled in its travel between die blocks by retaining ledges or shoulder screws. When using the retaining ledge construction, a recess is machined into the form blocks and a corresponding shoulder is machined on the pad. The ledge shoulder should always be made strong enough to withstand the pressure applied by either springs or an air cushion. The size of the shoulder to be used varies according to the size and metal thickness of the stamping. A good rule is to have the height of the ledge 1.5 times the width, as shown in Figure 8-29. Always design pad ledges with a radius in the corner. When using shoulder screws to control the travel of the pad, the die shoe must be thick enough to permit sufficient travel.

The pressure pad should always travel so that it extends slightly above the die blocks. This will

Figure 8-29. *Pressure-pad design.*

ensure uniform parts because pressure locks the part between the punch and pad faces before the actual forming takes place. The amount of travel in the pad depends on the height of the form die. It is not always necessary to travel the full height; in many cases, half of the die's form height is sufficient. For accurate work, allow the punch to give the part a definite set at the bottom of the stroke. When a stamping must have sides square with the bottom after forming, the corner radius should be set. This is done by designing the die blocks with the correct radius, *A*, shown in *Figure 8-30*. The pressure pad is made to match the height of the die block's radius edge. The punch radius, *C*, is made slightly smaller, approximately 10% less than the die-block radius.

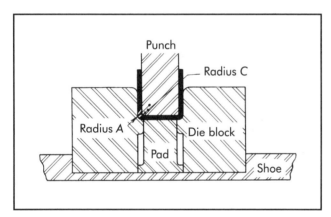

Figure 8-30. *Radii considerations for form die design.*

It may be necessary to machine a slight relief angle on the side of the punch to allow a slight overbending of the side being formed. This ensures that the sides of the formed part will be square with the base after forming.

### Rubber and Polyurethane

A rubber- or polyurethane-forming die often is used in press-brake operations, as illustrated in *Figure 8-31*. This reduces tooling costs, as only the male forming punch has to be made. This method is well-adapted to forming angles, channels, and radii. Another advantage is that the rubber die leaves no visible marks on the part, which is especially important when bending prefinished material.

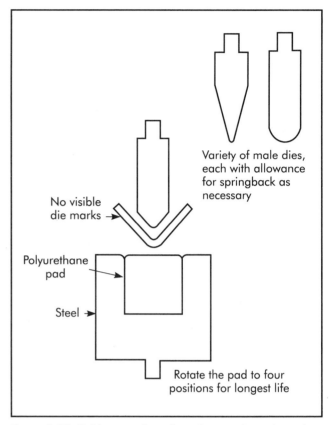

Figure 8-31. *Rubber or polyurethane forming die and punches.*

## EMBOSSING

In an *embossing* operation, a shallow surface detail is formed by displacing metal between two opposing mated tool surfaces. One surface has the depression, the other the projection. The metal is stretched slightly, rather than being compressed. Embossing is used for various purposes, the most common being the stiffening of the bottom of a pan or container; the embossing is designed to follow the outside profile of the part. A round canister may have an embossed circle or raised grooves of various widths or panels. When a can or box is square or rectangular, embossments follow the contours. Embossments often are ribs or crosses stamped in the metal to help make a section of a blank stronger by stiffening. An embossing die can be a male and female set of lettering dies or a profile of one of various shapes.

The construction of die blocks for an embossing operation depends on the size and shape of the form, and the accuracy and flatness required. When embossing simple shapes, such as stiffening ribs, it is not necessary to strike the bottom of the relief. The metal stretches over the punch and across the two radius edges of a die clearance.

The female die opening has the same width of the rib or embossing punch, and a slight radius is added to the edges of the opening to allow metal to flow freely. The punch is made slightly smaller than the required metal thickness per side so that it does not strike along this area. By avoiding coining or hard marks, the pressure required to stamp the embossing is reduced and die wear is much less.

A small embossment often is used as a spot-weld projection nib. *Spot-welding* is a process where metal is electrically welded together under controlled pressure and current. These nibs are helpful to ensure a highly uniform current flow and sound spot-welds.

## BEADING AND CURLING

In *beading* and *curling* operations, the edges of the metal are formed into a roll or curl. This is done to strengthen the part or to produce a better-looking product with a protective edge. Curls are used in the manufacturing of hinges, pots, pans, and other items. The size of the curl should be governed by the thickness of the metal; it should have a radius at least twice the metal thickness. To make good curls and beads, the material must be ductile; otherwise, it will not roll and will cause flaws in the metal. If the metal is too hard, the curls will become flat instead of round. If possible, the burr edge of the blank should be the inside edge of the curl. This location facilitates metal flow and helps keep the die radius from wearing or galling. In making curls and beads, a starting radius is always helpful. A starting radius is shown in *Figure 8-32*.

The curling radius of the die always must be smoothly polished and free of tool marks. Any groove or roughness will tend to back up the metal while it is rolling and cause defective curls. The inside surface of the blank must be held positively in line with the inside curling radius of the punch (*Figure 8-33*).

Figure 8-32. *Starting curl radius.*

Figure 8-33. *Curling punch design.*

When curling or beading pots, pans, cans, or pails, wires may be rolled inside the curls. This will make them stronger. The wire is made to the contour of the pan and placed on a spring pad. When the curling die closes, the edge of the pan is forced to curl around the wire, as shown in *Figure 8-34*. If the wire is omitted, the tooling is much the same, except the curled edge is termed a false wire curl.

Figure 8-34. *Curling die design.*

## TWISTERS AND BENDERS

The successful manufacturing of fans requires uniform blades formed alike and of equal weight. A hub spider must provide a precise location for the blade attachment holes in relationship to the hub shaft insert. Steel sheeting, by the very nature of the rolling process, cannot be made without slight variations in the way it bends with and across the rolling direction or grain of the sheet. Since the spider has a critical role in the balance and overall performance of the assembled fan, it must be made with precise bend angles in spite of the variation across the steel sheet and even larger differences from coil to coil. An example of a hub spider press is shown in *Figure 8-35*.

Figure 8-36. *A view of an individual spider bending station. The die is on the bench and held partly open by eight self-contained nitrogen cylinders.* (Courtesy Modern Die Systems, Inc.)

Figure 8-35. *Plan view of lower die with fan hub to be formed in place. Note the two locating pins that hold the part on location. Each of the spider legs rest on one half of a cam-actuated bending device.* (Courtesy Modern Die Systems, Inc.)

## Rotary Bender Operation

When the die closes with the flat spider in place (*Figure 8-36*), a rotary motion of the lower and upper die begins forming. All eight of the forming stations ride on a circular plate supported by eight nitrogen cylinders and guided by four guide pins and bushings. *Figure 8-36* is a view of an individual spider bending station. As the die closes, the cam lever arm contacts the fine adjustment screw resulting in a rotary motion of the lower and upper die forming halves.

The die is operated in a hydraulic press, ensuring a smooth, controlled closing. As the die closes, the plate floating on nitrogen cylinders descends until the lever arms meet the round

contact pads on each adjustment screw. Thus, the adjustment of each screw controls the angle to which each corresponding spider leg is bent. The amount of adjustment needed for variations in material is determined by a gaging station, which feeds the output of precision dial indicators to a computer. The computer provides a statistical process control (SPC) record and tells the operator how much adjustment is required to stay in the middle of the upper and lower SPC limits.

Once a computer interface has been attached, it can be used to measure parts and adjust the tool during production using servomotors to adjust the screws. One such computerized gage with 0.00005-in. (1.3-μm) resolution digital indicators can measure the angles of each spider leg to five decimal places faster than the operator can load and cycle the next part. This makes 100% inspection possible. The computer interface for the gage and die displays X-bar charts of the angle data for true "real-time" quality control. Quality discrepancies can be dealt with quickly.

## HOLE FLANGING OR EXTRUDING

*Hole flanging* or *extruding* is the forming or stretching of a flange around a hole in sheetmetal. The shape of the flange can vary according to part requirements. Flanges are made as countersunk, burred, or dimpled holes.

When countersunk-shaped, extruded holes are made in steel, it is necessary to coin the metal around the upper face and beveled sides to set the material. The holes also are made about 0.005 in. (0.13 mm) deeper than the required height of the rivet or screw head, which allows the metal compression that occurs when squeezing the rivet in place. A section of a die for this purpose is shown in *Figure 8-37*. The hole can be pierced before it is placed in the countersinking die, or it can be formed and pierced in a single operation.

As shown in *Figure 8-37*, the sheet is placed over the pilot diameter, *A*, which locates it centrally in the die. The die body, 1, descends and forces the metal down around the flange surface of the punch. Spring pressure strips the part from the punch and releases the formed part from the die.

*Figure 8-38* shows a two-step punch, 1, which first punches the hole in the part and then forces the metal around to the countersunk shape of the die block, 2. The hole punched by this method always is somewhat smaller than the size of the hole in the finished part. Spring pressure is used to strip the finished part from the punch. A shedder pin should be provided in the piercing point of the punch to remove the slug.

Although *Figures 8-37* and *8-38* show countersunk holes formed from the bottom up, they can be just as easily formed from the top down depending on part-design requirements.

Figure 8-37. *A forming die used to produce countersunk holes.*

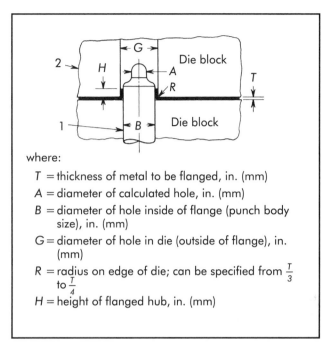

where:

$T$ = thickness of metal to be flanged, in. (mm)

$A$ = diameter of calculated hole, in. (mm)

$B$ = diameter of hole inside of flange (punch body size), in. (mm)

$G$ = diameter of hole in die (outside of flange), in. (mm)

$R$ = radius on edge of die; can be specified from $\frac{T}{3}$ to $\frac{T}{4}$

$H$ = height of flanged hub, in. (mm)

Figure 8-38. *A die for punching and countersinking a hole.*

## 90° Hole Flanging

The size of the pierced hole for a 90° hole flange can be calculated, but should never be used until it has been proven correct by using the same tools that will be employed in the die. To calculate the hole size, the same principles are employed as when finding a 90° bend.

When flanges are stretched more than 2-1/2 times the metal thickness in height, the wall can split. This can be prevented to some extent by burring the edge around the hole before the extruding operation. In this situation, the side of the hole having the burr is important. Coining the hole edges is a good way to strengthen the hole edge for flanging. This is an especially attractive option in progressive die operations, which are discussed later.

Forming a flange around a previously pierced hole at a bend angle of 90° (the most common operation) is nothing more than the formation of a stretch flange at that angle.

One manufacturer has standardized flange widths (*Figure 8-39*, dimension *H*) and starting hole diameter (Figure 8-39, dimension *J*) for holes that are to be threaded. Using taps that form the metal by displacement rather than cutting the threads will result in a much stronger female thread. The following are equations for low-carbon steel stamping stock:

$$B = A + \frac{5T}{4} \text{ when } T < 0.045 \text{ in. } (1.14 \text{ mm})$$
(8-9)

$$B = A + T \text{ when } T > 0.045 \text{ in. } (1.14 \text{ mm})$$
(8-10)

$$H = T \text{ when } T < 0.035 \text{ in. } (0.89 \text{ mm}) \quad (8\text{-}11)$$

$$H = \frac{4T}{5} \text{ when } T = 0.035 \text{ to } 0.050 \text{ in.}$$
$$(0.89 \text{ to } 1.27 \text{ mm}) \quad (8\text{-}12)$$

$$H = \frac{3T}{5} \text{ when } T > 0.050 \text{ in. } (1.27 \text{ mm}) \quad (8\text{-}13)$$

$$R = \frac{T}{4} \text{ when } T < 0.045 \text{ in. } (1.14 \text{ mm}) \quad (8\text{-}14)$$

$$R = \frac{T}{3} \text{ when } T > 0.045 \text{ in. } (1.14 \text{ mm}) \quad (8\text{-}15)$$

$$J = \frac{\sqrt{TB^2 + 4TA^2 + 4HA^2 - 4HB^2}}{9T}$$
(8-16)

where:

B = diameter of hole inside of flange (punch body size), in. (mm)
A = diameter of calculated hole, in. (mm)
T = thickness of metal to be flanged, in. (mm)
H = height of flanged hub, in. (mm)
R = radius on edge of die; can be specified

from $\frac{T}{3}$ to $\frac{T}{4}$

J = starting hole diameter, in. (mm)

Figure 8-39. Hole flange design.

The radius, P, on the nose of the punch should be blended into the body diameter, eliminating any sharpness that could cause the metal to score as it passes over it. The radius on the body, B, or hole-sizing portion of the punch, must be as large as possible, and smooth. The portion between the A and B diameters of the punch should have a radius C as large as possible (Figure 8-40a). When using the single-station method (Figure 8-40b), controlling the length of the flange is more difficult (Figure 8-39, H).

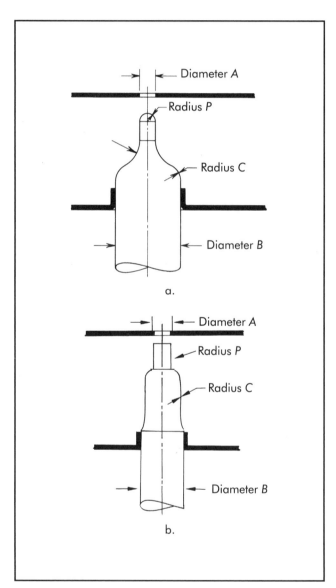

Figure 8-40. (a) Two-station and (b) single-station flanging punch design.

## COMPOUND DIES

A compound die usually refers to a one-station die designed around a common vertical centerline where two or more operations are completed during a single-press stroke. Usually, only cutting operations are done, such as combined blanking and piercing.

A common characteristic of compound-die design is inverted construction, with the blanking die on the upper-die shoe and the blanking punch on the lower-die shoe. The pierced slugs fall out through the lower-die shoe. The part or finished blank is retained in the female die, which is mounted on the upper shoe.

### Compound Blank and Piercing Dies

Compound dies are widely used to produce pierced blanks to close dimensional and flatness tolerances. Generally, the sheet material is lifted off the blanking punch by a spring-actuated stripper, which may be provided with guides to feed the material. If hand-fed, a stop is provided to position the strip for the next stroke.

The blank normally remains in the upper die, and is usually removed by a positive knockout at the top of the press stroke. Ejection of the blank from the die by spring-loaded or positive knockout occurs at the top of the stroke. Because of this feature, the die does not require angular die clearance. Not providing angular die clearance simplifies die construction and ensures constant blank size throughout the life of the die.

A compound die for making a washer is shown in *Figure 8-41*. The center hole is cut and outer diameter trimmed in a single-die station in one press stroke. The material is 0.015-in. (0.38-mm) cold-rolled steel strip. A piercing punch is attached to an upper-die shoe. The blanking punch is attached to a lower-die shoe. In this design, the piercing punch contacts the material slightly ahead of the blanking die. The part is stripped from both the blanking die and piercing punch by a positive knockout. The blanked strip is lifted off the blanking punch by a spring-loaded pressure pad.

### Part Removal

A potential disadvantage of compound dies is that the part must be removed from the upper die at the top of each stroke. The part usually is knocked out at the top of the stroke by means of a press-actuated knockout bar.

In the case of small parts, once knocked out of the upper die they may be ejected by a timed blast of air. Larger parts can be removed by means of a shuttle unloader that enters the die opening as the ram ascends. The press ram normally drives the unloader, although air, hydraulic, or servomotor-driven units may be used. Accomplishing part removal during each press stroke may limit the speed of the operation. For low-production jobs, manual removal with appropriate safeguarding precautions may suffice.

## STAMPING ANALYSIS

Forming flat sheet metal into complex, radically deformed stampings may appear to involve skills and processes that are more art than science. Modern stamping design and development techniques permit the product designer to work with manufacturing and tooling engineers to design parts that can be manufactured with certainty. A complex stamping is shown in *Figure 8-42*.

Stamping designs should be based on the data of successful prior designs and analytical formability methods. Uncertainty concerning the manufacturability of complex stampings often results in added expense and delays, such as:

- trial production on temporary tooling to prove process feasibility;
- delays in marketing the product while the process or product design is changed;
- specifying more operations than needed as a safety factor, and
- choosing alternative processes and materials such as molded plastics.

### Computerized Techniques

Simple computer software programs are available to ensure that proposed stamping designs

Figure 8-41. *A compound blanking and piercing die used to produce a washer. Dies of this type are widely used to produce accurate flat blanks. (Courtesy Cousino Metal Products Company)*

a = bend
b = bend and straighten
c = cup drawing
d = dome or biaxial stretch
p = plane-strain stretch

Figure 8-42. *A complex stamping is produced by a variety of forming operations, each of which may be analyzed separately. (Courtesy American Iron and Steel Institute)*

can be manufactured with certainty. Using these programs avoids costly trial-and-error guesswork. Software is available to analyze the amount and type of deformation in a stamping design. Computer-aided analysis ties in nicely with computer-aided design (CAD) design of stampings. The analysis should be applied early in the product-

design process. The CAD math data, which describes the part, is used for computerized formability analysis. Computerized analysis falls into several categories, such as:

- simple sectional analysis programs;
- general analysis programs that fully model the part, typically based on finite-element analysis, and
- programs that analyze the stamping, based on the type of deformation occurring in an individual area.

### Sectional Analysis Programs

Sectional analysis is good for identifying and troubleshooting a number of simple forming conditions. Computer programs are useful for determining the amount of strain present in a specific area of a stamping so anticipated design problems can be checked easily.

A moderately priced PC has sufficient computational capacity to run sectional-analysis programs to determine strain conditions. Estimating the effect of surface friction on metal movement is a useful feature of nearly all computerized formability analysis programs.

## General Analysis Programs

General analysis programs are required to completely model a part using the finite-element or finite-difference methods. Stamping the whole part is simulated in three dimensions with a single computer program. Many complex interactions occur during the stamping simulation. General analysis programs require a lot of computing power and time to calculate the interaction of many complex variables occurring throughout the forming process (Tharrett 1987).

## Simplifying Analysis

A simplified approach is to break down complete stampings into local regions to be analyzed individually. Following this approach, a stamping is analyzed as individual zones that interact in a predictable manner as they are formed. Some good programs include an expert-systems approach based on a library of successful designs.

## Circle Grid Analysis

Circle grid analysis (CGA) is a powerful process control tool. Data on essential areas of the stamping near the forming limit should be checked periodically to determine the effect of die wear on formability. Should a production stamping process start to experience problems, a blank of the material can be quickly gridded and analyzed. The CGA results can be compared with historical data for the part and steel formability specifications.

### Measuring Deformation

The CGA technique permits measurement of the deformation that occurs when forming stampings. First, a grid is stenciled on the surface of the blank by dye transfer or electrochemical etching. This grid deforms with the blank and allows accurate calculations of the strain or deformation that occurred during the stamping operation.

### Press Shop Applications

If a part always runs well within the safety zone, often a less costly steel or lubricant can be used. If only a few areas on the stamping are close to failure, a blankholder improvement or minor product change often will ensure the manufacturability of the product.

The CGA system is excellent for training apprentices. By making tooling, lubricant, and material changes, and then observing the metal deformation changes, cause-and-effect patterns can be readily discerned.

*Figure 8-43* illustrates a bumper jack hook. The grid of circles placed on the blank shows different types of deformation on the stamping. Note that most of the circles are deformed very little, while a few circles, especially one near the lip, show pronounced elliptical patterns. A stamping of this type will often fracture at the location where the edge is most severely stretched (Keeler 1969, 1986).

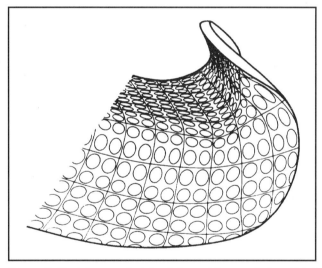

Figure 8-43. *A bumper jack hook formed from a circle-gridded blank.* (Courtesy National Steel Corporation)

### Measuring Forming Severity

The distribution of stretch is useful information by itself. Knowing the location of high-stretch concentrations and direction of the maximum stretch often is sufficient to suggest solutions to forming problems. However, CGA uses a numerical rating system for the deformation of the circles.

The system of rating forming severity is based on measuring the deformation of the circles and plotting the measurements on a graph. Grid measurements are easily made with transparent Mylar® tape imprinted with a calibrated scale (*Figure 8-44*). The tape is flexible and can be laid around a radius or tucked into a tight corner. The calibration of the tape eliminates any need to calculate stretch. The tape is used to measure the major (length) axis of the ellipse first, then rotated to measure the minor (width) axis.

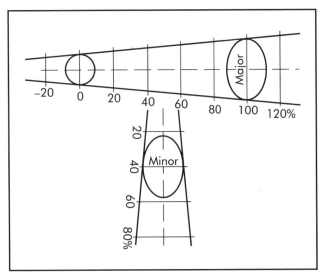

Figure 8-44. *Circle deformation is measured with a Mylar® tape overlay. (Courtesy National Steel Corporation)*

Many combinations of major and minor circle deformations can be found on different stampings. *Figure 8-45* illustrates five different types of deformation. *Figures 8-45a* and *8-45b* detail large major elongation, while the minor stretch is nega-

tive. These types of circle deformations are observed in the sidewalls of drawn cups and the corner sidewalls of rectangular drawn shells. This combined compression and elongation indicates the metal has been subjected to circumferential compression and tensile stretching as it is pulled toward and over the draw radius.

*Figure 8-45c* is special. Here the minor stretch component is zero. This is called *plane strain*. This stretch condition is found over edge radii or across character lines. Another special case is shown in *Figure 8-45e*, where both the major and minor axes of the ellipse are equal—the circle becomes a larger circle. This is called *balanced biaxial stretch*.

## Plotting the Measurements

Due to the variety of combinations, a method for plotting them on a single graph is necessary. The plotting technique used in *Figure 8-46* allows both the major and minor stretch for each circle to be plotted as a single point. The major stretch is plotted on the vertical axis, while the minor stretch is plotted on the horizontal axis. Circles that plot on the left side of the diagram

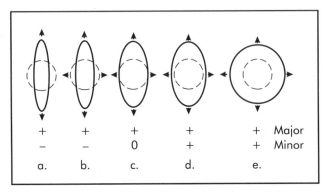

Figure 8-45. *Examples of deformed circles.* (Courtesy National Steel Corporation)

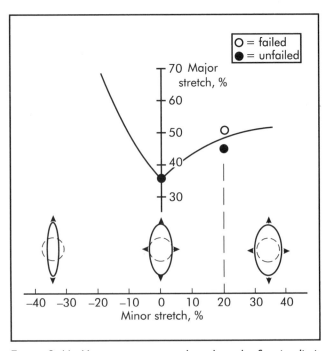

Figure 8-46. *Measurements are plotted on the forming limit diagram (FLD). (Courtesy National Steel Corporation)*

have negative minor stretch, while circles that plot on the right side of the diagram have positive minor stretch.

Three of the ellipses from *Figure 8-45* are plotted in *Figure 8-46*. Note that the case of plane stretch (*Figure 8-45c*) is plotted on the vertical axis.

*Figure 8-47* illustrates an asymmetrical V-shaped curve, which is the forming limit. Circles plotted below this curve show no evidence of necking or fracture, while those above it fail. A graph developed in this way is called a *forming limit diagram* (FLD). The point where the FLD intersects the major stretch axis is called $FLD_o$. Here, only plane strain deformation is occurring. To initially develop this diagram, many samples of failed versus not failed circles from the same material must be plotted.

The shape of the FLD (*Figure 8-47*) is constant for most low-alloy sheet steel used in automotive, appliance, agricultural, container, and similar industries. It illustrates how the FLD curve can raise or lower for different steel sheets. The level of the FLD—as specified by $FLD_o$—is a characteristic of the sheet steel. For example, a thinner sheet of steel would have a lower FLD than a

thicker sheet of steel. In addition, higher-strength steel would have a lower FLD than lower-strength steel. Thus, the location of the curve can be described by specifying the intersection of the curve's $FLD_o$ with the minor stretch axis.

## DEEP DRAWING OF CUPS

*Figure 8-48* illustrates the forces involved in deep drawing a metal cup. It is important to note that all of the force required for drawing is transmitted by the draw punch to the bottom of the cup.

The cup-drawing process starts with a flat round blank. The blank is subjected to radial tension and circumferential compression. The metal thickens as it flows toward the draw radius. Deep drawing is unique because of the deformation state of metal restrained by the blankholder.

*Figure 8-48* is an example of deep-cup drawing in which very little deformation occurs over the bottom of the punch. Nearly all deformation occurs in the metal restrained by the blankholder.

### Metal Flow

In general, metal flow in deep-cup drawing may be summarized as:

Figure 8-47. *Different steels move the* FLD$_o$ *point.* (Courtesy National Steel Corporation)

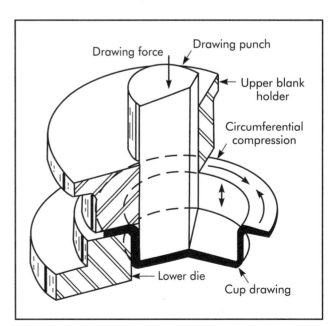

Figure 8-48. *A simple drawing die.*

- little or no metal deformation takes place in the blank area that forms the bottom of the cup;
- the metal flow occurring during the forming of the cup wall uniformly increases with cup depth, and
- the metal flow at the periphery of the blank involves an increase in metal thickness caused by circumferential compression.

## Success Factors

The success of a drawing operation depends on several factors, including:

- the formability of the material being drawn;
- limiting the drawing punch force to a lower value than that which will fracture the shell wall, and
- adjustment of the blankholder force to prevent wrinkles without excessively retarding metal flow.

## Causes of Failure

The maximum force requirement for the drawing process is limited by tensile failure of the material in the sidewall. As this limit is approached, the metal will neck or thin excessively in a localized area near the punch radius.

Many complex interactions occur during the cup-drawing process. The actual force required depends on the cross-sectional area of the cup wall and yield strength of the material as it is worked. Should the process fail, some or all of the following factors may be the root causes:

- the ductility or drawability of the stock may be too low;
- the blankholder force may be too high;
- scoring or galling may be present on die surfaces;
- the blankholder geometry and draw radius may not provide for metal thickening and smooth flow into the die cavity;
- there may be an incorrect or insufficient amount of drawing lubricant;
- the depth of draw or percentage of blank reduction may be too great, and/or
- one or more redrawing operations may be necessary to obtain the desired depth of draw.

Annealing may be required between redrawing operations, especially when using materials that work-harden rapidly.

### Draw Radius

The blankholder draw radius should be approximately four to six times the metal thickness for most applications. It has a large effect on the punch force required to pull the metal into the draw cavity. As the metal passes over the radius, it is bent and then straightened to form the sidewall of the drawn cup. If the radius is too small, it can lead to fracture because more force is required to pull the metal over a small radius than a larger one. Also, it will more severely strain the metal, increasing work-hardening. This, in turn, requires more force to draw the part.

There is little reduction of drawing force achieved by making the draw radius larger than six times metal thickness. Approaching a draw radius of 10 times metal thickness may result in puckering of the metal as it flows over the draw radius. Severe puckering can result in folded metal, which can lock up metal movement and result in fractures.

In cases where all of the metal on the blankholder is to be drawn into the cavity to form a straight-walled shell without a flange, a large radius may result in folded metal as blankholder control ceases.

### Measuring Thickness

Necking failures, such as those shown in *Figure 8-49*, are preceded by localized thinning, which may not be visible in the part. However, the onset of a necking failure can be detected by

Figure 8-49. *Localized thinning (necking) at the punch radius indicates a failure of the drawing process.*

measuring the metal thickness with an ultrasonic thickness-measuring device.

**Application of an ultrasonic thickness gage.** The ultrasonic thickness gage is a modern stamping analysis tool that is very useful for online process tracking, troubleshooting, and control. It features a portable control box, which provides thickness readout, and a probe having an ultrasonic transducer.

The principle of operation is much like that of a sonic or sonar depth finder used as a navigational aid by boaters. In its simplest form, a sonic depth finder measures the time between a sonic pulse sent out by a transducer attached to the boat's hull and the arrival of the return echo. The speed of sound in water is a known constant. The delay time between sending the sonic pulse and return echo can be easily converted and displayed in units of depth such as feet, meters, or fathoms. Ultrasonic metal thickness gages work in much the same way. An ultrasonic pulse is sent out by a handheld transducer, which attaches to the control unit.

The return echo time is very short, since the speed of sound in steel is in excess of 16,000 ft/sec (4,877 m/sec). For stamping thickness measurements, the operating frequency is approximately 15 MHz. A plastic delay line is used between the transducer and the point of contact, with the workpiece being measured so a faint return echo can be detected. Manufacturer-supplied compliant media is placed between the transducer and plastic delay line. This media requires periodic renewal to ensure accuracy.

A 0.125-in. (3.18-mm) diameter transducer is considered better than larger sizes (for example, 0.250 in. [6.35 mm]), especially for use on curved surfaces. A compliant media is needed between the transducer face and metal workpiece. The best readily available material meeting this requirement is common Johnson & Johnson K-Y® brand surgical jelly—a product made to strict quality requirements for uniformity and purity. Be sure to wipe this substance off the part after measurement because it can cause rust.

**Tracking thinning.** Stampings that are severely drawn or formed usually have one or more areas where thinning is apt to occur. These areas are spots for regular thickness checks. The areas where a necking failure or fracture is apt to occur on a stamping can be predicted with CGA during the development and die tryout period for new stampings. The failure locations on an existing stamping become well-known to pressroom personnel.

Once the areas found to thin leading to failure are identified, regular checks should be made with an ultrasonic thickness gage. It is helpful to chart the thinning trends in each area. This permits corrective action to be taken before a necking failure becomes visible. Causal factors for a pronounced increase in thinning include:

- excessive blankholder force;
- material problems such as a lack of ductility or drawability;
- material too thin, and
- scored die surfaces.

## Single-action Draw Dies

The simplest type of draw die is one with a punch and die. Each component may be designed in one piece without a shoe by incorporating features for attaching them to the ram and bolster plate of the press. *Figure 8-50a* shows a simple type of draw die in which the precut blank is placed in the recess on top of the die, and the punch descends, pushing the cup through the die. As the punch ascends, the cup is stripped from the punch by the counterbore in the bottom of the die. The top edge of the shell expands slightly to make this possible. The punch has an air vent to eliminate suction that would hold the cup on the punch and damage it when stripped from the punch.

The method by which the blank is held in position is important, because successful drawing is somewhat dependent on the proper control of blankholder pressure. A simple form of drawing die with a rigid flat blankholder for use with 0.0897 in. (2.278 mm) and heavier stock is shown in *Figure 8-50b*. When the punch comes into contact with the stock, it will be drawn into the die without allowing wrinkles to form.

Another type of drawing die for use in a single-action press is shown in *Figure 8-51*. This die is a plain, single-action type where the punch

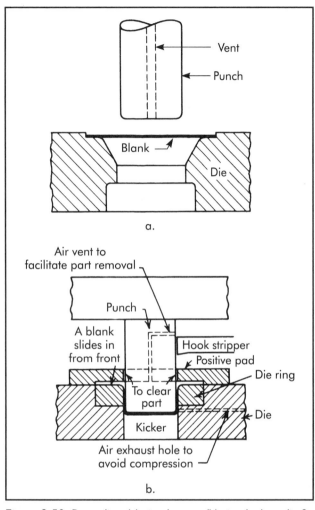

Figure 8-50. *Draw dies:* (a) *simple type;* (b) *simple draw die for heavy stock.*

pushes the metal blank into the die using a spring-loaded pressure pad to control the metal flow. The cup either drops through the die or is stripped off the punch by a pressure pad. The figure shows the pressure pad extending over the nest, which acts as a spacer and is ground to such a thickness that an even and proper pressure is exerted on the blank at all times. If the spring pressure pad is used without the spacer, the more the springs are depressed, and the greater the pressure exerted on the blank, thereby limiting the depth of draw. Because of the limited pressures obtainable, this type of die should be used with light-gage stock and shallow depths.

A single-action die for drawing flanged parts, having a spring-loaded pressure pad and stripper, is shown in *Figure 8-52.* The stripper also may be used to form slight indentations or re-entrant curves in the bottom of a cup, with or without a flange. Draw tools in which the pressure pad is attached to the punch are suitable only for shallow draws. The pressure cannot be easily adjusted, and the short springs tend to build up pressure too quickly for deep draws. This type of die is often constructed in an inverted position with the punch fastened to the lower portion of the die.

Deep draws may be made on single-action dies where the pressure on the blankholder is more

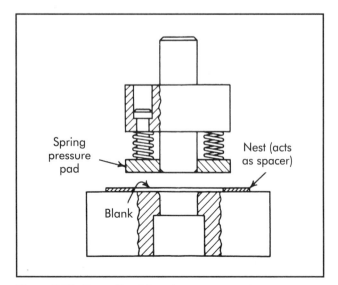

Figure 8-51. *Draw die with spring pressure pad.*

Figure 8-52. *A draw die with spring pressure pad and stripper.*

evenly controlled by a die cushion or pad attached to the bed of the press. The typical construction of such a die is shown in *Figure 8-53*. This is an inverted die with the punch on the die's lower portion.

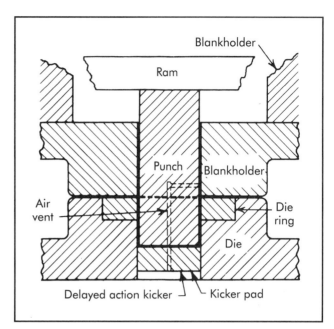

Figure 8-54. *A typical double-action cylindrical draw die.*

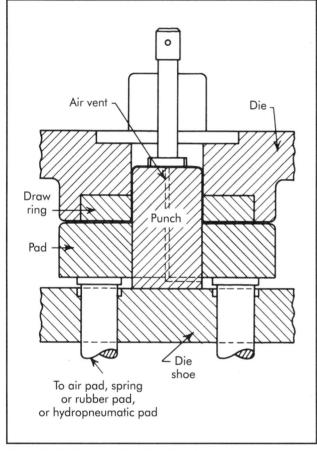

Figure 8-53. *Cross section of an inverted draw die for a single-action press; die is attached to the ram; punch and pressure pad are on the lower shoe.*

## Double-action Draw Dies

In dies designed for use in a double-action press, the blankholder is fastened to the outer ram, which descends first and grips the blank. Then the punch, which is fastened to the inner ram, descends, forming the part. These dies may be a push-through type, or the parts may be ejected from the die with a knockout attached to the die cushion or by means of a delayed action kicker. *Figure 8-54* shows a cross-section of a typical double-action draw die.

## Development of Blanks

The development of the approximate blank size should be done first to determine the size of a blank to produce the shell to the required depth, then to resolve how many draws will be necessary to produce the shell. This is configured by the ratio of the blank size to the shell size. Various methods have been developed to determine the size of blanks for drawn shells. These methods are based on mathematics alone, the use of graphic layouts, and a combination of both. Most of these methods are for use on symmetrical shells.

It is rarely possible to compute any blank size to close accuracy or maintain perfectly uniform height of shells in production because the thickening and thinning of the wall varies with the completeness of annealing. The height of ironed shells varies with commercial variations in sheet thickness, and the top edge varies from square to irregular, usually with four pronounced high spots resulting from the effect of the direction of the metal's crystalline structure. Thorough annealing should largely remove the directional effect. For all these reasons, it is ordinarily necessary to figure the blank sufficiently large to permit a sizing and trimming operation.

The drawing tools should be made first; then, the blank size should be determined by trial before the blanking die is made. There are times, however, when the metal to produce the product is not immediately available from stock and must be ordered at the same time the tools are ordered. This situation makes it necessary to estimate the blank size as closely as possible by formula or graphically to know what sizes to order.

## Blank Diameters

The following equations may be used to calculate the blank size for cylindrical shells of relatively thin metal. The ratio of the shell diameter to the corner radius $(d/r)$ can affect the blank diameter and should be taken into consideration. When $d/r$ is 20 or more:

$$D = \sqrt{d^2 + 4dh} \qquad (8\text{-}17)$$

When $d/r$ is between 15 and 20:

$$D = \sqrt{d^2 + 4dh - 0.5r} \qquad (8\text{-}18)$$

When $d/r$ is between 10 and 15:

$$D = \sqrt{d^2 + 4dh - r} \qquad (8\text{-}19)$$

When $d/r$ is below 10:

$$D = \sqrt{(d - 2r)^2 + 4d(h - r) + 2\pi r(d - 0.7r)} \quad (8\text{-}20)$$

where:

$D$ = blank diameter, in. (mm)
$d$ = shell diameter, in. (mm)
$h$ = shell height, in. (mm)
$r$ = corner radius, in. (mm)

The previous equations are based on the assumption that the surface area of the blank is equal to the surface area of the finished shell.

In cases where the shell wall is to be ironed thinner than the shell bottom, the volume of metal in the blank must equal the volume of metal in the finished shell. Where reduction of wall thickness is considerable, as in brass shell cases, the final blank size is developed by trial. A tentative blank size for an ironed shell can be obtained from:

$$D = \sqrt{d^2 + 4dh \frac{t}{T}} \qquad (8\text{-}21)$$

where:

$t$ = wall thickness, in. (mm)
$T$ = bottom thickness, in. (mm)

## Reduction Factors

After the approximate blank size has been determined, the next step is to estimate the number of draws to produce the shell and the best reduction rate per draw. For diameter reduction, the area of metal held between the blankholding faces must be reasonably proportional to the area on which the punch is pressing, since there is a limit to the amount of metal that can be made to flow in one operation. The greater the difference between blank and shell diameters, the greater the area that must be made to flow, and therefore the higher the stress required to make it flow. General practice has established that, for the first draw, the area of the blank should not be more than 3.5–4 times the cross-sectional area of the punch.

One important factor in the success or failure of a drawing operation is the thickness ratio, or the relation of metal thickness to the blank or previous shell diameter. This ratio is expressed as $t/D$. As this ratio decreases, the tendency of wrinkling increases, requiring more blankholding pressure to control the flow properly and prevent wrinkles from starting. The top limit of about 48% seems to be substantiated by practice for single-action first draws. The 30% limit for double-action redraws is dictated by practice and is modified by corner radii, friction, and the angle of blank-holding faces with respect to the shell wall. Because of strain-hardening stresses set up in the metal, third and subsequent draws should not exceed 20% reduction without an annealing operation.

## Force

### Drawing Force

The force applied to a punch necessary to draw a shell is equal to the product of the cross-sectional area and yield strength of the metal as shown in the following:

$$P = \pi d t s \left( \frac{D}{d} - C \right) \qquad (8\text{-}22)$$

where:

P = force, lbf (kN)
d = shell diameter, in. (mm)
t = wall thickness, in. (mm)
s = yield strength, psi (kPa)
D = blank diameter, in. (mm)
C = constant of 0.6 and 0.7 to cover friction and bending

### Blankholder Force

The amount of blankholder force required to prevent wrinkles and puckers is largely determined by trial-and-error. The force required to hold a blank flat for a cylindrical draw varies from very little to a maximum of about one-third or more of the drawing force. During cylindrical draws, the force is uniform and balanced at all points around the periphery because the amount of flow at all points is the same. On rectangular and irregularly shaped shells, the amount of flow around the periphery is not uniform; therefore, the force required also varies.

## Draw Die Design

The first step in planning a die for a part is not the development of the die but the blank for the drawn shell or cup. *Figure 8-55* shows the example part discussed, and *Figure 8-56* shows the die that will make the part.

Figure 8-55. *A drawn shell, the basis for blank development and draw die design (Figure 8-56).*

Figure 8-56. *Draw die for producing the shell of* Figure 8-55.

Since the ratio $d/r$ is more than 20, the diameter of the blank $D$ is, according to *Equation 8-17*:

$$D = \sqrt{(2.75)^2 + (4 \times 2.75 \times 1.5)}$$
$$= 4.9 \text{ in. } (124.5 \text{ mm})$$

The area of this blank is slightly less than four times the cross-sectional area of the punch (approximately 4.90 in.² [31.6 cm²]). This blank can be drawn into a shell of the dimensions specified in one draw because the ratio of its area to that of the punch nose is approximately 4:1.

The length of the draw radius (radius of the toroidal zone at the entrance to the die) must be determined before any other die design details are considered. The radius of the draw die should be kept as large as possible to aid in the flow, but if it is too large, the metal will be released by the blankholder before the draw is completed and wrinkle. When the radius is too small, the material will rupture as it goes over the radius or against the face of the punch. *Table 8-2* gives the practical drawing radii for certain stock thicknesses. The values in this table are based on a radius of approximately four times the stock thickness. In some cases, the radius may vary from four to six times the stock thickness. The length of the draw radius from *Table 8-2* is 0.563 in. (14.28 mm).

The required drawing force is determined with *Equation 8-22*:

$$P = 20 \text{ tons } (178 \text{ kN})$$

Table 8-2. Practical drawing radii
for certain thicknesses of stock

| Thickness of Stock, in. (mm) | Drawing Radius, in. (mm) |
|---|---|
| 1/64 (0.4) | 1/16 (1.6) |
| 1/32 (0.8) | 1/8 (3.2) |
| 3/64 (1.2) | 3/16 (4.8) |
| 1/16 (1.6) | 1/4 (6.4) |
| 5/64 (2.0) | 3/8 (9.5) |
| 3/32 (2.4) | 7/16 (11.1) |
| 1/8 (3.2) | 9/16 (14.3) |

Blankholder force is largely determined by trial and error; an allowable range is from a few pounds up to one-third of the drawing force; approximately 6 tons (53 kN) is more than adequate. An ordinary 30-ton (267-kN), open-backed inclinable press has enough capacity for the total possible maximum force of 26 tons (231 kN) required for the operation.

The blank is nested in a semicircular plate. Its thickness is established by the optimum blankholder pressure on the blank during drawing of the shell to specifications. The unit pressure on the blank becomes greater as it is pushed into the die. Shimming the nest plate will vary the pressure on both nest and blank, and results in a critical pressure on the latter without the production of defective shells.

The blankholder is a 0.625-in. (15.88-mm) thick circular plate suspended from the punchholder by six stripper bolts equally spaced around its circumference. Die springs fit around the bolts and are retained by pockets counter-bored in the punchholder and blankholder. In addition to holding the blank, the blankholder strips the shell from the punch.

The draw clearance (space between the punch and die) from *Table 8-3* should be from 0.138–0.140 in. (3.50–3.56 mm). The inside diameter (ID) of the die is 1.638 in. (41.61 mm) for tryout. If there is too much ironing of the shell, this ID can be increased up to approximately 1.640 in. (41.66 mm). The outside diameter (OD) of the draw punch is the same as the ID of the drawn shell.

The methods of determining die dimensions and hold-down size apply to the same elements in a drawing die, but generally may be increased in approximate proportion to the press forces applied and stresses in die blocks and blankholders.

## Draw Rings

The draw ring of many draw dies is the die itself, made of tool steel with the cavity edge forming a spherical zone over which the metal is drawn. In many large drawing dies, the die is not made in one piece but has an inserted draw ring of tool steel that is replaceable.

## Materials

The selection of material for a draw punch and a draw die is determined in the same manner as any other tool—the number of parts to be drawn and their ultimate unit cost. For low production, material cost is a major factor. If less than 100 parts are to be drawn, a plastic or zinc-alloy material is satisfactory.

For production of a few hundred parts, a plain cast-iron draw ring may be suitable. Cast alloyed irons that are flame hardenable provide excellent wear resistance. Cast-iron punches and dies can be hard-chrome plated or ion nitrided for long production runs. Tool steels should be used for long runs of parts in small dies. The choice of tool steel is based on wear, strength, etc. Carbide punches and dies or carbide inserts for these tools may prove to be the most economical material for extremely large runs when drawing abrasive materials. Material selection for blankholders, knockouts, and other parts of a draw die is based on these same factors.

## Lubricants

The functional requirements of lubricants for drawing operations are much more severe than for shafts and bearings. Possible corrosion of compounds on metals and the ease of its removal from drawn metal parts must be considered. Handbooks and manufacturers of various metals should be consulted for lubrication recommended for shallow- and deep-drawing operations.

Table 8-3. Draw clearance

| Blank Thickness, in. (mm) | First Draws | Redraws | Sizing Draw* |
|---|---|---|---|
| Up to 0.015 (0.38) | 1.07*t* to 1.09*t* | 1.08*t* to 1.1*t* | 1.04*t* to 1.05*t* |
| 0.016 to 0.050 (0.41 to 1.27) | 1.08*t* to 1.1*t* | 1.09*t* to 1.12*t* | 1.05*t* to 1.06*t* |
| 0.051 to 0.125 (1.30 to 3.18) | 1.1*t* to 1.12*t* | 1.12*t* to 1.14*t* | 1.07*t* to 1.09*t* |
| 0.136 (3.45) and up | 1.12*t* to 1.14*t* | 1.15*t* to 1.2*t* | 1.08*t* to 1.1*t* |

\* Used for straight-sided shells where diameter or wall thickness is important, or where it is necessary to improve the surface finish to reduce finishing costs.
*t* = thickness of the original blank

The following lubrication data for drawing mild steel are included as a guide.

Mild operations:

1. mineral oil of medium-heavy to heavy viscosity;
2. soap solutions (0.03–2%, high-titer soap);
3. fat, fatty-oil, or fatty and mineral-oil emulsions in soap-base emulsions; and/or

Medium operations:

1. fat or oil in soap-base emulsions containing finely divided fillers, such as whiting or lithopone;
2. fat or oil in soap-base emulsions containing sulfurized oils;
3. fat or oil in soap-base emulsions with fillers and sulfurized oils;
4. dissimilar metals deposited on steel plus emulsion lubricant or scrap solution;
5. rust or phosphate deposits plus emulsion lubricants or soap solution; and/or
6. dried soap film.

Severe operations:

1. dried soap or wax film, with light rust, phosphate, or dissimilar metal coatings;
2. sulfide or phosphate coatings plus emulsions with finely divided fillers and sometimes sulfurized oils;
3. emulsions or lubricants containing sulfur as combination filler and sulfide former; and/or
4. oil-base sulfurized blends containing finely divided fillers.

## PROGRESSIVE DIES

A progressive die performs a series of fundamental sheet-metal operations at two or more stations during each press stroke to develop a workpiece as strip stock moves through a die. Each working station performs one or more distinct die operations, but the strip must move from the first through each succeeding station to produce a complete part. One or more idle stations may be incorporated in the die, not to perform work on the metal but to locate the strip, facilitate interstation strip travel, provide maximum-size die sections, or simplify their construction.

The linear travel of the strip stock at each press stroke is called the progression, advance, or pitch, and is equal to the interstation distance. The unwanted parts of the strip are cut out as it advances through the die, and one or more ribbons or tabs are left connected to each partially completed part to carry it through the stations of the die.

The types of operations performed in a progressive die could be done in individual dies as separate operations. However, this would require hand feeding, positioning, and often unloading. In a progressive die, the part remains connected to the stock strip, which is fed through the die with automatic feeds and positioned by pilots with speed and accuracy.

### Selection

The selection of any multi-operation tool, such as a progressive die, is justified by the principle that the number of operations achieved with one handling of the stock is more economical than production by a series of single-operation dies and a number of handlings for each single die. Where total production requirements are high, particularly if production rates are large, total handling costs (man-hours) saved by progressive fabrication compared with a series of single operations are frequently greater than the cost of the progressive die.

The fabrication of parts with a progressive die under the above-mentioned production conditions is further indicated when:

- stock material is not so thin that it cannot be piloted or so thick that there are stock-straightening problems;
- overall size of die (functions of part size and strip length) is not too large for available presses, and
- total press capacity required is available.

## Strip Development

Individual operations performed in a progressive die often are relatively simple, but when combined in several stations, the most practical and economical strip design for optimum operation of the die often becomes difficult to devise. The sequence of operations on a strip and details of each operation must be carefully developed to produce good parts. As the final sequence of operations is developed, the following items should be considered:

- Pierce piloting holes and piloting notches in the first station. Other holes may be pierced that will not be affected by subsequent noncutting operations.
- Develop a blank for drawing or forming operations for free movement of metal.
- Distribute pierced areas over several stations if they are close together or close to the edge of the die opening.
- Analyze the shape of blanked areas in the strip for division into simple shapes so punches of simple contours may partially cut an area at one station and cut out remaining areas in later stations. This may suggest the use of commercially available punch shapes.

- Use idle stations to strengthen die blocks, stripper plates, and punch retainers, and to facilitate strip movement.
- Determine whether the direction of the strip grain will hinder or facilitate an operation.
- Plan the forming or drawing operations either in an upward or downward direction, whichever will ensure the best die design and strip movement.
- The shape of the finished part may dictate that the cutoff operation should precede the last noncutting operation.
- Design adequate carrier strips or tabs.
- Check strip layout for minimum scrap; use a multiple layout if feasible.
- Locate cutting and forming areas to provide uniform loading of the press slide.
- Design the strip so the scrap and part can be ejected without interference.

*Figure 8-57* shows the use of one die station instead of two to maintain a close-toleranced dimension. If two stations were used, the variation in the location of the stock guides and cutting punches could make it difficult to hold the ±0.001 in. (±0.03 mm) tolerance.

The strip development for shallow and deep drawing in progressive dies must allow movement of metal without affecting the positioning of the part in each successive station. *Figure 8-58* shows various types of cutouts and typical distortions

Figure 8-57. *Use of one station versus two to hold a close tolerance.*

to the carrier strips as the cup-shaped parts are formed and then blanked out of the strip. Piercing and lancing of the strip around the periphery of the part as shown in *a*, leaving one or two tabs connected to the carrier strip, is a commonly used method. The semicircular lancing as shown in *b* is used for shallow draws. This type of relief is used for deeper draws and places an extra strain on the metal in the tab, causing it to tear. The carrier strip is distorted to provide stock for the draw. A popular cutout for deep draws is shown in *c*. The double-lanced relief suspends the blank on narrow ribbons, and no distortion takes place in the carrier strips. Two sets of single, rounded lanced reliefs of slightly different diameters are placed diametrically opposite each other to produce the ribbon suspension. The hourglass cutout in *d* is an economical method of making the blank for shallow draws. The connection to the carrier strips is wide, and a deep draw would cause considerable distortion. An hourglass cutout for deep draws is shown in *e*, which provides a narrow tab connecting the carrier strip to the blank. The cupping operations narrow the width of the strip as the metal is drawn into the cup shape.

The hourglass cutout may be made in two stations by piercing two separated triangular cutouts in one station, and lancing or notching the material between them in a second station. The cutouts shown in *Figures 8-58f* and *g* provide an expansion-type carrier ribbon that tends to straighten out when the draw is performed. These cutouts are made in two stations to allow for stronger die construction. Satisfactory multiple layouts may be designed with most of the reliefs by using a longitudinal lance or slitting station to divide the wide strip into narrower strips as the stock advances. The I-shaped relief cutout in *h* is a modified hourglass cutout used for relatively wide strips from which rectangular or oblong shapes are produced.

Straight slots or lances crosswise of the stock are sometimes used on very shallow draws or where the forming is in the central portion of the blank. On deeper draws, this type of relief tends to tear out the carrier strips or cause excessive distortion in the blank and is not satisfactory.

The design of carrier strips for cup drawing in progressive dies tends to be a trial-and-error process. Software programs can give a good approximation of the carrier strip. Another successful method is to use beeswax of the same thickness as the metal to be formed. This material can be forced over wooden forms by hand to simulate the flow of steel in a die. The final tool operation and part shape can only be proven in a full-size section mock up of the tool, or in the actual finished tool. Often, station-to-station distances must be adjusted to accommodate slight pitch changes that occur as the metal is formed.

## Stock Positioning

Of prime importance in strip development is the positioning of stock in each station. The stock must be positioned accurately in each station so the operation can be done in the proper location. A commonly used method of stock positioning is to incorporate pilots in the die.

There are two methods of piloting in dies: direct and indirect. *Direct piloting* consists of piloting in holes punched in the part at a previous station. *Indirect piloting* consists of piercing holes in the scrap-strip and locating these holes with pilots at later operations. Direct piloting is the ideal method for locating the part in subsequent die operations. Unfortunately, ideal conditions may not exist and, in such cases, indirect piloting must be used to achieve the desired results of part accuracy and high production speeds.

There are two advantages of locating pilots in the scrap material area—they are not readily affected by workpiece change, and size and location are not as limited. Disadvantages of locating pilots in the scrap section are that the material width and lead may increase, and scrap-strip carriers distort on certain types of operations and make subsequent station use impossible.

How to pilot is an arbitrary decision that the tool designer must make. It is impossible to give definite rules and formulas because the material and hardness of the stock influence the decision. However, in the indirect piloting method, it is possible to use pilots of greater diameter than if holes in the part are used for piloting, such as in

Figure 8-58. *Cutout reliefs for progressive draws: (a) lanced outline; (b) circular lance; (c) double-lance suspension; (d) hourglass cutout; (e) cutout providing expansion-type carrier ribbon for circular draws; (f and g) cutout providing expansion-type carrier ribbon for rectangular draws; (h) I-shaped relief for rectangular draws.*

*Figure 8-59a.* The greater the diameter of the pilot, the less chance there is of distortion of either the strip or the pilot. In addition, small-diameter pilots introduce the possibility of broken pilots.

When holes in the part are held to close tolerances (*Figure 8-59b*), it is possible for the pilots to affect the hole size in their effort to move the strip to the proper location. When holes in the part are too close to the edges (*Figure 8-59c*), the weak outer portions of the part are likely to distort on contact with the pilots instead of the strip moving to the correct location. Care should be taken when planning a progressive die to avoid expen-

Figure 8-58. *(continued)*

sive die alterations and the possibility of scrap parts.

Just what constitutes a condition where the edge of the hole is too close to the edge of the part is, like many aspects of design, a matter of personal judgment. Many designers prefer at least twice the stock thickness, although with good stock control one stock thickness or less has been successfully achieved. A similar problem exists when part holes are located in a weak portion of the inside area of the part (*Figure 8-59d*). Here, there is a possibility of the part's buckling before the pilots can position the stock strip. In this case, it is advisable to pilot in the scrap strip.

To achieve accurate part location, the pilots must be placed as far apart as possible. When the holes in the workpiece are close together, as in *Figure 8-59e*, holes in the scrap strip should

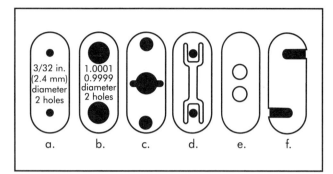

Figure 8-59. *Part conditions that require indirect pilots: (a) small holes; (b) close-tolerance holes; (c) holes too near edge; (d) hole in fragile areas of the part; (e) holes too close together; (f) slots in parts (Paquin 1962).*

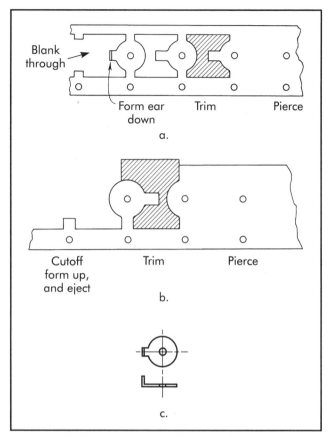

Figure 8-60. *Alternate strip developments for a workpiece.*

be used for piloting. A second method is to place a pilot in one hole in one station, then in the same hole in a succeeding station. The feasibility of the second method depends on the availability of an additional die station.

When slots are punched in the blank parallel to the stock movement (*Figure 8-59e*), the slots are not suitable piloting holes. Therefore, indirect pilots must be used.

## Disposition of Scrap Strip

A strip development is illustrated in *Figure 8-60a* using pierce, trim, form, and blank-through operations and carriers on both sides of the strip. The workpiece is dropped through the die, while the carrier bars continue to the scrap cutters to be cut into short lengths. The dropping of the workpiece through the die is the most desirable method of part ejection, but cannot always be obtained. Cutting the scrap into small sections simplifies material handling and produces a greater dollar return when sold as scrap metal. *Figure 8-60b* shows an alternate strip development with one side carrier. The workpiece is pierced, trimmed, cut off, and formed on a pad with air or gravity ejection, and the carrier bar is cut into short pieces by the scrap cutter. Remember that if a part is to be ejected, as this one is, the double-carrier bar design in *Figure 8-60a* should be avoided because the part may become trapped in these bars and cause die damage.

The design of the part in *Figure 8-60c* requires that the carrier be outside the part configura-

tion. This necessitates the use of stock wider than the part width plus the normal trimming allowance. The part shown in *Figure 8-61a* can be made of stock the same width as the part.

*Figure 8-61d* illustrates how the strip is pierced, trimmed, and the part cut off and formed. A slug-type cutoff punch is used and the flange is formed downward. The part then is ejected by an air jet or by gravity. This arrangement often is referred to as a scrapless development since no carrier strips remain after the part is cut from the strip.

*Figure 8-61e* shows strip development for the same part using a shear-type cutoff. The flange is formed upward as the combination cutoff and form punch descends. A spring-loaded pad supports the workpiece during forming and assists in ejecting the part from the die. The progression of this type of development is shortened by the width of the cutoff slug.

*Figure 8-62* shows another development in which the stock is the same as the developed

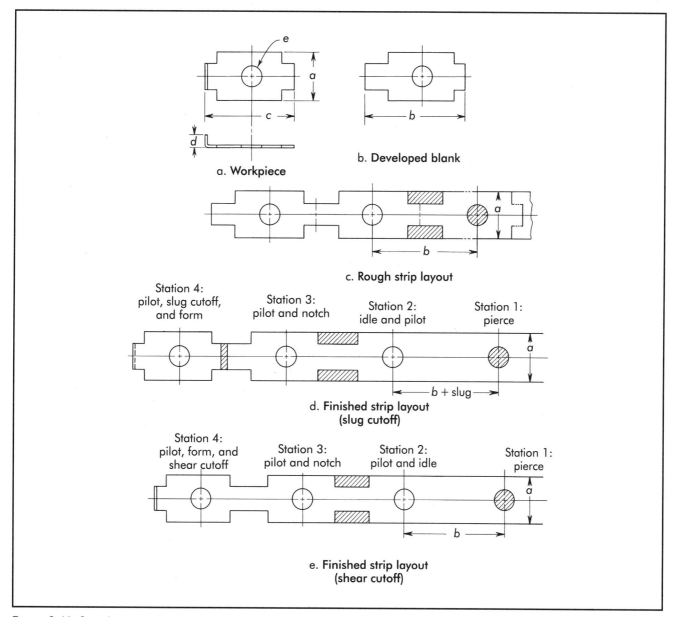

a. Workpiece

b. Developed blank

c. Rough strip layout

Station 4:
pilot, slug cutoff,
and form

Station 3:
pilot and notch

Station 2:
idle and pilot

Station 1:
pierce

d. Finished strip layout
(slug cutoff)

Station 4:
pilot, form, and
shear cutoff

Station 3:
pilot and notch

Station 2:
pilot and idle

Station 1:
pierce

e. Finished strip layout
(shear cutoff)

Figure 8-61. *Scrapless strip development.*

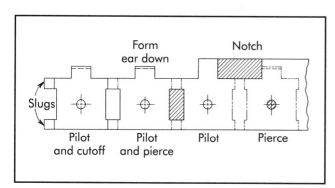

Figure 8-62. *Four-station strip development.*

width of the workpiece. The strip is pierced in station 1; piloted and notched in station 2; and piloted, pierced, and formed in station 3.

## Die Design

The die elements used in progressive dies, such as punches, stops, pilots, strippers, die buttons, punch guide bushings, die sets, guideposts, and guidepost bushings, are of similar design to those used in other types of dies.

A progressive die should be heavily constructed to withstand the repeated shock and continuous runs to which it is subject. Precision or anti-friction guideposts and bushings should be used to maintain accuracy. The stripper plates (if spring-loaded and movable), when also serving as guides for punches, should engage guide pins before contacting the strip stock. Lifters should be provided in die cavities to lift up or eject the formed parts, and carrier rails or pins should be provided to support and guide the strip when it is being moved to the next station. A positive ejection method should be provided at the last station. Where practical, punches should contain shedder or oil-seal-breaker pins to aid in the disposal of the slug. Adequate piloting should be provided to ensure proper location of the strip as it advances through the die.

*Figure 8-61a* shows a workpiece to be blanked and formed at high-production volume in a progressive die. The first step in die design is the development of the blank (*Figure 8-61b*). The grain of coiled metal normally is parallel to the length of the strip. Where possible, forming operations are performed perpendicular to the grain to forestall any tendency to fracture. The workpiece therefore is sketched with the bend line perpendicular to the strip.

The $a$ dimension appears to coincide with stock width. A comparison is made between the exact dimension with tolerance and available purchased coil stock. If the $a$ dimension is not within the slitting tolerance of the purchased stock, the part must be blanked from a wider strip and perhaps shaved to the finished dimension. In this case, the $a$ dimension is within the slitting tolerance, the slit edge will be satisfactory, and no work is needed to obtain the dimension.

The $b$ dimension is developed by adding the $c$ dimension, $d$ dimension, and a bend allowance. Tolerances must be carefully considered in developing the $b$ dimension, which will determine whether it can be achieved with a single shear or whether shaving will be required. In this case, the $b$ dimension is not critical and is, for the moment, accepted as the station-to-station distance.

With the $a$ and $b$ dimensions established, a centerline is drawn and the template of the blank traced along the centerline (*Figure 8-61c*). The areas from which stock must be removed are clearly indicated. The straight edges of the stock appear acceptable for guidance. The absence of scrap stock on either side of the workpiece determines that the strip must be piloted by registering workpiece features. Hole $e$ appears most suitable for piloting. Hole dimensions are examined to determine whether the hole can be pierced conventionally (hole diameter not less than 1-1/2 times stock thickness), or if a secondary operation may be needed. The hole dimensions in this case are not critical, so the hole can be pierced in the first station and used as a pilot hole in subsequent stations. The first station now can be drawn, which will include a stock guidance method that establishes the height of the punch and the die. The punch location can be transposed to the other stations where it will indicate the presence of a bullet-nosed pilot.

As the strip enters the second die station, it will register on the pilot. The next operation required on the workpiece is the notching of the edges. The workpiece dimensions are such that the notching should be performed by one stroke in one station. A preliminary layout discloses, however, that a notching operation in the second die station would be close to the piercing operation of the first station. The second station is tentatively designated as an idle station.

Idle stations are of great value. They enable the designer to distribute the workload uniformly throughout the length of the die. Individual die details consequently can be less complex and easier to build and maintain. Idle stations also permit later changes in workpiece design by providing space for added operations.

The third die station can now be drawn to include a pilot and the notching punches.

The stock strip will enter the fourth die station and register on the pilot. Two operations are required to finish the workpiece: the leading edge must be formed 90°, and the workpiece must be separated from the strip. Once the workpiece has been separated no further operations are normally possible, but the workpiece must be separated for forming. The two operations can, however, be performed simultaneously in one station, just as the two sides were notched in one station. Two methods may be considered. If the stock is to be separated by removing a slug, the width of the slug must be added to the $b$ dimension, and the station-to-station distance of the entire layout must be correspondingly amended. This method, however, would permit a second punch to simply wipe down (form) and thereby complete the part without changing the elevation of the stock. If the separation is to be performed by direct shear without a slug, the workpiece must change elevation. The die station would be spring-loaded to permit the shear action, and the forming action would take place at the depressed level.

*Figure 8-63* shows the proposed die, including stock elevation, stripper pads, and the ejection method. *Figure 8-63*(A-A) shows the die construction for a shear-type cutoff.

## Force Determination

The force required for each piercing station is computed. Bending forces are covered by the methods discussed earlier in this chapter. The total workload will be the sum of the force of each piercing station and bending through their points of application. In addition, the spring, die cushion, or nitrogen forces are determined. The sum of all of the forces required for each operation, together with these forces, is the theoretical maximum force required. Note that since the operations typically do not occur at the same time as the die closes, the total force figure may be conservative.

If the workload has been properly distributed throughout the length of the die, no rocking action should occur. If the required force is concentrated in any one station or area of the die, the planned operations should be shifted to achieve balance. When reasonable balance is ensured, the forces may be added to determine the required press capacity.

An error often made when balancing forces, even with the aid of a correctly installed tonnage monitor, is the sequence of operations that occur as the die closes. If there is a high cutting load on one end of the die and an embossing or coining load on the other, the ram will be deflected

even though the loads are equal because they do not occur simultaneously.

## Press Selection

The press selected must be of adequate capacity for the planned operation and must be able to withstand the overloads that may be accidentally encountered. It must be equipped with a precise stock feed mechanism. The bolster plate must not deflect excessively under the planned load, even after it has been weakened by milling an opening for slug disposal. The die sets are commercially available in a wide variety of sizes. Some shops prefer to make their own die sets. The quality of the die set chosen is influenced by the speed of the operation, accuracy required, and anticipated production during the life of the die.

## Examples

The part to be made is a small clip, *Figure 8-64*, made of 0.008-in. (0.20-mm) thick phosphor bronze, eight numbers hard. Because of the delivery date for the parts and anticipated total production, it was decided to pierce and blank the part in a progressive die, then form it in a separate die.

The product designer indicated that the grain direction for optimum part performance is to be perpendicular to the bends. The grain direction favors the bends shown on the part and will decrease the tendency to crack at the bend lines.

The pilots were located in the scrap area of the strip to:

- allow for any future design changes;
- avoid thin walls on the blanking punch if pilots were inserted in the holes of the part, and
- take the opportunity to use larger-diameter pilots, since the part holes are only 0.0655 in. ±0.0015 in. (1.664 mm ±0.038 mm) in diameter.

The perimeter of the blank and holes to be pierced is about 2.54 in. (64.5 mm). Multiplying the length of cut by the stock thickness of 0.008 in. (0.20 mm) and a shear stress of 80,000 psi (552 MPa) gives a shear load of approximately 1,600 lb

Figure 8-63. *Die for part of* Figure 8-61.

(726 kg). Therefore, the size of press will be more dependent on the die area than the capacity.

*Figure 8-65* shows the strip development for the part. In the first station, the two 0.067/0.064-in. (1.70/1.63-mm) diameter holes are pierced, the tip of the blank dimpled, and a 0.094-in. (2.39-mm) diameter pilot hole is pierced. When the full-length legs are required, station 2 is a piloting station only. When one or both of the legs are to be cut off, punches are inserted in this station.

In station 3, the strip is piloted and two rectangular holes are pierced. In station 4, the strip is piloted while the part is blanked through the die.

There are clearance spots on the die blocks for the dimple, thus permitting the strip to be flat

on the die blocks. The die blocks are made in three sections for ease of machining, as shown in *Figure 8-66*. A channel-type stripper, 1, was selected for this die because of its more reasonable cost, and less design and build time. The forming die in the second operation would remove some of the curvature caused by the blanking operation. Hardened guide bushings, 2, were used for the pilot-hole piercing punch, 3, and the punches, 4, for the 0.067/0.064 in. (1.70/1.63-mm) diameter holes. Guide bushings were used with these punches since they were small in diameter and for the most important holes in the part.

The punches, 5, for cutting off the legs were backed up with a socket setscrew, 6. When the

.420

⊄

.322

Calculated model blank

Note: tool to be made
so that one or both
terminal legs may be
cut off per dotted lines

Figure 8-64. *Phosphor bronze clip blanked in a progressive die.*

punches are not to be used, screws are backed off a few turns and the tight fit between the punch and punch holder holds them away from the stock. This process is called "gagging" the punch.

A 5.00 × 5.00-in. (12.7 × 12.7-cm), two-post precision die set was selected for the die. The lower shoe was 1.50 in. (38.1 mm) thick and the upper shoe was 1.25 in. (31.8 mm) thick with a 1.5625-in. (39.688-mm) diameter by 2.125-in. (53.98-mm) long shank.

The stripper was extended beyond the die set close to the push-type roll feed. This was to support the stock and avoid buckling and subsequent damage to the die.

*Figure 8-67* illustrates the die layout for a small, irregularly shaped stamping with eight small holes and one large hole. For accurate positioning of the strip, pilots are provided in each station. There are six stations in the die with progression of 1.718 in. (43.64 mm) between stations.

In station 1, the eight holes, the 0.1875-in. (4.763-mm) diameter pilot hole, and a 0.302 × 0.259 in. (7.67 × 6.58 mm) hole are pierced. In station 2, a 0.50-in. (12.7-mm) diameter hole, two 0.062 × 0.068 in. (1.57 × 1.73 mm) holes, and a 0.095 × 0.379 in. (2.41 × 9.63 mm) slot are pierced. In station 3, a slot is pierced symmetrically about the centerline. The strip is notched and slotted in the next station. Station 5 is an idle station because of the length of the notching punch in the previous station. The last station, 6, incorporates a slug-type cutoff punch. This punch trims the left-hand side of the blank piloted in station 5 and the right-hand side of the blank in station 6.

For ease of grinding, a die was made in five segments as shown in *Figure 8-68*. The die set was made specially with diagonally placed guideposts. One post is 1.50 in. (38.1 mm) in diameter, while the other is 1.25 in. (31.8 mm) in diameter. Guideposts of two different diameters were used to prevent placing the upper shoe on the lower shoe incorrectly and damaging the punches.

Die buttons in the lower die steels were used for each of the piercing punches. This feature allows changing high-wear items without the need to fabricate a completely new lower die section. The die also incorporated guide bushings in the channel stripper for the piercing punches because of their relatively small diameter and the location precision specified in the part design.

Figure 8-65. *Strip development for part of* Figure 8-64.

Figure 8-66. *Progressive die for part in* Figure 8-64.

Figure 8-67. *Die layout for a small, irregularly shaped stamping.*

Figure 8-68. *Details of die blocks for die of* Figure 8-67.

## EXTRUSION

Impact extrusion, also known as cold extrusion or cold forging, is closely allied to coining, sizing, and forging operations. The operations generally are performed in hydraulic or mechanical presses. The press applies sufficient pressure to cause plastic flow of the workpiece material (metal) and form the metal to a desired shape. A metal slug is placed in a stationary die cavity into which a punch is driven by the press action. The metal is extruded upward around the punch, downward through an orifice, or in any direction to fill the cavity between the punch and die. The finished part shape is determined by the shape of the punch and die.

Impact extrusion dies are classified as open or closed. In open dies, the metal does not completely fill the die cavity. In closed dies, the material fills out the details of the die. To compensate for variations in slug volume, it is desirable to relieve closed dies and thus minimize excessive pressures.

Product design is influenced by methods of manufacturing, such as machining from the solid, drawing, spinning, stamping, or casting. Impact extrusion may combine more manufacturing processes into a single operation than any other metalworking method. Cold extrusion can result in excellent surface finish and accurate size, and can improve the mechanical and physical properties of the workpiece material. Cold-extruded parts may have smooth surfaces ranging between 30–100 μin. (0.8–2.5 μm). Very close tolerances can be achieved by cold extrusion. Owing to severe cold working of the metal, a fine, dense grain structure is developed parallel to the direction of metal flow. These continuous flow lines increase the fatigue resistance of the material.

### Pressure

*Table 8-4* lists the pressures required to extrude common metals. The pressures required depend on the alloy and its microstructure, restriction to flow, severity of work hardening, and lubricant used. With high speed and restricted flow, the press load may suddenly increase to three times the anticipated tonnage requirement. Pressures have

**Table 8-4. Extrusion pressures for common metals**

| Material | Pressure, tons/in.$^2$/(MPa) |
|---|---|
| Pure aluminum, extrusion grade | 40.700 (561) |
| Brass (soft) | 30.500 (421) |
| Copper (soft) | 25.700 (354) |
| Steel C1010, extrusion grade | 50.165 (692) |
| Steel C1020, spheroidized | 60.200 (830) |

been recorded up to 165 tons/in.$^2$ (2,275 MPa). Owing to the high pressures required, presses must be carefully selected and dies must be properly sized to resist extreme loads.

The pressures required for extrusion also depend on the percentage of reduction in area. When the percentage of reduction of area for parts made of various aluminum alloys is known, the required pressure may be taken from *Figure 8-69*. Although the alloys mentioned usually are extruded at room temperature, a reduction in press pressure can be achieved by extruding at elevated temperatures. Part wall thickness is a close relation and a factor in establishing reduction of area. The increasing affect on punch pressure as the wall becomes thinner is illustrated in *Figure 8-70*.

Figure 8-69. *The effect of reduction in area to the punch load in extruding.*

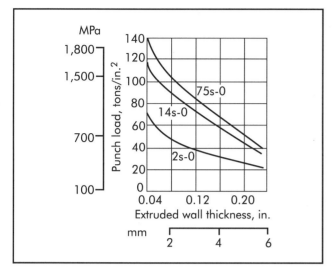

Figure 8-70. *The effect of extruded wall thickness on the punch load.*

0.58% chromium, 0.11% carbon, and 0.36% manganese, and 0.03% each sulfur and phosphorus.

## Lubrication

Correct lubrication reduces the pressures required and makes possible the cold extrusion of steel. Ordinary die lubricants break down because of the high pressures and the excessive surface heat generated by the plastic flow of material. A bonded steel-to-phosphate layer and bonded phosphate-to-lubricant layer are satisfactory to prevent metal-to-metal welding or pickup. The workpiece usually is phosphatized before extrusion. If the workpiece is subjected to several extrusion operations, it usually must be annealed and phosphatized between operations.

## Slugs

Because the slugs used for impact extrusion are cut from commercial rod or bar, it is desirable to coin them to a desired size and shape for better die-fitting characteristics before extrusion. Using coiled wire stock, headers can be used to

The relationship between reduction of area and extrusion pressures for a series of plain carbon steels is shown in *Figure 8-71*. The steels referred to in the different curves have a carbon content in the range of 0.05–0.50%, and less than 0.03% each of sulfur and phosphorus. Steel 11 contains

Figure 8-71. *Extrusion pressures: reduction relationship for the forward extrusion of a series of steels with carbon contents in the range 0.005–0.50%.*

cut and coin or upset as a continuous high-speed operation. *Figure 8-72* illustrates slug coining and upsetting. *Figure 8-73* illustrates how a profiled slug fits the die cavity and how the material flows when pressure is applied to cause plastic flow. Voids in the die cavity are filled as the slug collapses under initial pressure. Profiled slugs thus cushion the impact of the punch and allow higher ram velocity. Plastic flow then continues through small orifices between the diepot and the punch.

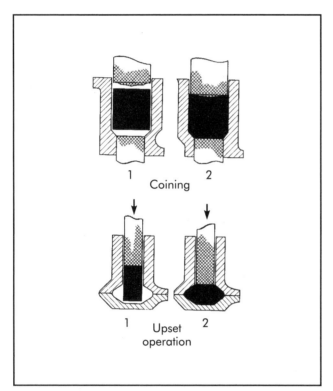

Figure 8-72. *Slug coining and upsetting.* (Courtesy American Machinist)

Figure 8-73. *Profiled slugs.* (Courtesy American Machinist)

## Methods

Basic methods of impact extrusion are backward, forward, and combination methods. In backward extrusion (*Figure 8-74*), metal flows in a direction opposite to the direction of punch movement. The punch speed can be from 7–14 in./sec (178–356 mm/sec). As the punch strikes the slug, the heavy pressure causes the metal to flow through the orifice created by the punch and die, and forms the sidewall of the part by extrusion. In forward extrusion (*Figure 8-75*), plastic flow of metal takes place in the same direction as punch travel. The orifice in this case is formed between the extension on the punch and the opening through the die. To prevent reverse metal flow, the body of the punch seals off the top of the die. The finished part is ejected backward after the punch is withdrawn from the part.

The design of a punch and die with double orifices to permit the plastic flow of metal in for-

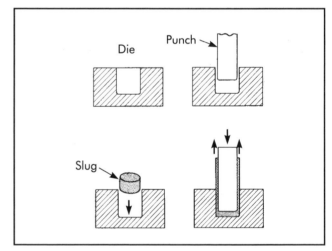

Figure 8-74. *Backward extrusion.* (Courtesy Design Engineering)

Figure 8-75. *Forward extrusion.* (Courtesy Design Engineering)

ward and reverse extrusion simultaneously will produce parts by combination impact extrusion. *Figure 8-76* illustrates examples of combination extrusion.

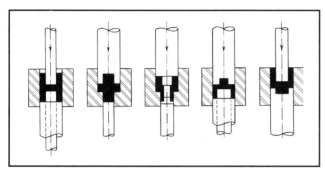

Figure 8-76. *Combination extrusion.* (Courtesy American Machinist)

## Die Design

### Coining Dies

In backward extruding dies, the punch always is smaller in diameter than the die cavity. This gives a clearance between punch and die equaling the desired wall thickness of the part to be produced. The punch is loaded as a column. To minimize punch failure, it is desirable to coin the slugs to a close fit in diameter to ensure concentricity. *Figure 8-77* illustrates a coining die to prepare a slug for backward extrusion. Coining the slug to fit the diepot and coining the upper end to fit

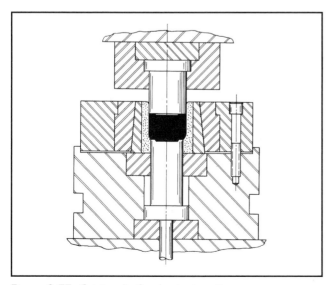

Figure 8-77. *Coining die for slug preparation.*

and guide the free end of the punch will minimize punch breakage of the extruding die.

### Backward Extrusion Dies

A typical backward extrusion die is shown in *Figure 8-78*. The use of a carbide die cavity will minimize wear due to excessive pressures. The carbide insert is shrunk into a tapered holder. The holder has a 1° side taper that pre-stresses the carbide insert to minimize expansion and fatigue failure. The inserts are well-supported on hardened blocks. The extruding punch is guided by a spring-loaded guide plate, which in turn is positioned by a tapered piloting ring on the lower die. Ejection of the finished part from the die is by cushion or pressure cylinder. *Figure 8-79* illustrates a backward extrusion die with an unusual punch penetration ratio of 5:1 made possible with a modified flat-end punch profile.

### Forward Extrusion Dies

*Figure 8-80* is an example of a typical forward extrusion die in which the metal flows in the same direction as the punch, but at a greater rate due

Figure 8-78. *Backward extrusion die. Note that the centering ring for the punch and the carbide die cavity are preloaded in shrink rings and supported on toughened load-distributing steels.* (Courtesy E. W. Bliss Company)

Figure 8-79. *Backward extrusion die.* (Courtesy American Machinist)

Figure 8-80. *Forward extrusion die.*

to change in the cross-sectional area. A lower carbide guide ring is added to maintain straightness. The nest above the upper-carbide-guide ring serves as a guide for the punch during the operation.

## Combination Extrusion

*Figure 8-81* illustrates another forward extrusion die in which the punch creates an orifice through which the metal flows. The extruding pressure is applied through the punch guide sleeve.

Figure 8-81. *Combination extrusion die.*

A typical combination forward and backward extrusion die is shown in *Figure 8-82*. In this die, the two-piece pressure anvil acts as a bottom ex-

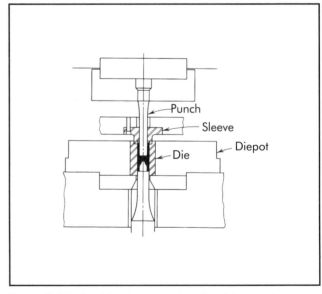

Figure 8-82. *Combination extrusion die with two-piece pressure anvil.* (Courtesy American Machinist)

truding punch and shedder. The upper extruding punch is guided by a spring-loaded guide plate, into which the guide sleeve is mounted. To maintain concentricity between the punch and die, the punch guide sleeve is centered into the die insert.

## Punch Design

The most important feature of punch design is end profile. A punch with a flat end face and corner radius not over 0.020 in. (0.51 mm) can penetrate three times its diameter in steel, four to six times its diameter in aluminum. A punch with a bullet-shaped nose or a steep angle will cut through the phosphate-coat lubricant quicker than a flat-end punch. When the lubricant is displaced in extrusion, severe galling and wear of the punch will take place. The punch must be free of grinding marks and requires a 4 μin. (0.10 μm) finish, lapped in the direction of metal flow. The punch should be made of hardened tool steel or carbide. In some backward extrusion dies, a shoulder is provided on the punch to square up the metal as it meets the shoulder.

## Pressure Anvil Design

The function of the pressure anvil is to form the base of the diepot, act as a bottom extruding punch, and act as a shedder unit to eject the finished part. Heat treatment and surface finish requirements are the same for pressure anvils as they are for punches.

## Diepot Design

To resist diepot bursting pressure, the tool steel or carbide die ring is shrunk into the shrink ring or die shoe. The die shoe normally is in compression. A shrink fit of 0.0040 in./in. (0.004 mm/mm) of diameter of the insert is desirable. Material, heat treatment, and finish requirements of the diepot are the same as for the punch. The recommended material for shrink rings is hot-worked alloy tool steel that is hardened to Rockwell hardness (HRC) 50–52. A two-piece diepot insert is sometimes used for complex workpiece shapes.

## Punch Guide Design

The guide ring minimizes the column loading on the punch by guiding it above the diepot. The spring-loaded guide sleeve pilots the punch into the diepot and maintains concentricity between them. The guide ring can also act as a stripper. The proper use of guide sleeves permits higher penetration ratios.

## BIBLIOGRAPHY

Committee of Sheet Steel Producers. 1984. *Sheet Steel Formability*, August 1. Washington, DC: American Iron and Steel Institute.

Keeler, S. 1986. *Circle Grid Analysis (CGA)*. Livonia, MI: National Steel Corporation Product Application Center.

——. 1969. *From Stretch to Draw*. Technical Paper MF69-513. Dearborn, MI: Society of Manufacturing Engineers.

Paquin, J.R. 1962. *Die Design Fundamentals*. New York: Industrial Press.

Smith, D. 2001. *Die Maintenance Handbook*. Dearborn, MI: Society of Manufacturing Engineers.

——. 1994. *Fundamentals of Pressworking*. Dearborn, MI: Society of Manufacturing Engineers.

——. 1991. *Quick Die Change*. Chapter 29, "Control the Process with Waveform Signature Analysis." Dearborn, MI: Society of Manufacturing Engineers.

——. 1990. *Die Design Handbook*. Section 3, "Die Engineering—Planning and Design." Section 6, "Shear Action in Metal Cutting." Dearborn, MI: Society of Manufacturing Engineers.

Tharrett, M. 1987. *Computer-aided Formability Analysis*. Die and Pressworking Tooling Clinic, August 25-27. Dearborn, MI: Society of Manufacturing Engineers.

Wick, C., et al., eds. 1984. *Tool and Manufacturing Engineers Handbook*, 4th edition. Volume 2, *Forming*. Dearborn, MI: Society of Manufacturing Engineers.

## REVIEW QUESTIONS

1. What is the most common pressworking operation?
2. How is clearance typically determined between the punch and die for die-cutting operations?
3. What is meant by "center of pressure?"
4. How is the theoretical peak cutting force determined?
5. How are cutting forces reduced during press operations?
6. What must occur to form a bend in a finished part?
7. How is the size of a curl governed? What are the limits of a curl?
8. What method of forming is used to produce medallions?
9. What are the advantages of using a rubber or polyurethane forming die?
10. What usually defines a compound die?
11. What is the function of an ultrasonic thickness gage?
12. What considerations need to be given in selecting lubricants for drawing operations?
13. What defines a progressive die operation?
14. Name the basic methods of impact extrusion.

# Tool Design for Inspection and Gaging

A drawing of a workpiece may be used as a guide for its manufacture only if it completely defines its limits of size, shape, and composition with dimensions and specifications. Tolerances may be given (directly or indirectly) for every dimension or specification.

*Tolerance* is the total amount by which a dimension may vary from a specified size. When providing tolerances, the designer recognizes that no two workpieces, distances, or quantities can be exactly alike. The designer specifies an ideal, unattainable condition (nominal dimension), then states what degree of error can be tolerated.

Each dimension on a workpiece must be specified as being between two limits. No proper dimension can ever be given as a single fixed value because a single value is unattainable. The designer guards against this pitfall in several ways. Critical dimensions on a drawing have the tolerance given as part of the dimension (for example, 0.125 ±0.002 in. [3.18 ±0.05 mm]). Less critical dimensions are provided tolerance by a general note in the title block (for example, unless otherwise specified, all dimensions are ±0.010 in. [±0.25 mm]). Tolerance also may be specified indirectly in the bill of material. If the designer calls for the use of purchased material or parts, and does not further specify tolerance, it must be assumed that the vendor's manufacturing tolerances are acceptable.

When establishing tolerances, the designer may consider a number of factors, including function, appearance, and cost. The workpiece will have an intended function, and the specified tol-erance must be compatible with it. Appearance is sometimes a factor because a workpiece dimension with no actual function may have aesthetic requirements among others. If for no other reason than the cost of materials, the size of a workpiece must be limited.

Cost can be the governing factor when deciding tolerances. As tolerances become smaller, the cost of meeting them rapidly increases. The designer, in concert with other departments, often must weigh the need for a close tolerance against its cost along with the capability of the machines that will produce the workpiece. If the function of the workpiece requires extremely close tolerances to ensure proper mating allowances, it may be less expensive to combine easily held tolerances with selective assembly. It may be possible in this way to obtain the necessary mating allowances for the specific function without increasing the cost with close tolerances.

In a conventional drawing, the workpiece is shown in a fixed relationship to reference planes at right angles to each other. The location of each element of the workpiece is defined by stating its distance (dimension) from each reference plane. Each dimension has a tolerance.

The reference planes in a single view may be compared with the $X$ and $Y$ axes of the Cartesian coordinate system. The location of the element is anywhere within a rectangular area formed by the minimum and maximum values of the $X$ and $Y$ dimensions as established by the tolerances.

Engineering intent often can be expressed more exactly by stating the nominal location of the element and how far from the true position (TP) the element may vary. In this case, a positional tolerance applied to the nominal location takes the place of the *X* and *Y* tolerances as shown in *Figure 9-1*. The workpiece designer's intent can be further shown by modern geometric dimensioning and tolerancing methods (see Appendix A).

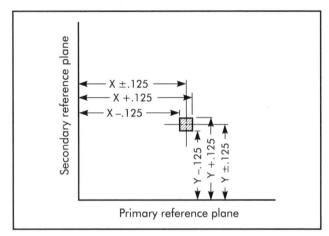

Figure 9-1. *Coordinate dimensioning and positioning tolerance illustrate the element. Dimension may be located anywhere within the shaded rectangle.*

## CONVERSION CHARTS

Since most machine movements are made in two directions at 90° to one another, *X* and *Y*, the chart shown in *Figure 9-2* may be used to determine the two-sided tolerance approximations of positional tolerance diameters. Mathematically, they also can be approximated by multiplying the position tolerance $\varnothing$ by 0.7. For example, 0.7 × 0.010 = 0.0070 ÷ 2 in. = ±0.00350 in. (0.7 × 0.35 = 0.245 ÷ 2 = ±0.1225 mm).

The chart in *Figure 9-3* may be used to determine whether an actual hole center (as measured on an open setup) is within the positional tolerance specified on the drawing, and the actual amount of difference (radius) from TP. For example, on a drawing, a hole is located by dimensions labeled TP with a positional tolerance. The location of the actual center of the hole on the part is determined by coordinate measurements. The difference in the actual measurements and corresponding TP dimensions are located on the chart using the values in the upper and right-

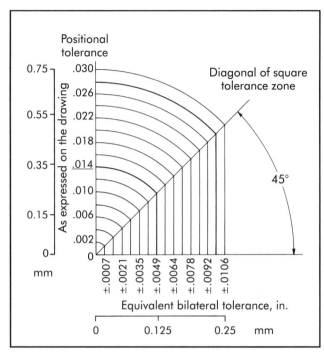

Figure 9-2. *Conversion chart: positional tolerance to bilateral.* (Courtesy Sandia Corp.)

hand borders. It can then be shown whether the location is within the required positional tolerance limits using values in the left-hand border.

To determine the actual difference from TP, a radius may be drawn through the point, intersecting the 45° diagonal line. The amount of difference may now be read from the scale on the 45° line using the values in the lower border. Mathematically, it also can be approximated by multiplying the total ± tolerance by 1.4. For example, 1.4 × 0.007 in. = 0.010 in. (±0.0035 in.) (1.4 × 0.20 mm = 0.28 mm [±0.100 mm]).

## GAGING PRINCIPLES

Since it is not possible to produce many parts with exactly the same dimensions, working tolerances are necessary. For the same reason, gage tolerances are necessary. Gage tolerance generally is determined from the amount of workpiece tolerance. A 10% rule is generally used for determining the amount of gage tolerance for fixed, limit-type working gages. When no gage tolerance is specified, the gagemaker will use 10% of the working tolerance as the gage tolerance for a

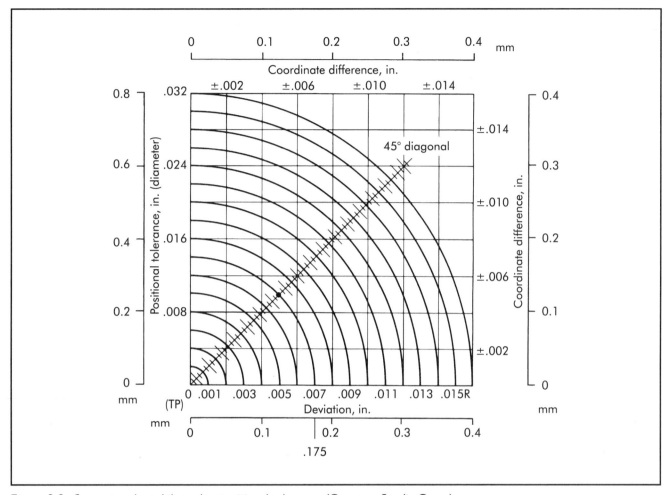

Figure 9-3. *Conversion chart: bilateral to positional tolerance. (Courtesy Sandia Corp.)*

working gage. Working gages are those used by production workers during manufacture.

The amount of tolerance on inspection gages—those used by the inspection department—is generally 5% of the work tolerance. Tolerance on master gages—those used for checking the accuracy of other gages—is generally 10% of the gage tolerance. Where tolerances are large, gages used by the inspection department are not different from the working gages.

Four classes of gagemakers' tolerances have been established by the American Gage Design Committee and are in general use. These four classes establish maximum variations for any desired gage size. The degree of accuracy needed determines the class of gage to be used. *Table 9-1* shows the four classes of gagemakers' tolerances.

- Class *XX* gages are precision-smoothed (lapped) to the very closest tolerances practicable. They are used primarily as master gages and for final close-tolerance inspection.
- Class *X* gages are precision-lapped to close tolerances. They are used for some types of master gage work and as close-tolerance inspection and working gages.
- Class *Y* gages are precision-lapped to slightly larger tolerances than Class *X* gages. They are used as inspection and working gages.
- Class *Z* gages are precision-lapped. They are used as working gages where part tolerances are large and the number of pieces to be gaged is small.

Going from Class *XX* to Class *Z*, tolerances become increasingly greater, and the gages are used

### Table 9-1. Standard gagemakers' tolerances, in. (mm)

| | | Class | | | |
|---|---|---|---|---|---|
| Above | To and Including | XX | X | Y | Z |
| .010 (0.25) | .825 (20.95) | .00002 (0.0005) | .00004 (0.0010) | .00007 (0.0018) | .00010 (0.0025) |
| .825 (20.95) | 1.510 (38.35) | .00003 (0.0008) | .00006 (0.0015) | .00009 (0.0023) | .00012 (0.0031) |
| 1.510 (38.35) | 2.510 (63.75) | .00004 (0.0010) | .00008 (0.0020) | .00012 (0.0031) | .00016 (0.0041) |
| 2.510 (63.75) | 4.510 (114.55) | .00005 (0.0013) | .00010 (0.0025) | .00015 (0.0038) | .00020 (0.0051) |
| 4.510 (114.55) | 6.510 (165.35) | .000065 (0.00165) | .00013 (0.0033) | .00019 (0.0048) | .00025 (0.0064) |
| 6.510 (165.35) | 9.010 (228.85) | .00008 (0.0020) | .00016 (0.0041) | .00024 (0.0061) | .00032 (0.00813) |
| 9.010 (228.85) | 12.010 (305.05) | .00010 (0.0025) | .00020 (0.0051) | .00030 (0.0076) | .00040 (0.0102) |

for inspecting parts having increasingly larger work tolerances.

To show the use of the 10% rule in connection with *Table 9-1*, assume a gagemaker is to choose the correct tolerance class for a working plug gage that is to be used on a 1.0000 in. (25.400 mm) diameter hole having a working tolerance of 0.0012 in. (0.030 mm). One-tenth of the work tolerance would indicate a gage tolerance of 0.00012 in. (0.0031 mm), as noted in the table, a class *Z* gage. If the work tolerance were only 0.0006 in. (0.015 mm) on the 1.0000 in. (25.400 mm) diameter hole, then a class *X* gage would be indicated, with a tolerance of 0.00006 in. (0.0015 mm). If the work tolerance, however, were 0.015 in. (0.38 mm), then the gage tolerance indicated by the 10% rule would be 0.0015 in. (0.381 mm). As this is larger than the maximum tolerance class, a class *Z* gage would be needed, and the gage tolerance would be 0.00012 in. (0.0031 mm).

The smaller the degree to which a gage tolerance must be held, the more expensive the gage becomes. Just as the production cost rises sharply as working tolerances are reduced, the cost of buying or manufacturing a gage is much higher if close tolerances are specified. Gage tolerances should be realistically applied from the work to be gaged.

## Allocation of Gage Tolerances

After deciding the tolerance for a specific gage, the direction (plus or minus) of that allowance must be decided. Two basic systems—the bilateral and unilateral—and the many variations to them, are used when making this decision. The final results of these allocation systems will differ greatly. The choice of the system to be used, modified, or unmodified must be determined by the product and the facilities for producing it. The objectives when choosing an allowance system should be the economical production of as near to 100% usable parts as possible, and the acceptance of good pieces and rejection of bad.

### The Bilateral System

In the bilateral system, the go and no-go gage tolerance zones are divided into two parts by the high and low limits of the workpiece tolerance zone. The division is illustrated by *Figure 9-4a*,

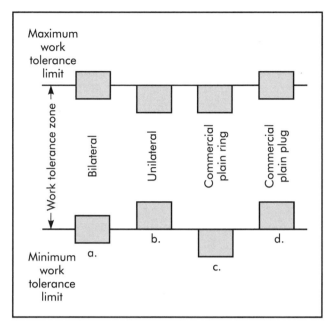

Figure 9-4. *Different systems of gage tolerance allocation.*

which shows the gray rectangles representing the gage tolerance zones are half plus and half minus in relation to the high or low limit of the work tolerance zone.

In *Table 9-1*, assume that the diameter of the hole to be gaged is 1.2500 ±0.0006 in. (31.750 ±0.015 mm). The total work tolerance in this case is 0.0012 in. (0.031 mm), since the hole size may vary from 1.2506 to 1.2494 in. (31.765 to 31.735 mm). Using 10% of the total work tolerance as the gage tolerance, the gage tolerance is then 0.00012 in. (0.0031 mm). From *Table 9-1*, this diameter would require a Class *Z* gage tolerance. The diameter on the go-plug gage for this example would be 1.2494 ±0.00006 in. (31.735 ±0.0015 mm), and the diameter of the no-go gage would be 1.2506 ±0.00006 in. (31.765 ±0.0015 mm).

One disadvantage of the bilateral system is that parts not within the working limits can pass inspection. Using the previous example, if the hole to be gaged is reamed to the low limit (1.2494 in. [31.735 mm]) and if the go-plug gage is at the low limit (1.24934 in. [31.7332 mm]), then the go-plug gage will enter the hole and the part will pass inspection, even though the diameter of the hole is outside the working tolerance zone. A part passed under these conditions would be very close to the working limit, and the tolerance on the

mating part should not be such as to prevent assembly. Plug gages using the bilateral system also could pass parts in which the holes are too large. A common misconception is that gages accept good parts and reject bad. With the bilateral system, parts also can be rejected as being outside the working limits when they are not.

### The Unilateral System

In the unilateral system (*Figure 9-4b*), the work tolerance zone includes the gage tolerance zone. This makes the work tolerance smaller by the sum of the gage tolerance, but guarantees that every part passed by such a gage, regardless of the amount of the gage size variation, will be within the work tolerance zone.

If the diameter of the hole is 1.2500 ±0.0006 in. (31.750 ±0.015 mm), again using 10% of the working tolerance as the gage tolerance, the go-gage diameter would be 1.24940 +0.00012 in. (31.7348 +0.0031 mm), and the no-go gage diameter 1.25060 –0.00012 in. (31.7652 –0.0031 mm).

The unilateral system of applying gage tolerance, like the bilateral system, may reject parts as being outside the working limits when they are not, but all parts passed using the unilateral system will be within the working limits. The unilateral system has found wider use in industry than the bilateral system for plain plug and ring gages.

One partial solution to the problem of gages rejecting parts within working limits is to use working gages with the largest unilateral gage tolerance practical, and inspection gages with the smallest unilateral gage tolerance practical. Thus, no piece can pass inspection that is outside tolerance, and the possibility of the inspection gage turning down acceptance work is reduced because of its small tolerance.

*Figure 9-4c* shows the commercial practice of allocating the plain ring gage tolerances negatively with reference to both the maximum and minimum limits of the workpiece tolerance. In another practice (*Figure 9-4d*), the no-go gage tolerance is divided by the maximum limit of the workpiece tolerance, and the go-gage tolerance is held within the minimum limit of the workpiece tolerance.

## GAGE WEAR ALLOWANCE

Perfect gages cannot be made. If one did exist, it would no longer be perfect after just one checking operation. Although the amount of gage wear during just one operation is difficult to determine, it is easy to measure the total wear of several checking operations. A gage can wear beyond usefulness unless some allowance for wear is built into it.

The wear allowance is an amount added to the nominal diameter of a go-plug and subtracted from that of a go-ring gage. It is depleted during the gage's life by wearing away the gage metal. Wear allowance is applied to the nominal gage diameter before the gage tolerance is applied.

The amount of wear allowance does not have to be decided in relation to work tolerance, although a small work tolerance can restrict wear allowances. When specifying a wear allowance, the material from which the gage and work are to be made, the quantity of work, and the type of gaging operation to be performed must be taken into consideration. It is important to establish a specific amount of wear allowance. When the gage has worn the established amount, it should be removed from service without question. This avoids any controversy as to whether a gage is still accurate.

One method, which uses a percentage of the working tolerance as the wear tolerance, can be explained with the following example: For a 1.500 ±0.0006 in. (38.10 ±0.015 mm) diameter hole, the working tolerance is 0.0012 in. (0.031 mm). The basic diameter of the go-plug gage would be 1.49940 in. (38.0848 mm). Using 5% of the working tolerance (0.00006 in. [0.0015 mm]) as the wear allowance, and adding this to the basic diameter, the new basic diameter then would be 1.49946 in. (38.0863 mm). The gagemaker's tolerance of 10% of the working tolerance (0.00012 in. [0.0031 mm]) then would be applied in a plus direction as allowed by the unilateral system, with a resultant go plug diameter of $1.49946^{+.00012}_{-.00000}$ ($38.0863^{+.0031}_{-.0000}$).

Some manufacturing companies do not build a wear allowance into their gages, but create a standard to determine when a gage has worn beyond its usefulness. The gage is allowed to wear a certain percentage above or below its basic size before being taken out of service. Gages should be inspected regularly for wear. No set policy for wear allowance or gage inspection is practical for all industries. In operations where an extremely high degree of accuracy must be maintained, the amount of allowable wear is smaller, and inspection must be more frequent than in operations where tolerances are greater.

A gagemaker normally makes the gage to provide maximum wear, even if no wear allowance is designed into the gage. That is, the go end of a plug gage is made to its high limit, and the go component of a ring gage is made to its low limit.

The no-go gage will slip in or over very few pieces and wear very little. As any wear on a no-go gage puts that gage farther within the product limits, no wear allowance is applied. When the no-go gage begins to reject work within acceptable limits, it must be retired.

## GAGE MATERIALS

For medium production runs, hardened alloy steel is used for the wear surfaces of gages. For higher-volume production runs, gage wear surfaces are usually chromium-plated. When a high degree of accuracy is needed, the production run is long, and/or wear is excessive, tungsten carbide contacts often are used on gages. Worn gaging surfaces can be ground down, chrome-plated, reground, lapped to size, and put back into service.

## GAGING POLICY

Gaging policy is the standardization of methods for determining gage tolerances and their allocation, and fixing wear allowance. It is a guide to determine when gages are required, when and how they are to be inspected, and what types of gages should be used.

There is no single policy in universal use. Gage users should have their own policy—the one best-suited to their work. A gage policy helps to eliminate controversy over the method of gaging, gage tolerances, allocation, and wear.

The policy should establish a general rule for determining the amount of gage tolerance. The 10% rule, used in conjunction with the four standard classes of gage tolerances, is widely used in industry.

The size and application of wear allowance is fixed in the gage policy. How often gages are inspected will depend upon how often they are used, what materials they check, and the closeness of tolerances. When a high degree of accuracy is needed, the gage policy must be strict.

## GAGE MEASUREMENT

A measurement compares an amount or length with a known standard. Dimensions are proven by end measurement; the end of a specified length is measured against the end of the standard; the last drop of a specified amount of liquid brings the fluid level up to a mark on the side of the vessel. Almost all dimensional measurement is end measurement. End measurement is determined by origin; that is, both the standard and the length or amount being measured must start at a common point.

Every workpiece related to primary, secondary, and tertiary datums must be measured by three planes located at right angles to each other. Every dimension has as its origin one of the three planes. Given a horizontal reference plane, two vertical reference planes can be constructed at right angles to each other, and any structure can be defined by dimensions starting at the reference planes.

## Surface Plate

The surface plate is used as the main horizontal reference plane. It is the primary tool of layout and inspection. Surface plates are usually made of cast iron or granite, with one surface finished extremely flat. When accuracy of 0.00001 in. (0.0003 mm) or finer is required, toolmakers' flats or optical flats are used instead. A number of accessories, such as squares, straight edges, parallels, angle irons, sine plates, leveling tables, height gages, length standards, and indicating equipment are used with the surface plate.

The workpiece being inspected is carefully placed with its primary reference plane on the plate surface. If the workpiece has elements extending below its established reference plane, it may be necessary to mount the workpiece on adjustable supports (jacks) or parallels to make the two reference planes parallel. Vertical reference planes are placed on the surface plate as required.

To avoid an increased chance of error, all dimensions on a part drawing should start from a reference point or plane rather than from the end of another dimension. If, however, a workpiece feature is measured from a point rather than a reference plane, its location should be proven in relation to that point rather than the reference plane. A typical surface plate setup is shown in *Figure 9-5*.

## Templates

A template represents a specified profile, or it may be a guide to the location of workpiece features with reference to a single plane. A straightedge may be used as a template to check straightness. To control or gage special shapes or contours in manufacturing, special templates are used for comparison by eye to ensure uniformity of individual parts. These templates are made from thin, easily machinable materials,

Figure 9-5. *Typical surface plate setup.*

some of which may be hardened later if production requirements demand longer use. *Figure 9-6* shows an application of the contour template to inspect a turned surface. Templates of this type are widely used in the sheet-metal industry, and where production is limited. Templates are satisfactory when the part tolerance will permit this type of inspection.

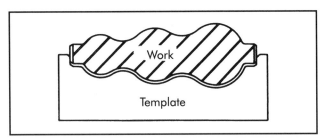

Figure 9-6. *Gaging the profile of a workpiece with a template.*

## GAGE TYPES

### Commercial Gages

To inspect or gage radii or fillets visually, a standard commercial type of template can be obtained from various gage manufacturers. These templates or gages are used by inspectors, layout personnel, toolmakers, diemakers, machinists, and pattern makers. The five different basic uses are shown in *Figure 9-7*. These gages are usually made to nominal sizes in increments of 1/64 in. for inch gages (0.5 mm for millimeter gages). Special gages of this type might readily be designed and made for specific jobs.

### Screw Pitch Gage

A screw pitch gage is used to determine the pitch of a screw. Individual gages can be obtained

Figure 9-7. *Commercial radius gage and applications: (a) inspection of an inside radius tangent to two perpendicular planes; (b) inspection of a groove; (c) inspection of an outside radius tangent to two perpendicular planes; (d) inspection of a ridge segment; (e) inspection of roundness and diameter of a shaft.*

for most of the commonly used thread forms and sizes. To determine the pitch, the gage is placed on the threaded portion as shown in *Figure 9-8*. A screw pitch gage is seldom designed for special uses, since it does not check thread size and will not give an adequate check on thread form for precision parts.

Figure 9-8. *Screw pitch gage and method of application.*

### Plug Gage

A *plug gage* is a fixed gage, usually made of two members. One member is called the go end, and the other the no-go or not-go end.

The actual design of most plug gages is standard, covered by American Gage Design (AGD) standards. However, there are many cases where a special plug gage must be designed.

A plug gage usually has two parts: the gaging member and a handle with the size (go or no-go) and gagemaker's tolerance marked on it. Generally, there are three types of AGD standard plug gages. First is the single-end plug gage. This type has two separate gage members (a go and a no-go), each having its own handle (*Figure 9-9a*). The second is called a double-end plug gage, which consists of two gage members (a go and a no-go) mounted on a single handle, with one gage member on each end (*Figure 9-9b*). The third type, called the progressive gage, consists of a single gage member mounted on a single handle (*Figure 9-9c*). The front two-thirds of the gage member is ground to the go size and the remaining portion is ground to the no-go size. The go and no-go sizes are put together in the same gage member.

Standard AGD plug gages generally have three methods of mounting gage members on the

Figure 9-9. *AGD cylindrical plug gages used to inspect the diameter of holes.*

handle. Smaller sizes are usually wire-type gages. The gage member is simply a straight blank or nib with no shoulder, taper, or threads. It is held in the handle by a setscrew or a collet chuck built into the handle, as shown in *Figure 9-10a*. The advantage of this type of mounting is that the gage members are reversible when one end becomes worn, thus increasing the life of the gage.

*Figure 9-10b* shows another type of mounting called the taper-lock design. The gage member is manufactured with a taper ground on one end. This taper fits into a tapered hole in the handle much the same as a taper-shank drill fits into a drill-press spindle. This type is not reversible.

The third type of mounting is called the trilock design. This design is usually for larger gages.

The gage member has a hole drilled through the center and is counterbored on both ends to receive a standard socket head screw. Three slots are milled radially in each end of the gage member. The gage is held to the handle by means of a socket head screw with the three slots engaging three lugs on the end of the handle as shown in *Figure 9-10c*. This type of gage is reversible.

Sometimes it is necessary to design special plug gages. There are many types, depending on the job requirement. A square, hexagonal, or octagonal hole requires a special plug gage. Internal splines require spline plugs designed in accordance with ANSI B92.1. Internal threads require thread plugs designed in accordance with ANSI B1.2, etc. *Figure 9-11* shows several special plug gages.

## Ring Gage

*Ring gages* are usually used in pairs, consisting of go and no-go members, as shown in *Figure 9-12*. They are fixed gages, and their design is covered by AGD standards.

In sizes up to 1.51 in. (38.4 mm), the design is a plain ring, knurled on the outside diameter

Figure 9-11. *Special plug gages used to inspect the profile or taper of holes. They check the individual feature elements but they do not check the boundary of perfect form.*

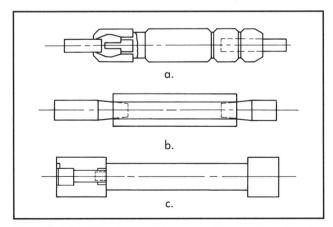

Figure 9-10. *AGD cylindrical plug gages:* (a) *reversible wire type,* (b) *taper-lock design* (c) *trilock design.*

Figure 9-12. *Ring gage set used to inspect the diameter of shafts.*

(OD), and lapped in the inside diameter (ID) to a close tolerance. The no-go member is identified by a groove around the OD of the gage.

In sizes up to 1.51 in. (38.4 mm), the gage has a flange to reduce weight and increase rigidity. The no-go in this type also is identified by a groove around the OD of the gage.

Special ring gages are occasionally required. One example would be for the inspection of a shaft having a keyway, where a key-shaped segment added to a ring gage would allow gaging of shaft size, key-slot width, and depth at the same time. External splines require spline rings designed in accordance with ASME B94.6-1995. External threads require thread rings designed in accordance with ASME B1.2-2001, etc. *Figure 9-13* shows several special ring gages.

## Snap Gage

A *snap gage* is a fixed gage arranged with inside measuring surfaces for calipering diameters, lengths, thicknesses, or widths. Snap gages come in a variety of types.

A plain adjustable snap gage is a complete external-caliper gage used for size control of plain external dimensions. It has an open frame with gaging members provided in both jaws. One or more pairs of gaging members can be set and locked to any predetermined size within the range of adjustment. The design of most types is covered by AGD standards. They are shown in *Figure 9-14.*

Figure 9-13. *Special ring gages to check profile or taper on parts.* (Courtesy Hemco)

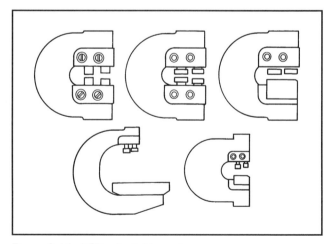

Figure 9-14. *AGD adjustable snap gages.*

A plain, solid, snap gage, shown in *Figure 9-15*, is another complete external-caliper gage used for controlling the size of plain external dimensions. It has an open frame and jaws, with the latter carrying gaging members in the form of fixed, parallel, nonadjustable anvils.

The thread-roll snap gage, shown in *Figure 9-16*, is a complete external-caliper gage employed to control the size of thread pitch diameter, lead of a thread, and thread form. It has an open frame and jaws in which gaging members are provided. One or more pairs of gaging members can be set and locked to a predetermined size within range of the thread to be checked. Gaging members vary with different diameters and pitch or thread.

A form-and-groove or blade-type snap gage, as shown in *Figure 9-17*, is a complete external-caliper gage used for controlling the size of grooves or checking close shoulder diameters. It has an open frame and adjustable blade anvils.

Many companies have master drawings of snap gages. The designer need only fill in the dimension needed to gage a part. These master drawings are then sent to gage manufacturers to fill the order. There are special cases when a single-purpose fixed-snap gage is desired. For example, the OD of a narrow groove (as shown in *Figure 9-18*) is checked with a special double-end snap gage designed to inspect the diameter.

At times, a snap gage may be better than a ring gage as an inspection tool. *Figure 9-19* illustrates how a ring gage may accept out-of-round workpieces that would be rejected by a snap gage.

Figure 9-15. *Plain snap gage.*

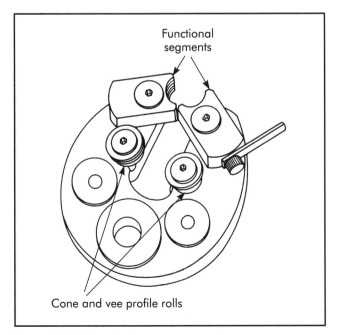

Figure 9-16. *Thread-roll snap gage.*

Figure 9-17. *Adjustable form-and-groove snap gage shown with typical anvil form modifications. (Courtesy Standard Gage Co.)*

Figure 9-18. *Special snap gage.*

## Flush-pin Gage

The *flush-pin gage* is a simple mechanical device used to measure linear dimensions. The important parts of the gage are the body and a sliding pin or plunger. The indicating device is a step, ground either on the plunger or on the flush-pin body, equal to the total tolerance of the dimension. When the gage is mounted on the workpiece, the position of the plunger can be checked visually or by fingernail touch.

*Figure 9-20* shows a slotted workpiece being checked with a flush-pin gage. Also shown are the relative positions of the plunger at the high and low limits of the depth. The flush-pin principle applied in this way is simple in operation, rugged and foolproof, does not require a master for presetting, and is economical when compared to micrometer or dial-gaging methods. The dimension also could be checked with a depth micrometer. However, the depth micrometer requires a greater degree of operator skill, and there is the possibility of misreading the instrument.

*Figure 9-21* shows an inspection fixture containing two flush-pin gages. Also shown is the workpiece being checked. The first dimension being checked ($X$) is the distance from the center of the spherical radius to a flat surface. Dimension $Y$ is between the center of the same spherical radius and over the outside of a roll (wire). A standard tooling ball in the base of the fixture locates the radius of the workpiece and provides the origin for both dimensions.

## Amplification and Magnification of Error

As changes in dimension become too small to be easily measured, it is necessary to amplify or magnify them prior to measurement. This can be done mechanically, pneumatically, optically, or electronically.

As tolerances become smaller, primary gaging methods are not precise enough to detect the degree of error. An inspector may, for example, find that qualification of a workpiece no longer depends on whether a gage enters a hole, but on how much pressure is needed to insert the gage or whether the gage enters with too little pressure. Inspection becomes more dependent on human judgment and is therefore less reliable.

A surface cannot be perfectly flat. Two parts cannot be exactly alike. Measured distances or quantities cannot be exactly equal. The inability of an inspection device to detect error proves only that the capability of the inspection device is limited. Error or deviation is always present and, unless specified or necessary limits are exceeded, are not in themselves a disqualifying factor. The designer specifies what degree of error can be allowed. The inspector confirms that the error in the workpiece is within tolerance. Accuracy of measurement is limited by the accuracy of the standard used for comparison and the skill of the person making the comparison.

## Indicating Gages and Comparators

*Indicating gages* and *comparators* magnify the amount a dimension deviates above or below a standard to which the gage is set. Most indicate in terms of actual units of measurement, but some show only whether a tolerance is within a given range. The ability to measure to 0.000001

Figure 9-19. *Gage comparison.*

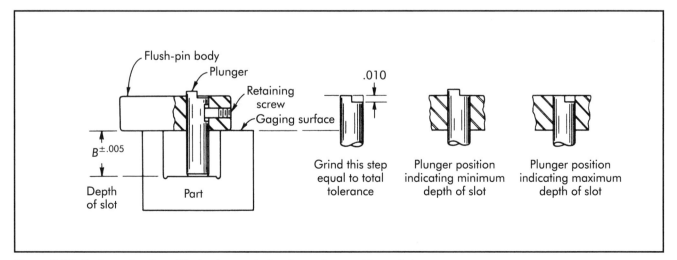

Figure 9-20. *Basic application of flush-pin gage indicating various positions of plunger.*

Figure 9-21. *Workpiece with flush-pin-type inspection fixture.*

in. (25 nanometers [nm]) depends upon magnification, resolution, accuracy of the setting gages, and staging of the workpiece and instrument. Graduations on a scale should be 0.06–0.10 in. (1.5–2.5 mm) apart to be clear. This requires magnification of 60,000–100,000× for a 0.000001 in. (25 nm) increment; less is needed, of course, for larger increments. Mechanical, air, electronic, and optical sensors and circuits are available for any magnification needed and will be described in the following sections. However, measurements have meaning and are repeatable only if based upon reliable standards, like gage blocks, and if the support of the workpiece and instrument is stable. An example is that of equipment to trace the roundness of cylindrical parts. Either the probe or part must be rotated on a spindle that must run true with an error much less than the increment to be measured.

## Dial Indicator

The *dial indicator* (*Figure 9-22*) is perhaps the most widely used instrument for precise measurement. Basically, it consists of a probe, rack, pinion, pointer, dial, and case. The probe, which is attached to the end of the rack, is placed on the workpiece. A change in workpiece size changes the position of the probe, which in turn moves the rack. The rack movement turns the pinion, which through a gear train causes the dial pointer to move. The graduated dial is calibrated for di-

Figure 9-22. *Dial indicators.* (Courtesy Federal Products Corp.)

rect reading of variation from the nominal dimension. Several amplification factors are involved.

Dial indicators are used for many kinds of measuring and gaging operations. They also serve to check machines and tools, alignments, and cutter runout. Dial indicators often are incorporated in special gages and measuring instruments.

Dial indicators are commercially available from many sources. Standard models vary greatly in size, amplification ratio, mounting facilities, and precision. Inexpensive models are available for production use as well as very precise models for use in the gage crib.

An indicator gage has one primary advantage over a fixed gage: it shows how much a workpiece is oversize or undersize. When using an indicator as part of a gaging device, a master block to the nominal dimension to be checked must be used to preset the indicator to zero. Then, in applying the gaging device, the variation from zero, the nominal dimension, is read from the dial scale.

Extremely precise dial indicators are part of the standard inspection equipment found in a gage crib, together with surface plates, parallels, V-blocks, and vernier height gages. *Figure 9-23* shows a dial indicator being used to inspect flatness of a workpiece mounted on gage blocks. *Figure 9-24* shows standard inspection tools being used to check the squareness of a workpiece. In checking squareness or runout, the indicator is zeroed with the probe placed on the workpiece. If the same standard components are used directly as a height gage, the gage first must be zeroed to a known standard, such as a gage block or master.

If inspection is frequently needed, it may be expensive to tie up quality control equipment. *Figure 9-25* shows a commercially available gaging fixture complete with surface, indicator, and stand. Attached centers and rolls for checking runout of centered or uncentered parts also are available.

Figure 9-24. *Squareness gaging fixture composed of standard inspection tools.*

Figure 9-25. *Gaging fixture with accessories.*

Figure 9-23. *Application of a dial indicator for inspecting flatness by placing the workpiece on gage blocks and checking full indicator movement (FIM).*

## Electric and Electronic Gages

Certain gages are called *electric limit gages* because they have the added feature of a rack stem that actuates precision switches. The switches connect lights or buzzers to show limits and may energize sorting and corrective devices.

An *electronic gage* gives a reading in proportion to the amount a stylus is displaced. It may also actuate switches electronically to control various functions. An example of an electronic gage and diagrams of the most common kinds of gage heads are shown in *Figure 9-26*. The variable inductance or inductance-bridge transducer

Figure 9-26. *Elements of electronic gages. Types of gage heads:* (a) *variable inductance;* (b) *variable transformer;* (c) *strain gage;* (d) *variable capacitance;* (e) *block diagram of typical electronic gage circuit;* (f) *one model of electronic gage.*

has an alternating current (AC) fed into two coils connected into a bridge circuit.

The reactance of each coil is changed as the position of the magnetic core is changed. This changes the output of the bridge circuit. The variable transformer or linear variable displacement transformer (LVDT) transducer has two opposed coils into which currents are induced from a primary coil. The net output depends on the displacement of the magnetic core. The deflection of a strain-gage transducer is sensed by the changes in length and resistance of strain gages on its surface. This is also a means for measuring forces. Displacement of a variable capacitance head changes the air gap between plates of a condenser connected in a bridge circuit. In every

case, AC is fed into the gage as depicted in (e). The output of the gage head circuit is amplified electronically and displayed on a dial or digital readout. The information from the gage may be recorded on tape or stored in a computer, a big advantage with this type of gaging.

Depending on capacity, range, resolution, quality, accessories, etc., electronic gages are priced from a little under $1,000 to many thousands of dollars. An electronic height gage like the one shown in (e) with an amplifier and digital display is at the low end of the price schedule. A digital-reading height gage like the instrument shown in (f) can be used for transferring height settings in increments of 0.00010 in. (0.0025 mm) with an accuracy of 0.000050 in. (0.00127 mm).

The instrument cost about $2,000 in the year 2000.

Electronic gages have several advantages: they are very sensitive (they commonly read to a few micrometers); output can be amplified as much as desired; a high-quality gage is quite stable; they generally can read either English or metric scales; a computer chip may be added to perform routine measurement algorithms; and they can be used as an absolute measuring device for thin pieces up to the range of the instrument. The amount of amplification can be switched easily, and three or four ranges are common for one instrument. Two or more heads may be connected to one amplifier to obtain sums or differences of dimensions, as for checking thickness, parallelism, etc. *Figure 9-27* shows electronic-gage application data for a wide variety of measurements.

## Air Gages

An *air gage* measures, compares, or checks dimensions by sensing the flow of air through the space between a gage head and workpiece surface. The gage head is applied to each workpiece in the same way, and the clearance between the two varies with the size of the piece. The amount the airflow is restricted depends upon the clearance. There are four basic types of air gage sensors shown in *Figure 9-28*. All have a controlled constant-pressure air supply. The back-pressure gage responds to the increase in pressure when the airflow is reduced. It can magnify from 1,000:1 to over 5,000:1, depending on range, but is somewhat slow because of the reaction of air to changing pressure. The differential gage is more sensitive. Air passes through this gage in one line to the gage head and in a parallel line to the atmosphere through a setting valve. The pressure between the two lines is measured. There is no time lag in the flow gage, where the rate of airflow raises an indicator in a tapered tube. The dimension is read from the position of the indicating float. This gage is simple, does not have a mechanism to wear, is free from hysteresis, and can amplify to over 500,000:1 without accessories.

A *venturi gage* measures the drop in pressure of the air flowing through a venturi tube. It combines the elements of the back-pressure and flow gages and is fast, but sacrifices simplicity. A few of the many kinds of gage heads and applications are also shown in *Figure 9-28*. An air gage is basically a comparator and must be set to a master for dimension or to two masters for limits. The common single gage head is the plug. Practically all inside and outside linear and geometric dimensions can be checked by air gaging. Air match gaging, depicted in *Figure 9-28i*, measures the clearance between two mating parts. This provides a means of controlling an operation to machine one part to a specified fit with the other. A multidimension gage has a set of cartridge or contact gage heads *(Figure 9-28h)* to check several dimensions on a part at the same time. The basic gage sensor can be used for a large variety of jobs, but a different gage head and setting master are needed for almost every job and size.

A major advantage of an air gage is that the gage head does not have to tightly fit the part. A clearance of up to 0.003 in. (0.08 mm) between the gage head and workpiece is permissible, even more in some cases. Thus no pressure is needed between the two to cause wear, and the gage head may have a large allowance for any wear that does occur. The flowing air helps keep surfaces clean. The lack of contact makes air gaging particularly suitable for checking against highly finished and soft surfaces. Because of its loose fit, an air gage is easy and quick to use. An inexperienced worker can measure the diameter of a hole to 0.000001 in. (25 nm) in a few seconds with an air gage; the same measurement to 0.001 in. (25 μm) with a vernier caliper by a skilled inspector may take up to 1 minute. The faster types of air gages are adequate for high-rate automatic gaging in production.

Air gaging is subject to varying shop air pressures, humidity, temperature, etc., which must be taken into account when designing these systems.

## Automatic Gaging Systems

As industrial processes are automated, gaging must keep pace. Automated gaging is performed in two general ways. One is in-process or on-the-machine control by continuous gaging of the work. The second way is post-process or after-the-machine gaging control. Here, the parts coming off

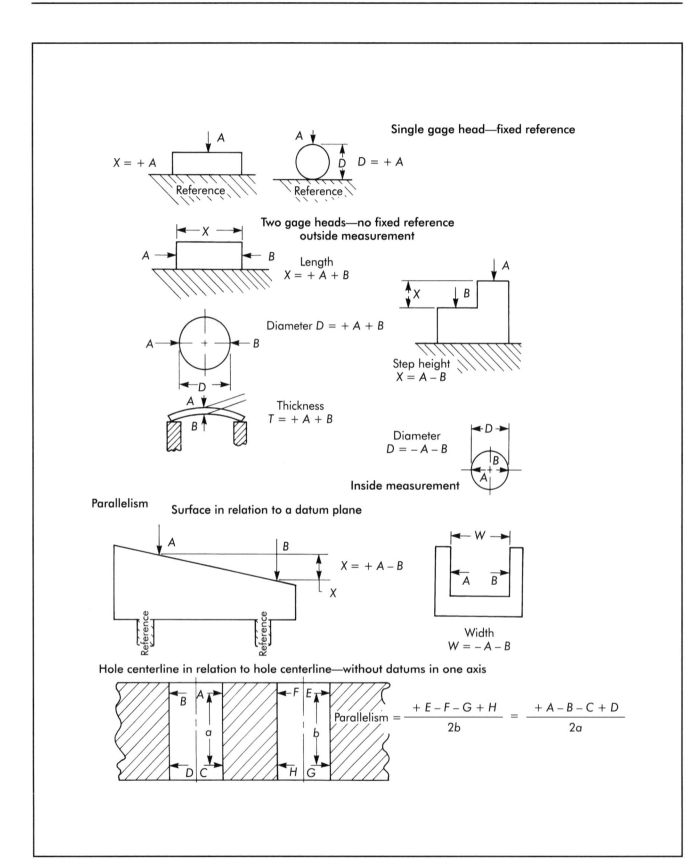

Figure 9-27. *Application data for electronic gaging.*

Figure 9-27. *continued.*

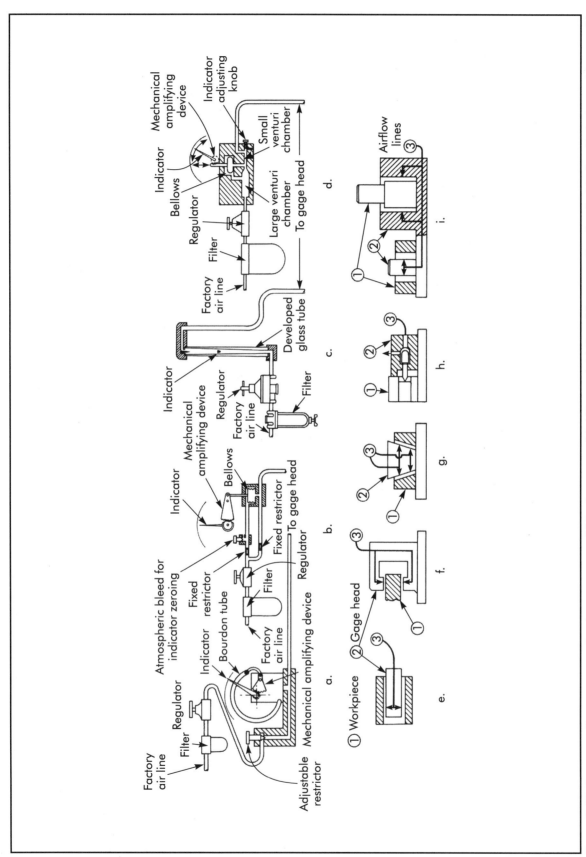

Figure 9-28. Diagrams of air-gage principles. Air gage sensors: (a) back pressure; (b) differential; (c) flow; (d) venturi. Gage heads: (e) plug; (f) thickness; (g) taper; (h) cartridge or contact; (i) matching.

the machine are passed through an automatic gage. A control unit responds to the gage to sort pieces by size and adjust or stop the machine if parts are found out of limits.

## Optical Projection Gaging

*Optical projection* is a method of measurement and gaging using a precision instrument known as an optical projector or comparator. Optical gaging is gaging by sight rather than feel or pressure. The measurement or gaging is performed by placing a workpiece in the path of a beam of light and in front of a magnifying-lens system, thereby projecting an enlarged silhouette shadow of the object upon a translucent screen as illustrated in *Figure 9-29*.

Measurement of a workpiece normally requires a chart gage with reference lines in two planes. In gaging applications, a precisely scaled layout of the contour of the part to be gaged, usually with tolerance limits, is drawn on the screen. *Figure 9-30* shows a workpiece with the chart gage used for its inspection.

Optical gages are almost unaffected by wear. There is no wear to a light beam, and any fixture wear can be compensated for by repositioning to the setting point. Little operator skill is required; no touching skill or sensitivity is needed. Dimensions can be changed on the screen easily and quickly. The chart provides exact duplication. Several dimensions can be checked at the same time, eliminating too much handling of the part.

Most optical projectors have standard magnification of 10× up to 100×. The effective area that can be projected through a given lens system can be determined by dividing the screen diameter by the magnification of the lens being used. For example, a comparator with a 14-in. (356-mm) diameter screen, using 10× magnification, will project a complete area of a 1.400-in. (35.56-mm) diameter specimen at one setting.

A chart gage is an accurately scribed, magnified, outline drawing of the workpiece to be gaged. It contains all the contours, dimensions, and tolerance limits necessary. The chart is made of glass or plastic. For short and quick checks, drafting paper can be used. However, paper should not be used consistently because it shrinks with changes of weather. Chart layout lines should be dark black, sharply defined, and from 0.006–0.020 in.

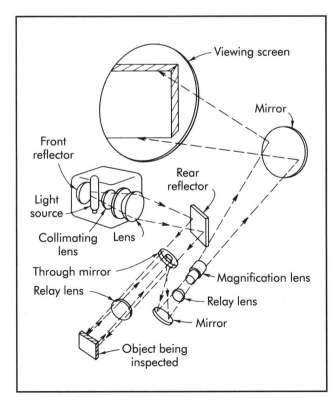

Figure 9-29. *Optical projection gaging principle.*

Figure 9-30. *Optical gage setup: (a) workpiece; (b) chart gage; (c) side view of holding fixture; (d) front view of holding fixture.*

(0.15–0.51 mm) wide for best legibility. Dimensions normally extend to the center of the lines. When maximum and minimum tolerance lines are used, the magnification should be high enough to maintain a minimum of 0.020 in. (0.51 mm) spacing between the lines. For closer tolerance checking, special lines or bridge arrangements, based on gaging to the edge of the lines, are often used. *Figure 9-31* shows two portions of a chart gage. The first portion (*a*) uses maximum and minimum tolerance lines. The second portion (*b*) has a close-tolerance bridge arrangement.

Workholding equipment includes vises for flat workpieces and staging centers for round workpieces, such as shafts and cylindrical parts with machined centers. V-blocks, in a range of sizes, are mounted on bases to clamp to projector worktable slots. Diameter capacities run from 0–5 in. (0–127 mm) in diameter.

## COORDINATE MEASURING MACHINES

The coordinate measuring machine (CMM) is a flexible measuring device capable of providing highly accurate dimensional position information along three mutually perpendicular axes. This instrument is widely used in manufacturing industries for the post-process inspection of a large variety of products and their components. It is also effectively used to check dimensions on a

variety of process tooling, including mold cavities, die assemblies, assembly fixtures, and other workholding or tool positioning devices. A system diagram of a measurement processor for automatic direct computer control (DCC) of a CMM is shown in *Figure 9-32*.

### Moving Bridge CMM

The two most common structural configurations for CMMs are the moving bridge and the cantilever. The basic elements and configuration of a typical moving-bridge type CMM are shown in *Figure 9-33*. The base or worktable of most CMMs is constructed of granite or some other ceramic material to provide a stable work locating surface and an integral guideway foundation for the superstructure. As indicated in *Figure 9-33*, the two vertical columns slide along precision guideways on the base to provide *Y*-axis movement. A traveling block on the bridge gives *X*-axis movement to the quill, and the quill travels vertically for a *Z*-axis coordinate. The moving elements along the axes are supported by air bearings to minimize sliding friction and compensate for any surface imperfections on the guideways.

Movement along the axes can be accomplished manually on some machines by light hand pressure or rotation of a handwheel. Movement on more expensive machines is accomplished by axis drive motors, sometimes with joystick control. Direct computer controlled CMMs are equipped with axis drive motors, which are program-controlled to automatically move the sensor element (probe) through a sequence of positions. To establish a reference point for coordinate measurement, the CMM and probe being used must be datumed. In the datuming process, the probe, or set of probes, is brought into contact with a calibrated sphere located on the worktable. The center of the sphere is then established as the origin of the *X-Y-Z* axes coordinate system.

The coordinate measuring machine shown in *Figure 9-34* has a measuring envelope of about 18 × 20 × 16 in. (0.5 × 0.5 × 0.4 m) and is equipped with a disengagable drive that enables the operator to toggle between manual and DCC. It has a granite worktable and the *X*-beam and

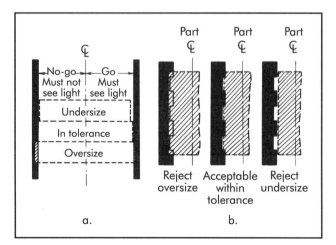

Figure 9-31. *Chart gage segments: (a) maximum and minimum tolerance lines; (b) close-tolerance bridge arrangement.*

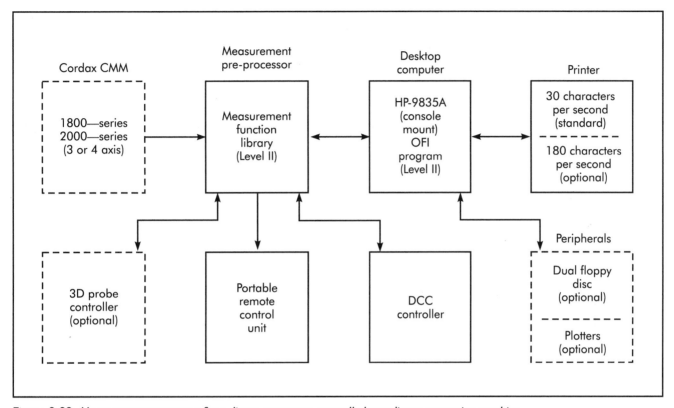

Figure 9-32. *Measurement processor for a direct, computer-controlled coordinate measuring machine.*

Figure 9-33. *Typical moving-bridge coordinate measuring machine configuration.*

Y-beam are made of an extruded aluminum to provide the rigidity and stability needed for accurate measuring. The measurement system has a readout resolution of 0.00004 in. (1 μm) and the DCC system can be programmed to accomplish 45–60 measurement points/min. A CMM of this size costs about $58,000 with computer support, joystick control, and an electronic touch probe. It can also be equipped with a videocamera for noncontact measurement.

## Contacting Probes

CMM measurements are taken by moving the small stylus (probe) attached to the end of the quill until it makes contact with the surface to be measured. The position of the probe is then observed on the axes readouts. On early CMMs, a rigid (hard) probe was used as the contacting element. The hard probe can lead to a variety of measurement errors, depending on the contact pressure applied, deflection of the stylus shank, etc. These errors are minimized by the use of a pressure-sensitive device called a touch-trigger probe.

Figure 9-34. *Coordinate measuring machine.* (Courtesy Brown & Sharpe Manufacturing Company)

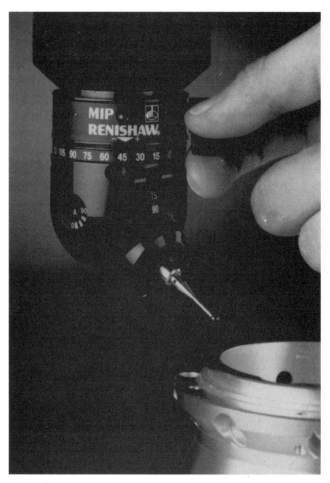

Figure 9-35. *Manual indexable probe.* (Courtesy Renishaw, Inc.)

The *touch trigger probe* permits hundreds of measurements to be made with repeatabilities in the 0.00001–0.00004 in. (0.25–1.0 μm) range. Basically, this type of probe operates as an extremely sensitive electrical switch that detects surface contact in three dimensions. The manual indexable touch-trigger probe shown in *Figure 9-35* can be used to point and probe without redatuming for each position measured. A probe of this type for use on manually operated CMMs costs about $4,000. Motorized probe heads are available at a cost of from $16,000–$19,000 for use on DCC CMMs.

## Noncontacting Sensors

Many industrial products and components that are not easily and suitably measured with surface-contacting devices may require the use of noncontact sensors or probes on CMMs to obtain the necessary inspection information. This may include two-dimensional parts such as circuit boards and very thin stamped parts; extremely small or miniaturized microelectronic devices and medical instrument parts; and very delicate, thin-walled products made of plastic or other light-weight materials.

There are several noncontacting sensor systems available that have been adapted either individually or in combination with the coordinate measuring process. All of these involve optical measuring techniques and include microscopes, lasers, and videocameras. The automated, three-dimensional, video-based, coordinate measuring system appears to offer many advantages because of its ability to accomplish multiple measurement points within a single video frame. Thus, it is often possible to obtain data from several hundred coordinate points with a video measuring system in the same time it would take to obtain a single-point measurement with a conventional touch-probe system. A large number of points on a part can be imaged with charged couple device (CCD) videocameras, depending on the optical view and array or pixel size of the camera. For example, one commercially available system has a field of view range from 0.02–0.30 in. (0.5–7.6 mm) with a 756 × 581 pixel array and can image the field of view at 30 video frames/second.

Video-based systems do not have the same throughput advantage for height ($Z$ axis) measurements as they do for two-dimensional ($X$-$Y$) measurements because of the time involved for focusing. On some machines, this disadvantage is overcome by integrating laser technology with the video system.

The multi-sensor coordinate measuring machine (MSCMM) shown in *Figure 9-36* incorporates three sensing technologies—optical, laser, and touch probe—for highly accurate noncontact and/or contact inspection tasks. The machine uses two quills or measuring heads to accomplish high-speed data acquisition within a four-axis ($X$, $Y$, $Z1$, $Z2$) configuration, thus permitting noncontact or contact inspection of virtually any part in a single setup.

The sensing head on the left quill in *Figure 9-36* contains an optical/laser sensor with a high-resolution CCD video and a coaxial laser. The videocamera has advanced image processing capabilities via its own microprocessor, enabling subpixel resolutions of 2 μin. (0.05 μm). The coaxial laser shares the same optical path as the video image and assists in focusing the CCD camera. This eliminates focusing errors associated with optical systems and increases the accuracy of $Z$ measurements. Single-point measurements can be obtained in less than 0.2 seconds, and high-speed laser scanning/digitizing can be accomplished at up to 5,000 points/second.

The right-hand quill of *Figure 9-36* contains the $Z2$-axis touch-probe sensor used to inspect features that are either out of sight of the optical/laser sensor or better-suited to be measured via the contact method.

The machine in *Figure 9-36* is being used to inspect a valve body with the $Z1$-axis optical/laser probe and a transmission case with the $Z2$-

Figure 9-36. *A multi-sensor coordinate measuring machine with optical, laser, and touch probes for noncontact and contact measurements.* (Courtesy WEGU, Inc.)

axis touch probe. The machine shown is a bench-top type with a maximum measuring range of about 35 in. (900 mm), about 32 in. (800 mm), and about 24 in. (600 mm) for the $X$, $Y$, and $Z1$ and $Z2$ axes respectively. Positioning of the moving elements is accomplished by backlash-free, recirculating-ball lead screws and computer-controlled direct current (DC) motors. Position information is provided by high-precision glass scales. A multi-sensor coordinate measuring machine of this type with Windows®-based computer control costs about $275,000.

## MEASURING WITH LIGHT RAYS

Light waves of any one kind are of invariable length and are the standards for ultimate measures of distance. Basically, all interferometers divide a light beam and send it along two or more paths. Then the beams are recombined and always show interference in some proportion to the differences between the lengths of the paths. One of the simplest illustrations of the phenomenon is the optical flat and a monochromatic light source of known wavelength.

The optical flat is a plane lens, usually a clear, fused quartz disk from about 2–10 in. (51–254 mm) in diameter and 0.5–1.0 in. (13–25 mm) thick. The faces of a flat are accurately polished to nearly true planes; some have surfaces within 0.000001 in. (25 nm) of true flatness.

Helium is commonly used in industry as a source of monochromatic or single-wavelength light because of its convenience. Although helium radiates a number of wavelengths of light, the portion emitted with a wavelength of 0.00002313 in. (587 nm) is so much stronger than the rest that other wavelengths are practically unnoticeable.

The principle of light-wave interference and the operation of the optical flat are illustrated in *Figure 9-37a*, wherein an optical flat is shown resting at a slight angle on a workpiece surface. Energy in the form of light waves is transmitted from a monochromatic light source to the optical flat. When a ray of light reaches the bottom surface of the flat, it is divided into two rays. One ray is reflected from the bottom of the flat toward the eye of the observer, while the other continues downward, is reflected, and loses one-half wavelength on striking the top of the workpiece. If the rays are in phase when they re-form, their energies reinforce each other, and they appear bright. If they are out-of-phase, their energies cancel and they are dark. This phenomenon produces a series of light and dark fringes or bands along the workpiece surface and the bottom of the flat, as illustrated in *Figure 9-37b*. The distance between the workpiece and the bottom surface of the optical flat at any point determines which effect takes place.

If the distance is equivalent to some whole number of half wavelengths of the monochromatic light, the reflected rays will be out-of-phase, thus producing dark bands. This condition exists at positions $X$ and $Z$ of *Figure 9-37a*. If the distance is equivalent to some odd number of quarter wavelengths of the light, the reflected rays will be in phase with each other and produce light bands. The light bands would be centered between the dark bands. Thus a light band would appear at position $Y$ in *Figure 9-37a*.

Since each dark band indicates a change of one-half wavelength in distance separating the work surface and flat, measurements are made very simply by counting the number of these bands and multiplying that number by one-half the wavelength of the light source. This procedure may be illustrated by the use of *Figure 9-37b*. There, the diameter of a steel ball is compared with a gage block of known height. Assume a monochromatic light source with a wavelength of 23.13 μin. (0.6 μm). From the four interference bands on the surface of the gage block, it is obvious that the difference in elevations of positions $A$ and $B$ on the flat is equal to $(4 \times 23.13)/2$ or 46.26 μin. ($[4 \times 0.6]/2$ or 1.2 μm). By simple proportion, the difference in elevations between points $A$ and $C$ is equal to $(46.26 \times 2.5)/0.5 = 231.3$ μin. ($[1.2 \times 64]/13 = 5.9$ μm). Thus, the diameter of the ball is $0.750 \times 0.0002313 = 0.0001734$ in. ($19.05 \times 0.005875 = 0.111919$ mm).

Optical flats are often used to test the flatness of surfaces. The presence of interference bands between the flat and the surface being tested is an indication that the surface is not parallel with the surface of the flat. The way dimensions are measured by interferometry can be explained by moving the optical flat of *Figure 9-37a* in a direction perpendicular to the face of the workpiece or mirror. It is assumed that the mirror is rigidly attached to a base, and the optical flat is firmly held on a true slide. As the optical flat moves, the distance between the flat and mirror changes along the line of traverse, and the fringes appear to glide across the face of the flat or mirror. The amount of movement is measured by counting the number of fringes and fraction of a fringe that pass a mark.

It is difficult to precisely superimpose a real optical flat on a mirror or the end of a piece to establish the end points of a dimension to be

Figure 9-37. (a) *Light-wave interference with an optical flat;* (b) *application of an optical flat;* (c) *diagram of an interferometer.*

measured. This difficulty is overcome in sophisticated instruments by placing the flat elsewhere and by optical means reflecting its image in the position relative to the mirror in *Figure 9-37a*. This creates interference bands that appear to lie on the face of and move with the workpiece or mirror. The image of the optical flat can be merged into the planes of the workpiece surfaces to establish beginning and end points of dimensions.

A simple interferometer for measuring movements of a machine tool slide to millionths of an inch (nanometers) is depicted in *Figure 9-37c*. A strong light beam from a laser is split by a half mirror. One component becomes the reference *R* and is reflected solely over the fixed machine base. The other part, *M*, travels to a reflector on the machine slide and is directed back to merge with ray *R* at the second beam splitter. Their result is split and directed to two photodetectors. The rays pass in and out of phase as the slide moves. The undulations are converted to pulses by an electronic circuit; each pulse stands for a slide movement equal to one-half the wavelength of the laser light. The signal at one photodetector leads the other according to the direction of movement.

When measurements are made to millionths of an inch (nanometers) by an interferometer, they are meaningful only if all causes of error are closely controlled. Among these are temperature, humidity, air pressure, oil films, impurities, and gravity. Consequently, a real interferometer is necessarily a highly refined and complex instrument; only its elements have been described here.

## Optical Tooling

Telescopes and accessories to establish precisely straight, parallel, perpendicular, or angled lines are called *optical tooling*. Two of many applications are shown in *Figure 9-38a, b*. One is to check the straightness and truth of the ways of a machine-tool bed at various places along the length. The other is to establish reference planes for measurements on a major aircraft or missile component. Such methods are especially necessary for large structures. Accuracy of one part in 200,000 is regularly realized; this means that a

point at a distance 100 in. (2.5 m) can be located within 0.0005 in. (13 μm). Common optical-tooling procedures are autocollimation, autoreflection, planizing, leveling, and plumbing.

### Autocollimation

*Autocollimation* is performed with a telescope having an internal light that projects a beam through the crosshairs to a target mirror as indicated in *Figure 9-38a*. If the mirror face is truly perpendicular to the line of sight, the cross-hair image will be reflected back on itself. The amount the reflected image deviates from the actual reticle image is an indication of the tilt in the target. A target may have a cross-line pattern for alignment with the line of sight. An autocollimated image is not clear for distances over 50 ft (15.2 m) and then a somewhat less accurate method must be used.

### Autoreflection

*Autoreflection* is performed with an optical flat containing a cross-line pattern mounted on the end of the illuminated telescope and focused to twice the distance of the target mirror. Then, if the mirror is perpendicular to the line of sight, the pattern of the flat is reflected in coincidence with the cross hairs in the telescope.

### Planizing

*Planizing* is composed of fixing planes at 90° with other planes or with a line of sight. This may be done from accurately placed rails on which transits are mounted in a tooling dock as indicated in *Figure 9-38b*. A transit is a telescope mounted to swing in a plane perpendicular to a horizontal axis. Square lines also may be established with an optical square or planizing prism mounted on or in front of a telescope as depicted in *Figure 9-38c*. Angles may be precisely set by autocollimating on the precisely located faces of an optical polygon as in *Figure 9-38d*.

### Leveling and Plumbing

*Leveling* establishes a horizontal line of sight or plane. This may be done with a telescope fitted with a precision spirit level to fix a horizon-

Figure 9-38. *Optical tooling: (a) autocollimation, or autoreflection, for checking the straightness of the ways on a machine-tool bed; (b) planizing to establish reference planes in a tooling dock; (c) establishing a line at right angles to a line of sight by means of an optical square; (d) collimating the faces of an optical polygon to check angles on an index table; (e) example of a plumbing operation to establish a vertical line; (f) use of a centering detector to establish alignment with the center of a laser beam.*

tal line of sight. A transit or sight level so set may be swiveled around a vertical axis to generate a horizontal plane. *Plumbing,* shown in *Figure 9-38e,* consists of autocollimating a telescope from the surface of a pool of mercury to establish a vertical axis.

## Laser Light Beam

An advanced step in optical tooling is the use of an intense light beam of a laser. A centering detector, shown in *Figure 9-38f,* has four photocells equally spaced (top and bottom and on each side) around a point. Their output is measured and becomes equalized when the device is centered with the beam. This provides a means to obtain alignment with a straight line. Squareness may be established by passing a laser beam through an optical square.

## GAGING METHODS

Many different part attributes can be measured or checked using a variety of the gaging devices previously described. The following is a discussion of a number of the most common methods used in gaging and measurement.

### Flatness

Various methods of checking flatness are available. Method selection depends on the accuracy required, size of the part, and time available to make the check. Flatness cannot be easily checked by functional methods.

One of the most widely used methods for checking flatness is direct contact. With this method, the part being checked is brought into direct contact with a reference plane of known flatness that has been covered with a thin coating of layout fluid (Prussian blue). The high spots on the part are indicated by the transfer of the blue dye. This method, however, does not lend itself to geometrics because the results are not measurable.

Another common method for checking flatness is with an indicator, stand, leveling device, and a surface plate of known flatness, as shown in *Figure 9-39*. With this method, the surface to be checked is first adjusted with the leveling device as shown by the indicator to establish the most accurate plane possible parallel to the surface plate. The entire surface then is explored with the indicator to determine the full indicator movement (FIM), which is the measure of flatness.

### Straightness

Straightness of surface elements can be checked by various methods, depending on the accuracy required and part size.

One common method of checking element straightness is with a straight-edge or surface

Figure 9-39. *Leveling a part.*

plate of known straightness and gage blocks or thickness stock. With this method, the part to be checked is supported at two places equidistant from the surface, and the differences from parallel are measured with gage blocks or an indicator set up to determine the FIM or equivalent. Cylindrical parts and narrow surfaces sometimes are placed directly on the surface plate and the differences checked with thickness stock. For very small tolerances, element straightness can be checked by placing a toolmaker's knife edge directly on the part as shown in *Figure 9-40*. Differences in element straightness are indicated by the presence and width of light gaps. Deciding element straightness by this method, however, generally does not provide measureable results.

A more practical check can be made with a measuring machine as shown in *Figure 9-41*. The tracking accuracy of these machines permits checking, with a high degree of resolution, the locations of many separate points along the surface, and provides numerical information on any out-of-straight condition. When these machines are computer-assisted, precision setups are not needed and the condition can be displayed directly.

Figure 9-40. *Checking surface element straightness with a toolmaker's straightedge.*

Figure 9-41. *Checking straightness while the part is held motionless.*

Large, accurate straightness measurements are made with precision levels, electronic levels, autocollimators, or alignment telescopes using the same methods described for checking flatness.

When straightness of an axis is specified maximum material condition (MMC), it can be checked with a functional gage. A gage to check the straightness of a center plane MMC would basically consist of two parallel surfaces 0.312 in. (7.92 mm) apart, through which the part must pass to be acceptable.

## Circularity

Although the common method for checking circularity uses a V-block or checking from centers, they are not recommended in this text. The many possible sources of error, such as lobing, angle of V-block, center misalignment, etc., contribute to an inaccurate reading of the conditions. Circularity should be checked with a precision rotating spindle, rotating table, or circular tracing instrument. Measurement is made by centering the part on the table, establishing an axis, and placing a stylus in contact with the surface of the circular cross-section. The stylus contacts the surface normal to the axis being examined to pick up, magnify, and display departures from roundness for determination of any out-of-round condition. These instruments can be equipped with auto-centering capability, eliminating the need to precisely align the part before checking. Properly equipped, these instruments can be used for the complete analysis of circular sections.

Along with any of the previously mentioned checks, elements of the surface must be measured separately to ensure that they are within the specified size tolerance and boundary of perfect form at MMC.

## Cylindricity

Cylindricity is checked with the same equipment used to check roundness, except that roundness readings must be taken at a number of sections along the entire length of the part and placed in a way to establish a common axis from which a tolerance zone can be established and measurements made to determine whether they fall within the tolerance zone specified. As with the circularity checks, elements of the cylindrical surface must be measured separately to ensure that they are within the specified size tolerance and boundary of perfect form at MMC.

## Profile of a Line or Surface

Many profiles are checked with contour gages having the opposite form of the nominal contour of the part, or limit gages representing the go and no-go sizes of the part as determined by the maximum and minimum material condition resulting from the compound effects of form and size tolerances. Neither of these methods is suitable for geometrics.

One suitable method is to compare the part with a master part conforming exactly with the basic dimensions of the part. A method to do this is shown in *Figure 9-42*.

Another method has been used to check cams and other profiles that can be rotated. Modern technology, however, now makes it possible to read these profiles directly with long-range digital reading sensors, compare readings with a perfect master in the memory of a microcomputer, and print any out-of-tolerance conditions along with their location.

*Figure 9-43* shows another generally accepted method that uses a fixture with several indicating devices mounted side by side in a plane containing the profile to be inspected. The indicators must be set with a master representing the basic profile. Conformance of the part is easily determined by reading part differences directly from the indicators or display unit.

Perhaps the best way to check small complex profiles is with an optical projector using one of the methods previously described. One note, however, is that low spots cannot be detected. Profiles also can be checked on an optical comparator with a contour projector tracer attachment as shown in *Figure 9-44*.

Figure 9-42. *Checking profile—indicator following a master profile.*

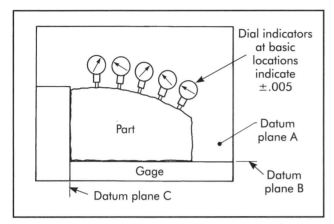

Figure 9-43. *Checking profile—fixed indicators set to master part.*

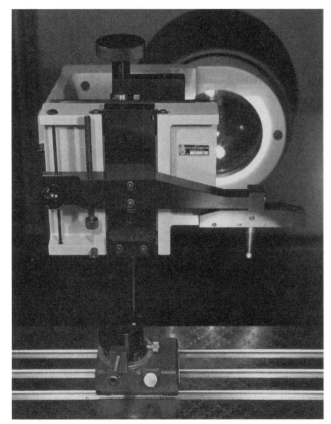

Figure 9-44. *Optical comparator with a contour projector tracer attachment. (Courtesy Optical Gaging Products, Inc.)*

## Perpendicularity (Squareness)

The most common method of checking perpendicularity of surfaces is by direct comparison with gages of known squareness, such as the precision square. To make this check, the square and the part are placed in contact with each other while resting on a surface plate. Out-of-squareness is determined by measuring the gap between the square and sections along the part with a thickness gage, as shown in *Figure 9-45.*

Another direct contact method uses a direct reading cylindrical square, with one face-off angle and dotted curves and graduations etched on the surface indicating the amount of out-of-squareness for the bounded area. In use, the part and the cylindrical square are placed on a surface plate and brought together while rotating the cylindrical square to produce the smallest light gap. The number of the topmost dotted curve in contact with the part surface indicates the squareness error.

Another widely used surface-plate method for checking perpendicularity is as follows: check

Figure 9-45. *Checking squareness with a square. (Courtesy L.S. Starrett Co.)*

whether two preselected points on the vertical surface of the part are in a common plane at right angles to the surface plate on which the part and gage are resting. Measurement is made by contacting the part, with the spherical base of the comparator square and the probe of the indicator mounted in the stand, as shown in *Figure 9-46*. Before the measurement is made, the indicator is set to zero with the aid of the cylindrical square shown in *Figure 9-46*. With this method, multiple checks must be made at varying heights to get an indication of the complete surface condition. The squareness gage, shown with its master in *Figure 9-47*, measures in a similar way except that the part surface can be completely indicated in a vertical manner due to the precision guideways built into the gage. Accuracy is maintained by comparing and adjusting the squareness to the master gage.

Large, accurate perpendicular measurements generally are made with an autocollimator, optical square, and reflector stand using a setup such

Figure 9-47. *Squareness gage and master. (Courtesy PMC Industries)*

Figure 9-46. *Transfer inspection of squareness using cylindrical and comparator squares.* (Courtesy Taft-Peirce)

as the one shown in *Figure 9-48*. Sometimes, special fixtures are used such as those shown in *Figure 9-49*, which permit the concerned feature to be searched with indicating depth gages. None of these methods, however, are applicable to the other examples shown.

Examples specifying perpendicularity, cylindrical size feature regardless of feature size (RFS), and noncylindrical feature RFS cannot be checked by functional gaging. They also would be difficult to check economically by other means, since a number of direct or differential measurements would have to be made, depending on the feature

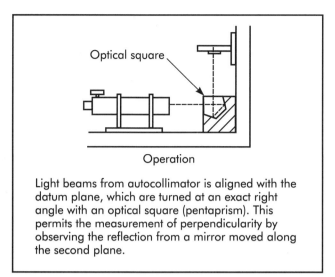

Operation

Light beams from autocollimator is aligned with the datum plane, which are turned at an exact right angle with an optical square (pentaprism). This permits the measurement of perpendicularity by observing the reflection from a mirror moved along the second plane.

Figure 9-48. *Checking perpendicularity with an autocollimator.*

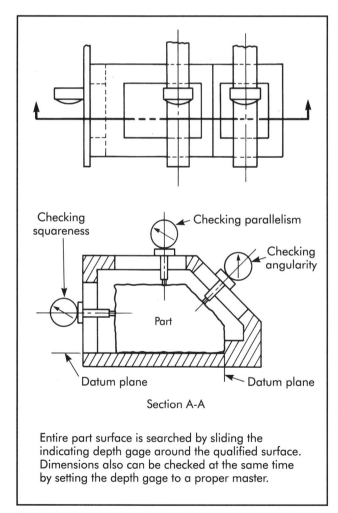

Section A-A

Entire part surface is searched by sliding the indicating depth gage around the qualified surface. Dimensions also can be checked at the same time by setting the depth gage to a proper master.

Figure 9-49. *Checking squareness, parallelism, and angularity with indicating depth gages.*

basic form errors, just to determine the attitude of the axis or median plane. For small quantities, the part could be checked by standard layout methods with the help of a simple staging fixture to quickly arrange the datum surface so the concerned feature was properly lined up with the surface plate to easily check with a height stand and indicator, as shown in *Figure 9-50*. For production quantities, fixtures generally are designed with air or electronic probes properly positioned and connected to determine the squareness of the axis or median plane to the datum surface.

The example specifying perpendicularity, cylindrical size feature, and MMC could be checked with a functional gage as shown in *Figure 9-51*.

## Angularity

Angularity tolerance cannot be checked with functional gaging. Angularity control can be called out at MMC or least material condition (LMC). RFS only applies to feature sizes.

Most short-run parts are checked for angularity with standard surface-plate methods. Sine plates or simple staging fixtures are used to place the datum surface of the concerned feature in proper alignment with the surface plate.

For production quantities, fixtures are generally designed with specifically located and interrelated air or electronic probes that determine, then display any out-of-tolerance condition of the feature angle being checked.

Figure 9-50. *Checking perpendicularity with a height stand and an indicator.* (Courtesy Scherr-Tumico)

Figure 9-51. *Functional gage to check perpendicularity, cylindrical size feature, and MMC.*

Large, accurate angularity measurements generally are made with an autocollimator or special fixtures and indicating depth gages in a manner similar to those shown for checking perpendicularity in *Figures 9-48* and *9-49*.

## Parallelism

Parallelism of a surface generally is checked by placing the datum of the part on a surface plate and searching for any out-of-parallel conditions with a height stand and indicator.

Many times, a part, because of its geometry, must be placed in a special fixture to check perpendicularity, angularity, location, etc. In these cases, a check for parallelism generally is incorporated in the same fixture shown in *Figure 9-49*.

When parallelism concerns cylindrical size features or datums, the inspection becomes more difficult, especially when the feature, datum, or both are RFS. Here again, standard layout methods using a surface plate, height stand, and indicator can be used for small quantities as long as the inspector understands the meaning expressed by the feature control symbol. *Figure 9-52* shows an inspection gage design approach that can be used to check the parallelism of these cylindrical size features.

## Runout

Checking runout on a part on which the datum axis is established by two part centers is relatively simple and easily understood. Low-production parts are generally mounted between the centers of a bench center and rotated 360° with a dial indicator (*Figure 9-22*) in contact with the surface to be checked. For high production, special fixtures with several air, electronic, or mechanical indicators are generally more practical.

Parts where the datum axis is established by two functional diameters generally are checked in the same manner, except that the datum is established by closing in on the datum features with expanding chuck-like devices, as shown in *Figure 9-53*, and taking the reading in the same manner as from centers. Another method widely used with electronic gaging allows the part to be mounted in any suitable fashion, while indicating the datums as well as the features to be checked. The readings then are adjusted electronically within the equipment and the runout can be displayed directly.

Methods for checking the runout on parts where the datum is established from a diameter, or a combination of datum diameter and functional face, are shown in *Figures 9-54*, *9-55*, and *9-56*. The cylindrical datums must be established by chuck-like devices or electronically as noted previously since runout always is RFS.

When checking runout, all features of the part also must be measured separately to ensure they are within the specified size limits and the boundary of perfect form at MMC.

## Position

When position tolerance is applied on an MMC basis, it allows functional gaging to be used.

*Figure 9-57* shows two identical parts containing clearance holes assembled with two ⌀ 0.50 in.

Figure 9-52. *Gage for checking parallelism—cylindrical size feature.*

Figure 9-53. *Hydraulically actuated arbors and chucks.*

Figure 9-54. *Checking runout from an OD datum.*

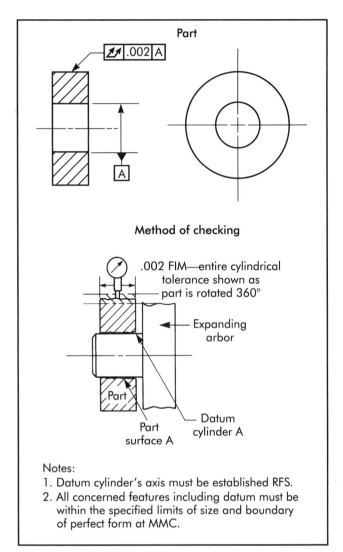

Figure 9-55. *Checking runout from an ID datum.*

(12.7 mm) bolts. Each part can be dimensioned and tol-eranced as shown, with MMC specified after the hole location tolerance. Also shown is a hole relation gage for each of the parts. Hole relation gages check hole-to-hole relationship, not hole location to some other part feature. The gage could contain fixed-pin gage features in place of the separate gage pins shown, which fit tightly into basically located bushings in the gage.

Locational and squareness tolerance are the actual differences in size between the gage pin and the clearance hole feature. Since feature size will vary from hole to hole, part to part, and process to process, the true tolerance is a variable since the ∅ 0.50 in. (12.7 mm) bolt or gage pin and the interchangeable design requirement is still a constant. The positional tolerance is ∅ 0.10 in. (2.5 mm).

Substituting RFS for the MMC callout would make gaging difficult, if not impossible, since the gage pin must always be 0.01 in. (2.5 mm) smaller than the actual hole size. Eleven or more gage pins would be required for each hole (∅ 0.500 [12.70 mm]), ∅ 0.501 in. [12.73 mm], etc.) so that, for instance, if a particular hole measured ∅ 0.512 in. (13.00 mm), the ∅ 0.502 in. (12.75 mm) gage would be used.

*Table 9-2* shows how the true tolerance varies with hole size. All tolerances greater than ∅ 0.01 in. (2.5 mm) are bonus tolerances and not specifically allowed in any system other than the positional tolerancing system.

*Figure 9-58* shows the tolerances and gages for the manufacture and assembly of two different parts. One part contains clearance holes and the other holes tapped for ∅ 0.50 in. (12.7 mm) bolts.

**Part**

**Method of checking**

.002 FIM—entire cylindrical tolerance shown as part is rotated 360°

Expanding arbor

Part surface A

Datum cylinder B

Datum surface A

Part surface B

Notes:
1. Datum 'A' must rest on arbor face and datum axis 'B' must be established RFS.
2. All concerned features including datum must be within the specified limits of size and boundary of perfect form at MMC.

Figure 9-56. *Checking runout from a functional face and ID datums.*

Also shown is the feature relation gage required for the tapped part. Gage thickness or bushing height must be at least the maximum thickness of the untapped part to guarantee the bolts will be properly located and square for assembly. The two go-thread gages simulate the bolts at assembly. Gage bushing size is determined by adding the ∅ 0.01 in. (2.5 mm) positional tolerance specified for the tapped features to the bolt or tap size. Stepped gage pins with go threads may be used to take advantage of standard bushing size as long

as a 0.01 in. (2.5 mm) difference in size is maintained between the gage bushing and that portion of the gage pin that lies within the bushing.

## BASIC DESIGN RULES FOR POSITIONALLY TOLERANCED PARTS

Two basic rules govern the design of gages for positionally toleranced parts. These principles apply regardless of the number of features that make up an interchangeable pattern.

- For parts with internal features, the nominal gage feature size is directly determined by subtracting the total positional tolerance specified at MMC from the specified MMC size of the feature to be gaged for location.
- For parts with external features, the nominal gage feature size is directly determined by adding the total positional tolerance specified at MMC to the specified MMC size of the feature to be gaged for location.

*Figure 9-59* shows a workpiece with four holes that must be located from the specific center (datum) of the workpiece regardless of the actual workpiece size (RFS). The design specification drawing includes the exact pickup points so the same center can be repeatedly found. Also shown is a hole relation gage that uses four dial indicators to determine and correctly position the datum for the hole-gaging operation.

*Figure 9-60* shows a similar workpiece with the less critical MMC requirement on the datum diameter requiring that the holes be located from the center of the datum diameter when the datum is at MMC. Also shown is the design of a gage to check the part functionally.

*Figure 9-61* shows another workpiece in which the holes never need to be exactly located from the center of a specified datum feature. A gage fit allowance has been specified that is directly reflected in the size of the datum gage feature since the gage is ∅ 0.995 in. (25.27 mm) and differs from the MMC size of the part (∅ 1.000 in. [25.40 mm]) by 0.005 in. (0.13 mm). A fairly large allowance could be specified if the datum was merely a convenient starting place for manufacturing.

Figure 9-57. *Two parts with clearance holes assembled with bolts.*

Table 9-2. Variable tolerances allowed by MMC

| Feature Diameter (Ø) | Positional and Squareness Tolerances Diameter (Ø) |
|---|---|
| Ø .510 (MMC or most critical size) | Ø .010 at MMC (tightest tolerance) |
| Ø .511 | Ø .011 |
| Ø .512 | Ø .012 |
| Ø .513 | Ø .013 |
| Ø .514 | Ø .014 |
| Ø .515 | Ø .015 |
| Ø .516 | Ø .016 |
| Ø .517 | Ø .017 |
| Ø .518 | Ø .018 |
| Ø .519 | Ø .019 |
| Ø .520 (LMC or least critical size) | Ø .020 at LMC (loosest tolerance) |

A pattern of interchangeable features (holes) is the most critical feature on the part shown in *Figure 9-62*, but it is not locationally critical in relation to any single datum feature. The 0.30 in. (7.6 mm) minimum breakout specification is the result of a stress analysis, and is an end-product requirement. No single datum is specified, and the 0.30 in. (7.6 mm) minimum specification can be readily gaged with a tubing micrometer or a fork gage as shown.

*Figure 9-63* shows a workpiece with seven holes. The specified positional tolerance includes the location and angularity tolerances for each radial hole. Also shown is the gage required for checking the part. In use, all seven gage pins must go through the part at one time. In designating the datum, if RFS callouts had been used instead of MMC for the diameter and width of the slot, the gage would be required to center on the two datum features. As a result, one gage pin could be used to individually qualify each hole in reference to the datum.

Position tolerance is often used to control the location of coaxial mating part features, as shown in *Figure 9-64*, along with the gages to check them.

*Figure 9-65* shows the method used to check bore alignment functionally when specified at MMC. Another common application of position tolerance MMC is on noncylindrical mating parts, such as those shown in *Figure 9-66*. Occasionally, a projected tolerance zone will be specified to prevent interference at assembly, as shown in *Figure 9-67*, along with its interpretation.

A case where it was desired to place all the usable size and location tolerances into the size limit by stating the position tolerance as zero (0.000) at MMC is shown in *Figure 9-68*, along with the functional gage to check the part. It should be noted that the gage eliminates the need for a go-plug-gage check.

## Concentricity

Since concentricity tolerance is always on an RFS basis, it requires that the datum axis be established by chuck-like devices that close in on the part, or by electronic means previously described under runout, which allows the part to be mounted in any suitable fashion while indicating the datums and the features to be checked. Results are electronically interpreted and any eccentricity is displayed directly on the readout. All features must be measured separately to ensure that the part conforms to specified size tolerances.

## Symmetry

Symmetry tolerances will be specified on noncylindrical parts on an RFS basis. Symmetry can be checked with standard surface-plate methods. However, for high production the multi-sensor-type electronic gaging methods mentioned previously for checking other RFS features are recommended.

## BIBLIOGRAPHY

ANSI B4.4M-1981 (R1994). *Inspection of Workpieces*. Washington, D.C.: American National Standards Institute.

ANSI/IEEE 268-1992. *Metric Practice*. Washington, D.C.: American National Standards Institute.

Figure 9-58. *Tolerances and gages for the manufacture and assembly of two different parts.*

Figure 9-59. *Workpiece with four holes that must be located from the specific center of the workpiece.*

Figure 9-60. "Shake" gage to check a less critical (MMC) OD datum feature.

Figure 9-61. *Gaging workpiece features not specifically located relative to the datum: (a) workpiece; (b) hole location (shake) gage.*

Figure 9-62. *Gaging an end product: (a) workpiece; (b) gage for checking required edge distance; (c) gage for checking hole location.*

Figure 9-63. *Gaging radial hole patterns: (a) workpiece; (b) hole location and squareness gage.*

Figure 9-64. *Gaging coaxial mating parts at MMC.*

ASME B1.2-2001. *Gages and Gaging for Unified Inch Screw Threads.* New York: American Society of Mechanical Engineers.

ASME B4.2-1999. *Preferred Metric Limits and Fits.* New York: American Society of Mechanical Engineers.

ASME Y14.38-1999. *Abbreviations and Acronyms Revision and Redesignation of ASME Y1.1-1989.* New York: American Society of Mechanical Engineers.

ASME Y14.7.2-1999. *Gear and Spline Drawing Standards—Part 2: Bevel and Hypoid Gears.* New York: American Society of Mechanical Engineers.

ASME Y14.3M-1999. *Multiview and Sectional View Drawings.* New York: American Society of Mechanical Engineers.

ASME Y14.2M-1998. *Line Conventions and Lettering.* Reaffirmed 1999. New York: American Society of Mechanical Engineers.

ASME Y14.6-1998. *Screw Thread Representation.* Reaffirmed 1999. New York: American Society of Mechanical Engineers.

ASME Y14.6aM-1998. *Screw Thread Representation (Metric Supplement).* Reaffirmed 1999. New York: American Society of Mechanical Engineers.

ASME Y14.7.1-1998. *Gear Drawing Standards— Part 1: For Spur, Helical, Double Helical, and Rack.* Reaffirmed 1999. New York: American Society of Mechanical Engineers.

ASME B89.3.1-1997. *Measurement of Out-of-roundness.* Reaffirmed 1999. New York: American Society of Mechanical Engineers.

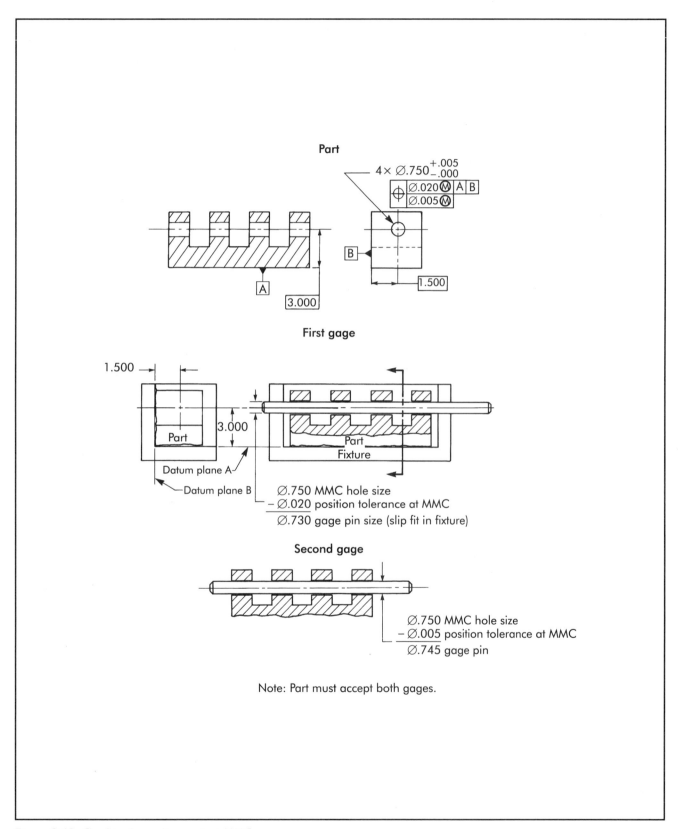

Figure 9-65. *Checking bore alignment at MMC.*

Figure 9-66. *Checking noncylindrical mating parts at MMC.*

Figure 9-67. *Projected tolerance zone.*

ASME Y14.8M-1996. *Castings and Forgings.* Reaffirmed 1999. New York: American Society of Mechanical Engineers.

ASME Y14.36M-1996. *Surface Texture Symbols.* Reaffirmed 1999. New York: American Society of Mechanical Engineers.

ASME B46.1-1995. *Surface Texture (Surface Roughness, Waviness, and Lay).* Reaffirmed 1999. New York: American Society of Mechanical Engineers.

ASME B94.6-1995. *Knurling.* Reaffirmed 1999. New York: American Society of Mechanical Engineers.

ASME Y14.1-1995. *Decimal Inch Drawing Sheet Size and Format.* Reaffirmed 1999. New York: American Society of Mechanical Engineers.

ASME B5.10-1994. *Machine Tapers—Self-holding and Steep Taper Series.* Reaffirmed 1999. New York: American Society of Mechanical Engineers.

ASME Y14.5.1M-1994. *Mathematical Definition of Dimensioning and Tolerancing Principles.* Reaffirmed 1999. New York: American Society of Mechanical Engineers.

ASME B94.11M-1993. *Twist Drills.* Reaffirmed 1999. New York: American Society of Mechanical Engineers.

## REVIEW QUESTIONS

1. Define tolerance.
2. Convert a bilateral tolerance of ±0.007 in. (±0.18 mm) to a tolerance of position.
3. Convert a tolerance of position of 0.014 in. (0.36 mm) diameter to a bilateral tolerance.
4. Using the unilateral system of tolerancing, specify the correct nominal sizes and the gagemaker's tolerance to be applied or go and no-go working-ring gages to check a shaft with a diameter of 0.8805 in. ±0.0003 in. (22.4 mm ±0.008 mm). Include 5% wear allowance.

Figure 9-68. *Positional tolerance of zero at MMC.*

5. What bilateral tolerance would be applied to a master gage used to set a working gage measuring a dimension with a tolerance of ±0.002 in. (±0.05 mm)?

6. Why is the unilateral system of gage-tolerance allocation preferred over the other systems?

7. What is the largest specimen diameter that could be completely viewed on a 30 in. (76.2 cm) optical comparator at 10× magnification? at 20×? at 31-1/4×? at 50×? at 62-1/2×? At 100×?

8. What gage material is used for a medium-production run? For higher-volume production? For situations requiring high accuracy or high wear applications?

9. What are the materials most used for surface plates?

10. What are the limitations of a screw-pitch gage?

11. What is the most widely used instrument for precise measurement?

12. What is the meaning of FIM?

13. What are the basic elements of electronic gages?

14. What are the two most common structural configurations for CMMs?

15. What is the primary application of optical flats?

# ◄ 10 ►

# Tool Design for Joining Processes

Joining processes generally fall into two classes: physical and mechanical. In the physical joining process, parts are made to join along their contacting surfaces through the application of heat or pressure, or both. Often, a filler material is added, with the edges losing their identity in a homogeneous mass. Mechanical joining ordinarily does not involve changes in composition of the workpiece material. The edges of the pieces being joined remain distinct.

Two pieces of wood nailed together are joined mechanically. The same two pieces of wood could be joined physically by an adhesive. At the exact center of the joint, only the adhesive would be found. The adhesive would penetrate into the pores of the wood for some distance, and the workpiece edges would no longer exist as true entities, but become a blend of wood and glue.

Joining processes may require tooling to hold the parts in correct relationship during joining. Another function of the tooling is to assist and control the joining process. Often, several parts may be joined mechanically and physically. Thus, two workpieces may be bolted together to ensure alignment during subsequent welding. At times, mechanical joining may be considered as the tooling method for final physical joining.

## TOOLING FOR PHYSICAL JOINING PROCESSES

Physical joining processes generally cannot be performed without tooling because the high temperatures required make manual positioning impractical. The tooling must hold the workpieces in correct relationship during joining, and assist and control the joining process by affording adequate support. Tooling used for hot processes must not only withstand the temperatures involved, but in many cases either accelerate or retard the flow of heat. Hot fixtures must be designed so their heat-expanded dimensions remain functional.

### Welding Fixtures

The purpose of a welding fixture is to hold the parts to be welded in the proper relationship both before and after welding. Typically, distortion is a by-product of the welding process due to the amount of heat input into the parts. *Figure 10-1* reflects the degree of heat input associated with a given welding process. Often, a fixture will maintain the proper part relationship during welding but distort after removal from the fixture. The degree and type of distortion can be a function of the welding process alone or in combination with the materials to be welded, the fixturing, or other weld-related processing such as preheating and post-weld heat treating. *Figure 10-2* shows typical types of distortion occurring during welding, and *Figure 10-3* the distortion potential based on material type. It is said that an imperfect fixture is required to make a perfect part. In other words, the fixture needs to be designed to misalign parts so distortion caused by welding brings the parts into the correct relationship to each other.

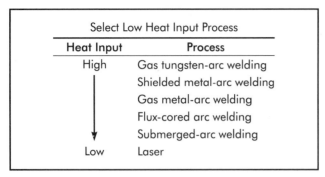

Figure 10-1. *Degree of heat input based on process.*

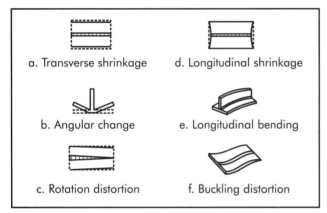

Figure 10-2. *Typical types of distortion occurring during welding.*

Figure 10-3. *Distortion potential based on material type.*

Good fixture design itself will largely determine product reliability. Major fixture design objectives, some basic and others special, include:

- holding the part in the most convenient position for welding;
- providing proper heat control of the weld zone;
- providing suitable clamping to reduce distortion;

- providing channels and outlets for welding atmosphere;
- providing access for the welding process, and
- providing for ease of operation, including part loading and unloading.

Other factors include:

- cost of fixture;
- size of the production run and rates;
- adaptability of available welding equipment;
- complexity of the weld;
- quality required in the weldment;
- welding process to be employed;
- conditions under which the welding will be performed;
- dimensional tolerances;
- material(s) to be welded;
- part-surface finish requirements (scratches, spatter, etc.), and
- coefficient of expansion and thermal conductivity of both workpiece and tool materials.

The tool designer may have to be familiar with gas-, arc-, and resistance-welding processes. Each process will require individual variations to the general design factors involved. For instance, heat dissipation is not a critical factor in some of the welding processes. Expansion is not a problem if outer ends of the workpiece are not restricted.

## Gas-welding Fixtures

The general design of a gas-welding fixture must take into consideration the heating and cooling conditions. Minimal heat loss from the welding area is required. If heat loss is too rapid, the weld may develop cracks. Heat loss by materials, particularly aluminum and copper, must be carefully controlled. To accomplish this, large fixture masses should not be placed close to the weld line. The contact area and clamps should be a minimum size consistent with the load transmitted through the contact point. When welding copper and aluminum, the minimum contact surface often permits excessive heat loss, and prevents good fixture welds. This necessitates tack welding the fixtured parts at points most distant from the fixture contact points, with the rest of the welding done out of the fixture. Excessive distortion may result with this method, and the part may subsequently need to be relieved of stress by proper heating and cooling techniques.

One of the simplest fixtures for gas welding is the gravity-type fixture shown in *Figure 10-4*. This design eliminates excess fixture material from the weld area to minimize heat loss, while providing sufficient support and locating points. The design also permits making welds in a horizontal position.

*Figure 10-5* shows another simple form of gas-welding fixture, which holds two flat sheets for joining. C-clamps hold the workpieces to steel support bars. Alignment is performed visually or with a straight edge. A heat barrier of alumina-ceramic fiber is placed between the workpieces and steel bars. Hold-down plates are used to keep workpieces flat and control distortion. If the parts to be welded have curved surfaces, the supporting bars and hold-down plates may be machined to match the part.

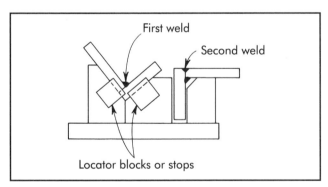

Figure 10-4. *Simple welding fixture using gravity to help locate parts.*

Figure 10-5. *Workpieces with simple fixturing for gas-welding operations.*

Simple parts may be properly located or positioned in a fixture visually. Positive location is desirable as the shape of the workpiece becomes complex or production rate increases. The same locating methods used in workholder design can be readily adapted for designing welding fixtures.

The selection of material for gas-welding fixtures is governed by four factors:

- part print tolerances;
- material heat resistance;
- heat-transfer qualities, and
- the fixture rigidity required to ensure workpiece alignment accuracy.

The fixture material should not be affected in the weld zone and should prevent rapid heat dissipation from the weld area. Commonly used fixture materials include cast iron, carbon steel, and stainless steel.

## Arc-welding Fixtures

Arc welding concentrates more heat at the weld line than gas welding. The fixtures for this process must provide support, alignment, and restraint on the parts, while permitting heat dissipation.

The most important consideration for arc welding is whether the parts will be manually or automatically welded. In manual welding, it is sufficient just to locate the parts in relationship to each other; the welder can find the weld joints and properly place the weld in location. For automatic welding, assuming no use of special sensors, the weld joints themselves must be fixtured in the correct location. Automation typically locates the anticipated welds in a defined place, whether the weld joint is supposed to be there or not.

Depending on the weld-joint configuration, either partial or full penetration welds are required. If full penetration is necessary, a backing bar often is necessary to support the weld pool. In other words, welding takes place against the backing bar directly, or the bar serves to provide shielding to the backside of the weld and pull heat out of the weld. The molten weld pool does not actually contact the backup bar in certain situations. If a backing bar is used, some of the more important design considerations for arc-welding fixtures are as follows:

- The fixture must exert enough force to prevent or minimize parts from moving out of alignment during the welding process. Too much restraint can lead to weld cracking since the residual stresses want to relieve themselves and the solidifying weld metal provides the path of least resistance. This force must be applied at the proper location by a clamp supported with a backing bar, which should be parallel to weld lines.
- Promote heat dissipation from the weld line.
- Support the molten weld, govern the weld contour.
- Protect the root of the weld from the atmosphere.

Backing bars usually are made from solid metal or ceramics. A simple backup could be a rectangular bar with a small groove directly under the weld. In use, the backup would be clamped against the part to make the weld root as airtight as possible. Some common shapes are shown in *Figure 10-6*. *Figure 10-7* shows a backing bar in position against a fixed workpiece.

The size of the backup bar depends on the metal thickness and material to be welded. A thin weldment requires larger backup to promote heat transfer from the weld. A material with greater heat-conducting ability requires less backup than that required for a comparable thickness of a poor conductor.

*Figure 10-8* shows backing bars designed for use with gas, which may be used to blast (*a*), flood

Figure 10-7. *Workpiece with simple fixturing for arc-welding operations.* (Courtesy Alloy Rods Division, Chemetron Corp.)

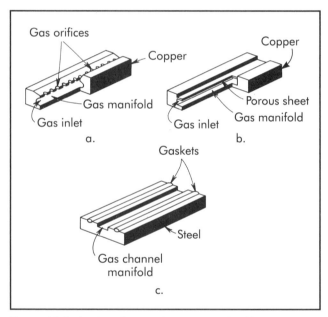

Figure 10-8. *Backing bars with provisions for (a) directed gas flow, (b) diffused gas flow, and (c) pressurized gas.*

(*b*), or concentrate in the weld area (*c*). Backup bars may be made of copper, stainless steel (used for tungsten inert gas), titanium, ceramic, or a combination of several metals (sandwich construction).

## RESISTANCE WELDING

One of the simplest and most economical processes for joining two or more metal parts is resistance welding. In resistance welding, fusion is produced by heat generated at the junction of the

Figure 10-6. *Typical backing bars.* (Courtesy Alloy Rods Division, Chemetron Corp.)

workpieces by local resistance to passage of large amounts of electric current and applied pressure. *Figure 10-9* illustrates the elements of a resistance-welding machine. Resistance-welding processes used for low-cost, high production are shown in *Figure 10-10* and include spot, projection, seam, pulsation, flash-butt, upset-butt, and cross-wire welding.

Other popular joining techniques include ultrasonic, high-frequency resistance, foil-seam, magnetic-force, percussion, friction, thermo-pressure, diffusion-bond, electro-slag, electron-beam, plasma-arc, and laser welding.

## Spot Welding

Spot welding, when compared to riveting (both produce a similar outcome), may be considerably faster and less expensive since there is no need to drill holes and insert rivets. *Figure 10-11* illustrates typical spot-welded joints and electrode shapes that can be produced on standard welders. *Figure 10-12* illustrates the principle of series welding where two welds are made with each stroke of the welder without any markings, indentations, or discoloration on one side of the assembly. *Figure 10-13* illustrates the principle of indirect welding ordinarily used where the

Figure 10-10. *Resistance-welding methods.*

Figure 10-11. *Typical spot-welded joints.*

Figure 10-9. *Elements of a typical resistance welder: (1) housing; (2) low-voltage, high-current transformer; (3) primary coils; (4) tap switch; (5) welding timer; (6) power interrupter; (7) foot switch; (8) secondary loop; (9) bands from electrodes to secondary; (10) electrodes; (11) cylinder that exerts pressure on work.*

Figure 10-12. *Assembly showing series-welded joint.*

Figure 10-13. *Spot-welded assembly showing a typical joint design for an indirect weld.*

welding current must pass through the side or ends of parts due to the design. The welding-electrode tip size directly affects the size and shear strength of the weld. Tip area is a function of the workpiece material gage. For thin sheets (up to 0.250 in. [6.35 mm]) the diameter of the electrode tip can be calculated by:

$$d = 0.1 + 2t \qquad (10\text{-}1)$$

where:

$d$ = diameter of the electrode tip, in. (mm)
$t$ = material thickness, in. (mm)

For thick material, use (0.25–2.00 in. [6.4–50.8 mm]):

$$d = t \qquad (10\text{-}2)$$

## Projection Welding

In projection welding, embossments or projections are formed on one or both workpieces for heat localization. Dimpled workpieces are placed between plain, large-area electrodes. Projection welding provides increased strength with reduced electrode maintenance.

## Seam Welding

In seam welding, the material to be welded passes between two rotating disk electrodes. As the current is turned on and off a continuous tight seam is produced.

## Pulsation Welding

In pulsation welding, the current is applied repeatedly to make a single weld while the pressure is applied. This process will produce a better weld for heavier material.

## Flash-butt Welding

In flash-butt welding, the work is clamped in dies, the current is turned on, and two joints are brought together by means of cam control to establish flashing and upsetting, followed by discontinuation of the welding current. *Figure 10-14* illustrates a flash welding fixture.

## Upset-butt Welding

Upset-butt welding differs from flash-butt welding in that pressure is continuous through the clamping dies after the welding current is applied. In this manner, heat is developed entirely from the resistance effect of the current.

## Function of Electrode Holders and Electrodes

In resistance welding, parts are positioned between electrodes that exert heavy pressure, conduct the current into the materials to be welded, and dissipate heat from the outer surface of the materials being welded. Holders and adapters are mounted in the machine so the electrode's position can be adjusted to suit a particular workpiece. Wherever possible, electrode tips should be water-cooled. *Figure 10-15* illustrates typical standard electrode tips. The design of welding electrodes and the material from which they are made are of great importance. For increased life, the design must provide sufficient strength with adequate heat conduction and cooling. Electrode nose shapes are illustrated in *Figure 10-16*.

## Resistance Welding Fixtures

There are two general types of fixtures for resistance welding. The first is a fixture for welding in a standard machine having a single electrode. The second is a fixture and machine designed as a single unit, usually to attain a high production rate.

Figure 10-14. *Flash-butt welding.*

Figure 10-15. *Typical standard electrode tips and operations.*

Certain design considerations apply to fixtures for resistance welding:

• keep all magnetic materials, particularly ferrous materials, out of the throat of the welding machine;

• insulate all gage pins, clamps, locators, index pins, etc.;

• protect all moving slides, bearings, index pins, adjustable screws, and any accurate locating devices from flash;

• give consideration to the ease of operation and protection of the operator;

• provide sufficient water cooling to prevent overheating, and

• bear in mind that stationary parts of the fixture and work are affected by the magnetic field of the machine.

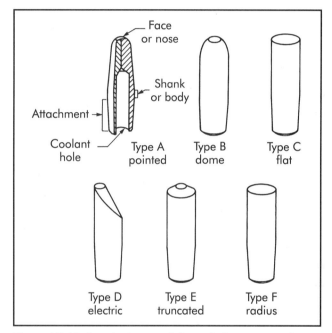

Figure 10-16. *Standard types of electrode face or nose shapes. (Type D was formerly called offset.)*

Workholder parts and clamp handles of nonmagnetic material will not be heated, distorted, or otherwise affected by the magnetic field.

Other considerations will affect the design of resistance welding fixtures and the machine if high production is required.

• A fixture loop or throat is the gap surrounded by the upper and lower arms or knees containing electrodes and the base of the machine that houses the transformer. This gap or loop is an intense magnetic field, within which any magnetic material will be affected. In some cases, materials have actually been known to melt or puddle. Power lost by unintentional heating of fixture material will decrease the welding current and lower the welding efficiency. This power loss may sometimes be used to advantage, for example, if the current is burning the parts to be welded, the addition of a magnetic material in the throat will increase the impedance, lower the maximum current, and halt the burning of parts.

- The throat of the machine should be as small as possible for the particular job.
- Welding electrodes should be easily and quickly replaceable.
- Water for cooling should be circulated as close to the tips as possible.
- Adjustment should be provided for electrode wear.
- Current-carrying members should run as close to the electrodes as possible, have a minimum number of connections or joints, and be of adequate cross-sectional area.
- Provide adjustment for electrode wear.
- Check the welding pressure application.
- Have knockout pins or strippers if there is a tendency of the electrode to stick to the electrode face. These may be leveraged or air-operated.

### General Design Considerations

Simple fixtures may locate a part visually with scribed lines as a guide. This is quite similar to locating parts for gas welding. For higher production, a quicker locating method is needed. A locating land may be incorporated in the fixture to accurately establish the edge position of the part to be welded (*Figure 10-17*). In some cases, set blocks may be used in place of a locating land (*Figure 10-18*).

When welding a variety of similar parts with different dimensions, set blocks have a distinct advantage over the land method of locating. With proper design, set blocks can be interchangeable to accommodate varying workpieces, and dowel pins may be used as locators (*Figure 10-19*). Other means of locating include V-blocks, adjustable clamps, rest buttons and pads, spring plungers, and magnets.

Figure 10-17. *Locating lands.*

Figure 10-18. *Set-block locators.*

Figure 10-19. *Dowel-pin locators.*

### Clamping Design Considerations

Clamps used in welding fixtures must hold parts in the proper position and prevent their movement due to alternate heating and cooling. Clamping pressure should not deform the parts to be joined. Clamps must be supported underneath the workpiece (*Figure 10-20*). Due to the heat involved, deflection could remain in the part.

Quick-acting and power-operated clamps are recommended to achieve fast loading and unloading. C-clamps may be used for low production volume. Power clamping systems may be direct-acting or work through lever systems (*Figure 10-21*). In heavier plate applications, urethane tip or spring-loaded clamp spindles are recommended to compensate for plate thickness variations.

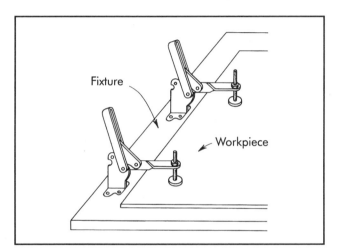

Figure 10-20. *Typical clamp installation with the fixture supporting the workpiece directly beneath the clamps.*

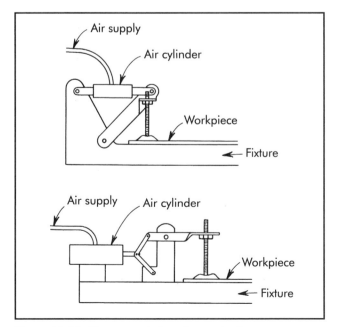

Figure 10-21. *Air-actuated clamping methods.*

## LASER WELDING FIXTURES

Because of the laser's high heat intensity, it can be used for a variety of processes, such as cutting, drilling, metal forming, heat-treating, cladding, brazing, and welding. There are many types of lasers, but the two mainly used in industry are $CO_2$ and neodymium: yttrium aluminum garnet (Nd:YAG) lasers. These lasers are capable of producing up to 6 kW (21,600 kJ) for the Nd:YAG and up to 50 kW (180,000 kJ) for the $CO_2$.

Lasers deliver energy in the form of light. This light is in the visible spectrum for many laser systems, but the $CO_2$ and Nd:YAG lasers produce light in the invisible, infrared region. Because of the large amounts of power needed to operate industrial lasers and the damaging characteristics of the emitted light to the eye and skin, great care must be taken when performing maintenance on a laser or working in the vicinity of the emitted laser beam.

When welding, power can be delivered in pulses or in a continuous wave operation and no contact is needed with the workpiece to create a welded assembly (*Figure 10-22*). The raw laser beam can be focused on a workpiece through a lens or parabolic mirror. Focusing the beam produces a very intense heat, which causes localized melting of the metal and thus the potential for high processing rates. Laser welding is capable of penetrating depths up to approximately 0.50-in (12.7 mm) in many steels and can be used simultaneously with other processes, such as gas metal-arc welding (GMAW) (or metal inert-gas-shielded arcwelding [MIG]) to create greater depths of penetration. This method, known as a hybrid process, also enables tailoring the weld to

Figure 10-22. *Laser welding.*

match desired mechanical properties with the addition of filler material.

The laser should be mounted in a firm structure to prevent vibration. The design should allow sufficient room between the worktable and focusing optic to accommodate various types of positioning devices.

The tooling used to position parts for laser welding is similar to other areas of welding. However, because of the lower overall heat input into the material with laser processing, lower thermally induced stresses (low distortion) is present and less massive, so less restricting fixtures can be used. In addition, laser tooling can be scaled down to accommodate smaller piece-parts. In numerical control operations where more than one fixture is used, working heights of fixtures must be controlled in relation to each other to prevent accidental defocusing of the laser beam. Accidental defocusing may cause less penetration in the weld than desired. Because laser energy is delivered to the workpiece without mechanical force, no fixturing is required on small piece-parts. Use of mechanical force for these applications could induce unwanted strains and deformations.

## TOOLING FOR SOLDERING AND BRAZING

Soldering and brazing differ from welding in several respects. The filler metal is nonferrous, usually lead, tin, copper, silver, or their alloys. The workpiece or base metal is not heated to the melting point during the operation. The added filler metal is melted and usually enters the joint by capillary action. Brazing and soldering are fundamentally the same processes. An arbitrary distinction is made between them based on the melting temperature of the filler metal. Soldering filler metals melt below 842° F (450° C) and brazing filler metals above 842° F (450° C).

The success of brazing and soldering depends on part cleanliness, temperature control, and the clearance between the surfaces to be joined. Cleanliness usually is obtained by introducing a flux that cleans, dissolves, and floats off any dirt or oxides. The flux also covers and protects the area by shielding it from oxidation during the process. It

may, to some extent, reduce the surface tension of molten metal to promote free flow. The worst contamination usually is due to oxidation during the process. This is more prevalent with brazing because of the higher process temperatures. Some brazing and soldering operations are conducted in a controlled atmosphere or vacuum.

Temperature control, although influenced somewhat by fixture design, is dependent primarily on the heat application method. In a simple low-production process, the worker may hold a torch closer to the workpiece for a longer period of time. In a very precise high-production process, the instrumentation of a controlled atmosphere-type furnace may be adjusted.

Clearance between the surfaces being joined helps determine the amount of capillary attraction, the thickness of the alloy film, and consequently the strength of the finished joint. The range of typical joint clearance is 0.002–0.015 in. (0.05–0.38 mm). The optimal joint clearance is dependent upon the particular filler metal alloy, base metal being joined, and joint configuration. Too large a clearance would lack sufficient capillary attraction, while too small a clearance may inhibit proper flow of some filler metals.

### General Considerations

Many soldering and brazing operations are conducted without special tooling. As with mechanical joining methods, many workholding devices can be used to conveniently present the faces or areas to be joined. An electrical connecting plug can be easily held in a vise while a number of wires are soldered to its terminals. In many high-production assembly operations, parts are manually mated with a preformed brazing ring or foil between them. Then they are placed directly on the continuous belt of a tunnel-type furnace or gas heater-ring fixture.

If the shape of the workpiece will not support itself in an upright or convenient position, a simple nesting fixture may be required. *Figure 10-23* shows a simple nesting fixture in which two workpieces and a brazing ring have been placed. The fixture can be mounted on a table while an operator applies heat with a hand torch. The same fixture could be mounted on a pow-

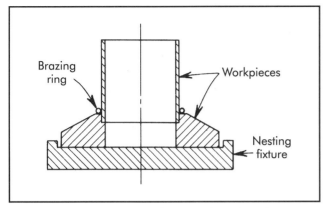

Figure 10-23. *Simple nesting fixture with work in place.*

ered rotating base in the flame path of a fixed torch while a feed mechanism introduces wire solder at a predetermined rate (*Figure 10-24*). The same fixture could be attached in quantity to the belt of a tunnel furnace. Or, a number of the fixtures could be attached to a rack for processing in a batch furnace.

Some brazing and soldering processes dictate more tooling considerations than others. For instance, hand soldering often requires no special tooling at all, while induction brazing may require significant tooling/fixture design and development. Nevertheless, several aspects are common to most all brazing and soldering tooling and/or fixtures.

• The tooling material must be stable at the intended process temperatures. Tooling and fixtures often are heated along with the parts being joined. Consequently, they must be made from materials with a melting temperature that is high enough, possess ad-

Figure 10-24. *Soldering machine using simple nesting fixture.*

equate hot strength, and not react with base metals or the process atmosphere unfavorably. For high-production processes, corrosion of tooling and fixtures can be a particularly serious concern.

• The process must account for the difference in coefficient of thermal expansion (CTE) of tooling and fixtures relative to that of the parts being joined. This is particularly true for fixturing that maintains critical tolerances. A significant CTE mismatch between the fixture and the workpiece may lead to excessive distortion of either one or both.

• Fixtures must allow adequate and uniform heating of the entire joint surface. If a fixture acts as a chill for one part of the joint, it may result in incomplete bonding and a defective joint. However, even the best fixture design always must be accompanied by proper procedure development to prevent this type of problem.

• The mass of any tooling or fixtures that will be heated during the process, even in part, should be minimized as best as possible to limit the amount of excess heat required. This is particularly true for high-production processes where the cost just for heating the tooling could be substantial. Low thermal-conductivity materials, such as ceramics, often are used to minimize the heat loss to and/or through a fixture component.

• Precautions must be taken to prevent the tooling or fixtures from being inadvertently bonded to the workpiece. Proper design considerations and/or using materials, such as ceramic, stainless steels, and titanium that are not as easily wet by conventional brazing and soldering alloys provide added insurance. It should be noted that in some cases, the filler metal may wet and flow well beyond the intended joint area.

As is the case with most joining processes, good tooling and fixtures cannot compensate for poor procedures development or selection of the wrong brazing or soldering process. For example, a fixture may appear to be acting as a chill on part of an assembly causing incomplete wetting in a torch brazing application. However, this may actually be the result of an inadequate heating rate due to poor torch placement, gas flow rates set too low, improper flame adjustment, or improper fuel-gas selection.

## TOOLING FOR INDUCTION BRAZING

*Figure 10-25* illustrates a nest-type fixture to hold mating workpieces within the field of an induction work coil. *Figure 10-26* shows the same fixture altered to permit use of an internal induction-heating coil. If the external coil is used, the fixture designer must provide some method of moving the fixture or coil while workpieces are loaded and unloaded.

Induction coils (inductors) provide a convenient and precise way of quickly and efficiently heating any selected area of an electrically con-

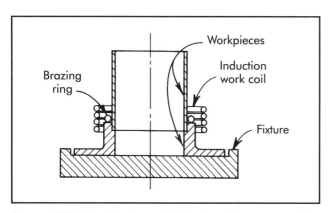

Figure 10-25. *Nesting fixture for brazing with an external inductor.*

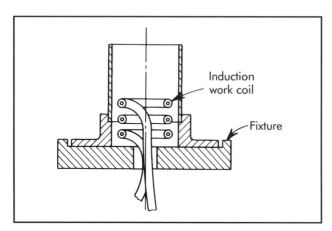

Figure 10-26. *Nesting fixture for brazing with an internal inductor.*

ductive part or assembly to a required depth to provide a brazed joint. Correct selection must be made regarding frequency, power density, heating time, and inductor design.

## Induction Heating Theory

The flow through an electrical conductor results in heating as the current meets resistance to flow. Thus, $I^2R$ losses ($I$ = current; $R$ = resistance) may be as low as current flowing through copper wire (resistance is low) or high when this same current reaches and flows through a heating element (resistance is high). This high loss is the resistance heating obtained with conventional electric heaters.

Induction heating is resistance heating with current flow meeting resistance. A workpiece heated in this manner has been made the secondary of a simple transformer, the primary being the inductor that generally surrounds the part and through which alternating current is flowing. The flow of alternating current in a primary induces a flow of current in the secondary by electromagnetic forces (magnetic flux). Alternating current tends to flow on the surface, and there is a relationship between the frequency of the alternating current, the depth to which it flows, and the diameter of stock that can be heated efficiently.

Considering a piece of plain carbon steel being heated for brazing or surface hardening, the depth at room temperature in which there is instantaneous flow of current is related only to the frequency:

$$D = 4/F \qquad (10\text{-}3)$$

where:

$D$ = depth, in. (mm)
$F$ = frequency in cycles/sec

The depth, $D$, increases with temperature. For heat to be generated, the $I^2R$ must have a time factor $t$ to become:

$$I^2Rt \qquad (10\text{-}4)$$

Additional depth results from the current following the path of least resistance (cooler underlying metal) and heat flow by conduction. This depth may be approximated by:

$$D = 0.0015t \qquad (10\text{-}5)$$

where:

$D$ = depth of heat penetration, in. (mm)
$t$ = time, sec

The higher the frequency for a given heating time, the shallower the depth of heat. The converse is true. For a given frequency, the depth of heat is also directly proportional to the time (square root).

An examination of *Equation 10-3* will provide an answer to the relationship between frequency and the diameter of stock that can be heated efficiently. Large diameters may be heated efficiently with low frequency. Sixty cycles are used efficiently on 10-in. (254-mm) diameter steel workpieces.

The theoretical depth of current penetration mentioned previously increases with temperature and must never exceed the radius of the stock being heated. For practical purposes, this depth, $D$, ($D = 4/F$ at room temperature) should be two to three times the radius for through-heating, and 10 times the radius for surface hardening.

Power density and heating time are closely related to the design of the inductor. The inductor design is integral to the success of any induction-heating application. The purpose of the inductor is to set up a magnetic flux pattern in which the work to be heated is correctly positioned so the required heating is accomplished (*Figure 10-27*). Coil design is influenced by the application, as is the selection of frequency, power density, and heating time.

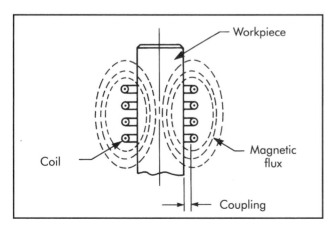

Figure 10-27. *Magnetic flux in induction heating.*

Regardless of the material used in the induction-brazing process, the inductor must be designed to heat the area of the joint sufficiently to cause the bonding material to flow. Temperatures can range from a few hundred degrees for soft solder and epoxies to 2,100° F (1,149° C) for copper. When attempting to heat heavy sections and avoid overheating of sharp corners on small or medium-size workpieces, the use of motor generators at 3,000 or 10,000 cycles is preferred.

Copper normally is used for coils with appropriate water cooling. This can be accomplished by using copper tubing, solid copper to which tubing has been attached, or solid copper that has been drilled out or machined to provide water passages.

An inductor should have the following characteristics:

- proper physical shape to surround the section to be heated;
- large enough diameter to permit loading and unloading or, in the case of a continuous operation, clear passage;
- proper support or rigidity to maintain its designed shape;
- insulation to avoid electrical breakdown (spacing or an air gap between turns and the workpiece usually is sufficient), and
- ability to withstand operating conditions (exposure to water, dirt, and flux must be taken into consideration).

The air gap or space between the work and the inside of the inductor should be kept low for efficiency. A gap of 0.125 in. (3.18 mm) is reasonable. For irregularly shaped sections where efficiency is not too important, gaps of 0.25 in. (6.4 mm) and greater can be tolerated.

The area to be heated determines the length or width of the inductor. For most joint designs on small parts, a single-turn is sufficient. On larger diameters, it is necessary to heat a wider band requiring either several turns or a wider inductor. An inductor that is too narrow will simply require a longer heating time to allow the heat flow to cover the area. An inductor that is too wide will heat more metal than necessary and be less efficient. A single-turn inductor should be no wider than the bore diameter; for greater coverage, multi-turn inductors are used.

The electrical characteristics of a high-frequency power source determine the number of turns in an inductor. Generally, the high-frequencies of a vacuum tube oscillator (200–3,000 kHz) usually require multi-turn coils, whereas the motor-generator sets (1–10 kHz) can operate into either multi-turn or single-turn coils through the use of variable-ratio transformers. Turns of the inductor should be kept as close as possible without touching. Typical inductors are illustrated in *Figure 10-28*.

## Number of Inductors

The number of inductors used at one time is, of course, influenced by production requirements. The number can be extended to as many as it may appear convenient to load. Instead of a large number, a continuous coil can be used, through which parts are moved progressively. Generally, this method is limited to a workpiece having a joint on or near its end. The determination of how many parts are to be processed at a time is related to heating time and power density. The time generally runs from 10 or 15 seconds on small parts to as long as two minutes on large parts where heat must penetrate. The power density is approximately 0.5–1.5 kW/in.$^2$ (279–837 kJ/cm$^2$) of surface area. An average value of 1 kW/in.$^2$ (558 kJ/cm$^2$) is recommended. The actual value depends on the ratio of surface area to volume.

The amount of energy (kilowatt seconds) is determined by the volume of metal to be heated and the temperature necessary to flow the bonding material. Since the exact amount of metal heated in a brazing operation depends on many factors, such as conductivity of the work and heating time, only a rough approximation can be attempted. The following procedure is recommended (using U.S. Customary units):

1. Estimate the volume of metal expected to be brought to brazing temperature.
2. Convert the volume to weight in pounds based on steel weighing 0.3000 lb/in.$^3$
3. For 400° F solder, convert to kilowatt seconds by multiplying by 100.

Figure 10-28. *Designs of induction-heating coils for brazing.*

4. For 1,200° F brazing material, convert to kilowatt seconds by multiplying by 300.
5. For 1,600° F brazing material, convert to kilowatt seconds by multiplying by 500.
6. For 2,000° F brazing material, convert to kilowatt seconds by multiplying by 700.

7. Estimate the surface area of the joint. Based on 1 kW/in.$^2$, divide the area into the preceding value of kilowatt seconds to determine the approximate brazing time in seconds.
8. Determine from production requirements (allow for handling each part) the number of parts to be processed at a time.
9. Total power required will depend on the above value and kilowatts per piece based on 1 kW/in.$^2$ of surface area.
10. Power available is determined by the equipment to be used. The maximum number of parts that can be processed at once can be approximated by dividing the kilowatt rating of the equipment by the surface area of each part in square inches (kW/in.$^2$).

## Material

A solid-type inductor is preferred because of its greater rigidity. Accidental contact while loading parts will not bend it out of shape. Less expensive coils can be made of copper tubing which, except for extremely low power (1 kW [3,600 kJ] per coil), should not be less than 0.19 in. (4.8 mm) in diameter. Both round and rectangular sections are used; the rectangular section is especially adaptable when a wide, single-turn is needed. Its wide flat area can be adjacent to the work for more uniform heating. A 1.00-in. (25.4-mm) wide inductor can be made of 0.25 × 1.00 in. (6.4 × 25.4 mm) tubing or from a wide, flat strip brazed to a round tube for cooling.

## Leads

The leads to the inductor should be kept as short as possible and close together. In some situations, coaxial leads may be necessary (450 kHz and up). If the leads are not a continuation or part of the inductor and a brazed joint is used, they must not reduce current flow to the inductor. Silver solder is recommended, and carefully prepared joints (even mitered) are essential. Insulating materials should be used between leads if they are not self-supporting. The connection to the power source will be determined by the design of the output transformer.

## Cooling

Inductors must have sufficient cooling to avoid overheating. Inductors for brazing operations can be made by forming a copper tube on a suitable mandrel. The tube must not collapse to restrict internal coolant flow.

Cooling water may be brought in from the output transformer through connectors carrying the current, or by separate insulated connections to a water supply.

## Coil Support

Solid inductors usually are self-supporting, as are single-turn inductors. Support of multi-turn inductors or an array of inductors can be accomplished with any nonmetallic material. Many plastic materials can be used. Coils can be attached by studs (preferably copper) brazed to the inductor. Use of copper or bronze-brass studs is highly recommended, as cooling water will cause rust problems with steel or any ferrous metal and hinder their removal.

## Internal Heating

When tubular sections are to be joined, it is often more convenient to heat from the inside. The factors of coil design essentially are the same as for inside diameter (ID) inductors, except that the leads must be brought to the coil axially instead of radially (*Figure 10-26*). Since heating from the ID is considerably less efficient (the magnetic flux is weak outside the confines of the coil), two to three times as much energy (kilowatt seconds [joules]) may be required. This can be obtained by using more power, or time, or both. When high production requirements preclude the use of a longer heating cycle, the ID coil efficiency can be improved by using an inside core that increases the magnetic flux outside of the coil. Basic design elements are as follows:

1. For vacuum-tube oscillators, compacted ferro-magnetic stock may be used. This material, which can be purchased in round bars, is then made a core inside the inductor. The turns can be held tightly on the surface or wound in grooves machined in the

core. No insulation is required, and one lead may be brought through a hole in the center of the core.

2. For motor generators, iron laminations are used. They are made from 0.007-in. (0.18-mm) thick strips of transformer steel, 0.25-in. (6.4-mm) wide, which has been cut to the correct coil length. The strips are wound together and inserted in a multi-turn coil to form a core. If a single-turn inductor is used, laminations are cut in a C-shape and stacked tightly around the turn, as shown in *Figure 10-29*.

The design of inductor coils for brazing operations is relatively simple. There should be no problem with electrical matching to the proper power source. The physical configuration is dictated by the shape and size of the area to be heated. The inductor is a primary winding for a transformer with the workpiece acting as a secondary. The coil need only be formed to surround or—for ID heating—be adjacent to the surface to be heated. A wide latitude of shape is permitted. A square part can be heated with an inductor formed on a square (*Figure 10-30*) or round mandrel. Irregularly shaped parts can be heated with round inductors. If a corner of a workpiece overheats, the inductor can be modified to merely provide a larger air gap at the corner.

Occasionally, a joint must be heated by proximity. An ID coil cannot be used as it would necessitate heating entirely too much metal. An inductor for heating such an area is shown in *Figure 10-31*. Such an inductor is insufficient and can be improved by use of laminations as shown. The C-shape is open toward the work.

Figure 10-30. *Inductor formed on a rectangular mandrel.*

Figure 10-31. *Inductor with C-shaped laminations added.*

Generally, an inductor should have the same contour as the area to be heated. With a reasonably uniform coupling, the part can be loaded with no difficulty. When there are large changes in diameter in the section to be heated, the larger diameter will tend to overheat unless some compensation is made. This can be done by having fewer turns per linear inch (millimeter) in that area, increasing the coupling, or using two turns in series (electrically) on the small diameter (*Figure 10-32*). The intensity of the magnetic field in which the surface finds itself should be as uniform as possible.

Figure 10-29. *Special-purpose inductor.*

Figure 10-32. *Inductor with loops in series.*

## TOOLING FOR THERMAL CUTTING

Three basic types of thermal-cutting systems are used in metal-fabricating operations: oxy-flame, plasma-arc, and laser.

Laser cutting is primarily used on different types of numerically controlled (NC) and computer-numerically controlled (CNC) machines. Machines equipped with lasers generally have a large working table area (60.00 × 50.00 in. [152.4 × 127.0 cm] nominal sheet size). The workpiece is held in position by several hydraulic holders or clamps. Many of the machines have nesting capabilities, which allow more than one part per material sheet.

Plasma-arc cutting can be used as a means by which irregular shapes can be cut. Plasma-arc cutting also is used on thermal machining centers. These machining centers carry one or more torches on a traveling bridge. The torches are guided by an optical scanner, which follows a paper template or can be operated by tape control or computer. The material to be cut is placed on water tables. The water table is used to control smoke, noise, and ultraviolet radiation. A dye in the water eliminates ultraviolet emissions.

Oxy-flame or oxyacetylene cutting machines employ the same method as plasma-arc machining centers, except that the cutting tables do not normally contain water. The cutting table contains a number of cross-supports on which the workpiece is set. These cross-supports are held

in position by a square or rectangular container into which the drop-outs may fall. After several cutting operations, the cross-supports must be replaced because they become damaged during the cutting operation.

Some fixturing is used on plasma-arc and oxy-flame machining centers. These fixtures are constructed to support the material to be cut, allowing for proper heat dissipation to prevent distortion. The fixtures use standard components. The key to designing fixtures for cutting operations is allowing clearance for cutting media below the piece-part. If there is insufficient clearance, fixture damage may result.

Added precautions must be taken when cutting high-alloy steels. This is due to the hardening ability of the workpiece at the line of cut, and suitable machining allowance or post-heat treatment must be considered.

## TOOLING FOR MECHANICAL JOINING PROCESSES

Many workpieces can be held for mechanical joining without tooling. A worker often can manually align two workpieces and insert a fastener. This method has several limitations. The workpiece must be small or light enough to be positioned manually, and the forces incurred in the joining process must be relatively small. The complexity of an assembly joined in this manner is limited by the number of components a worker can conveniently handle. An elementary workholder often can be used to advantage in even the most simple joining operation.

A universal-type vise can be of great value in mechanical joining. A primary workpiece can be held in any position while other workpieces are fastened to it. Several workpieces may be held in alignment between the jaws, while the worker applies a fastener. Clamping pressure can be used to counteract the joining forces, such as torque applied to a threaded fastener.

### Threaded Fasteners

Threaded fasteners are used for a wide variety of applications, and the bolt and nut are the most common. Tooling for threaded fasteners is as varied as their application. There are, however, design principles that apply to all cases.

A primary application of threaded fasteners is joining and holding parts together for load-carrying requirements, especially when disassembly and reassembly may be required. Typical assemblies for several types of threaded fasteners are illustrated in *Figure 10-33*. Threaded fasteners also are used extensively for assemblies subject to environmental conditions, such as high temperatures and corrosion.

Advantages of threaded fasteners include their commercial availability in a wide range of standard and special types, sizes, materials, and strengths, as well as the ability of the assembly to be disassembled easily. Extensive standardization efforts have made most threaded fasteners interchangeable.

### Bolts and Studs

Bolts are externally threaded fasteners generally assembled with nuts (*Figure 10-33a*). While most bolts have heads, some do not. The means of distinguishing between bolts and screws are discussed in ANSI Standard B18.2.1 ("Square and Hexagonal Bolts and Screws, Inch Series"). Studs are cylindrical rods threaded on one or both ends or throughout their lengths (*Figure 10-33c*).

Bolts with hexagonal heads, frequently called hex heads, are the most commonly used. These heads have a flat or indented top surface, six flat sides, and a flat bearing surface. The flat sides facilitate tightening the bolts with wrenches. Hex heads often are used on high-strength bolts and are easier to tighten than bolts with square heads. Generally, they are available in standard and special strength grades and requirements for specific applications.

Figure 10-33. *Typical assemblies using threaded fasteners: (a) bolt and nut; (b) cap screw; (c) stud.*

Round-head bolts have thin circular heads with rounded or flat top surfaces and flat bearing surfaces. When provided with an underhead configuration that locks into the joint material, round-head bolts resist rotation and are tightened by turning their mating nuts. Included in this classification, even though the configurations differ, are countersunk and T-head bolts. Variations of round-head bolts include those with square, ribbed, or finned necks on the shanks below the heads to prevent the fasteners from rotating in their holes.

Square-head bolts have square-shaped, external wrenching heads. They are available in two strength grades. Lag bolts, sometimes called lag screws, usually have square or hex heads, gimlet or cone points, and thin, sharp, coarse-pitch threads. They produce mating threads in wood or other resilient materials and are used in masonry with expanding anchors.

Battery bolts have square heads and are mostly stainless steel or lead- or tin-coated for clamping onto battery terminals. Fitting-up bolts have square heads and coarse-pitch, 60° stub threads. They are used for the preliminary assembly of structural steel components. T-bolts are square-head bolts used in the T-slots of machine tools.

Bent bolts are cylindrical rods having one end threaded and the other bent to various configurations. These include eyebolts, hook bolts, and J-bolts. Other bent bolts, such as U-bolts, have both ends threaded. The ends of bent bolts are usually square (as sheared).

Studs are unheaded, externally threaded fasteners. They are available with threads on one or both ends or continuously threaded. Studs with collars and threaded on one or both ends also are available. Heat-treated and/or plated studs are available to suit specific requirements. They are made with chamfered or dog-point ends.

An advantage of studs for some applications, such as the assembly of large and heavy components, is their usefulness as pilots to facilitate mating of components, which expedites automatic assembly. For many applications, studs provide fixed external threads, and nuts are the only components that must be assembled.

## Nuts

Nuts are internally threaded fasteners that fit on bolts, studs, screws, or other externally threaded fasteners for mechanically joining parts. They also serve for adjusting and transmitting motion or power in some applications, but they generally require special thread forms.

Hex and square nuts, sometimes referred to as full nuts, are the most common. Hex nuts are used for most general-purpose applications. Square machine-screw nuts usually are limited to light-duty and special assemblies. Regular and heavy square nuts often are used for bolted flange connections.

Single-thread nuts, sometimes called spring nuts, are formed by stamping a thread-engaging impression (arched prongs) in a flat piece of metal (*Figure 10-34*). These nuts generally are made from high-carbon spring steel (SAE 1050–1064), but are also available in corrosion-resistant steel, beryllium copper, and other metals.

Stamped nuts are hex fasteners stamped from spring steel or other metals, with prongs formed to engage mating threads. Like single-thread nuts, they rely on spring action for clamping and resistance to loosening, but they have more prongs to engage the threads on the mating fastener. Applications include replacement for full nuts in low-stress uses and as retaining nuts against full nuts (*Figure 10-35*). Stamped nuts are made with integral washers, in closed top or bottom styles, and as wingnuts.

Figure 10-34. *Single-thread nut.*

Figure 10-35. *Stamped nut applied and tightened after full nut is in place.*

## Screws

Screws are externally threaded fasteners capable of insertion into holes in assembled parts, mating with preformed internal threads, or cutting or forming their own threads. Because of their basic design, it is possible to use some screws, which are sometimes called bolts, in combination with nuts.

Screws are available in a wide variety of types and sizes to suit specific requirements for different applications. Major types discussed in this section include machine screws, cap screws, setscrews, sems (screw and washer assemblies), tapping screws, and captive screws.

***Machine screws.*** Machine screws usually are inserted into tapped holes, but are sometimes used with nuts. They are generally supplied with plain (as sheared) points, but for some special applications they are made with various types of points. Machine screws have slotted, recessed, or wrenching heads in a variety of styles and are usually made from steel, stainless steel, brass, or aluminum. Many machine screws are made from unhardened materials, but hardened screws are available.

***Cap screws.*** Cap screws are manufactured to close dimensional tolerances and designed for applications requiring high tensile strengths. The shanks of cap screws are not fully threaded to their heads. They are made with hex, socked, or fillister slotted heads (*Figure 10-36*). Low-head cap screws are available for applications having head clearance problems. Most cap screws are made from steel, stainless steel, brass, bronze, or aluminum alloy.

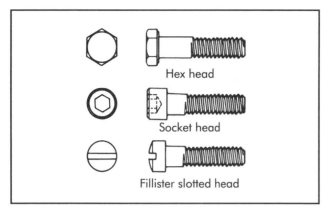

Figure 10-36. *Cap screws with various heads.*

**Setscrews.** Setscrews are hardened fasteners generally used to hold pulleys, gears, and other components on shafts. Hardness of the shaft is an important consideration in selecting a proper setscrew. They are available in various styles, with square-head, headless-slotted, hex-socket, and splined- (fluted) socket styles being the most common. Holding power is provided by compressive forces, with some setscrews providing additional resistance to rotation by penetrating their points into the shaft material.

**Sems.** Sems (screw and washer assemblies) is a generic term for preassembled screw and washer fasteners. The washer is placed on the screw blank prior to roll threading and becomes a permanent part of the assembly after roll threading but is free to rotate. Sems are available in various combinations of head styles and washer types. Washers commonly used include flat (plain), conical, spring, and toothed lock.

Sems are used extensively in the manufacture of automobiles and appliances and in other mass-production industries because they are suitable for automatic-assembly operations. These fasteners permit convenient and rapid assembly by eliminating the need for a separate washer-assembly operation. They also ensure the presence of the proper washer in each assembly and prevent the loss of washers during maintenance.

**Tapping screws.** Tapping screws will cut or form mating threads when driven into holes. Self-drilling, self-piercing, and special tapping screws are also available. They are made with slotted, recessed, or wrenching heads in various head

styles and with spaced (course) inch or metric threads. Tapping screws are mostly used in thin materials.

Advantages of tapping screws include rapid installation because nuts are not needed and access is required from only one side. Mating threads fit the screw threads closely, with no clearances necessary. Underhead serrations or nibs on some screws increase locking action and minimize thread stripout.

**Captive screws.** Captive screws remain attached to panels or assembly components after they have been disengaged from their mating parts. Advantages include fast assembly and disassembly, and prevention of damage to other assembly components caused by lost or loose screws.

## Threaded Assembly

The most common method of threaded assembly is to place two workpieces in their correct relative location, drill a hole through them, insert a bolt in the hole, and torque a nut onto the bolt. The hole may have been drilled in both pieces prior to mating. In many cases, no tooling will be required. If it is convenient for a worker to hold the workpiece together while inserting the bolt and adding and tightening the nut, then a simple workholder may be advantageous. The workholder must locate the workpiece so the holes are conveniently positioned, and it should support the workpiece against the torque and thrust loads imposed in tightening the fasteners.

One of the two workpieces being assembled may be tapped to receive the threaded fastener, or a self-tapping fastener may be used. Power tools may be used to drive the fasteners. The fixture design principles for nut runners are: (a) location—the fixture must align the workpiece precisely with the nut runner; and (b) support—the fixture must withstand the weight of the workpiece plus the thrust and torque loads imposed.

*Figure 10-37* shows a template nesting fixture resting on the table of a power screwdriver. It is designed to hold a metal cabinet while two side shields are assembled to it by 20 self-tapping

Figure 10-37. *Partial nest-type workholding fixture for assembly of a cabinet.*

screws. A round locating pin is attached to the machine table exactly under the driver. The fixture base is a template and has 10 radiused notches that can be manually held against the locating pin. When the components of the cabinet are placed in the fixture, each template notch is directly beneath a predrilled hole in the components, into which a screw is to be inserted. As each notch is held against the locating pin, the predrilled hole is placed directly beneath the driver and the machine cycled to insert and drive one screw. After 10 screws have been driven to secure one side shield, the cabinet is inverted and another 10 screws are driven to fasten the other shield.

The fixture design principles are the same. It must precisely locate the workpiece relative to the driver. In this case, 10 precise location points are involved. The fixture also must establish the exact height of the workpiece. The fixture must support the workpiece in resistance to the torque and thrust loads imposed.

For high production volumes, several machines may be arranged to simultaneously insert and drive screws into a single-fixtured workpiece.

## RIVETS

Perhaps the most widely used pin-type fastener is the rivet—a pin with a head on one end with the other being plastically deformed after insertion to prevent retraction. The riveting process is extremely varied, and may be used to assemble the parts of a timepiece or the structural members of a bridge. Rivet diameters may vary from perhaps 0.015–5.000 in. (0.38–127.00 mm). Holes may be drilled or pierced before or during the operation.

It is important that a riveting tool of the correct size and shape be used, and that excessive driving pressures be avoided. Excessive pressure in driving may result in:

- bulging of the edge of the piece being riveted;
- buckling or other distortion, particularly if thin material is used, and/or
- weakening or fracturing of metal near the hole.

*Figure 10-38* shows an L-shaped workpiece clamped in a fixture while a second workpiece, a channel, is being riveted to it. The portable riveting yoke literally squeezes the rivets to deform them. Holes are drilled and rivets are inserted prior to the operation. In practice, it would be

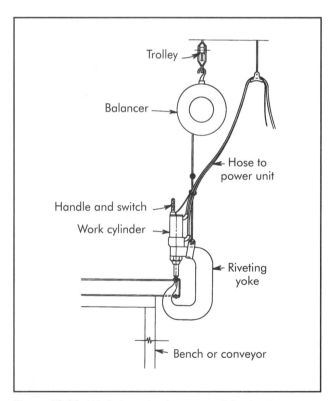

Figure 10-38. *Workpiece simply supported for riveting.*

necessary to employ stops or other means to locate the channel with reference to the primary workpiece.

*Figure 10-39* shows the sequence of a punching and riveting operation used in mass production. The two workpieces are placed between the tools, and the machine is cycled to automatically pierce a hole, insert a rivet, and then head the rivet.

## Eyeleting

Eyelets, like rivets, are used extensively as low-cost fasteners for light assembly work. For high-production operations, costs can be reduced to a minimum with the use of eyelet-attaching machines. These machines are small power presses with hopper-feed mechanisms. *Figure 10-40* illustrates typical tooling of an automatic eyelet-setting machine in loading and clinching positions.

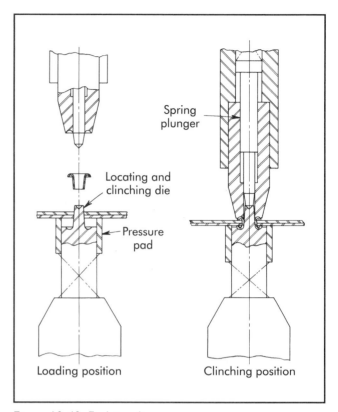

Figure 10-40. *Eyelet curling.*

## Tubular Riveting

A tubular rivet is a cross between a solid rivet and an eyelet. The straight end of the rivet has a center hole that permits this part of the rivet to clinch easily when struck by the contoured riveting tool. *Figure 10-41* illustrates tooling for a typical tubular riveting operation.

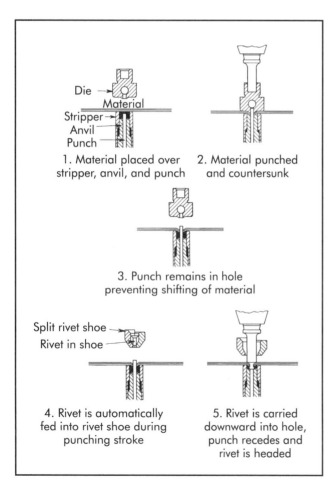

Figure 10-39. *Sequence of punching and riveting operations.*

Figure 10-41. *Tubular riveting.*

## Clinch Allowance

Clinch allowance is the space caused by the part of a rivet or eyelet that extends beyond the combined thickness of the assembly before the rivet or eyelet has been set. *Figure 10-42* shows the relation of rivet or eyelet length to material thickness and clinch allowance. The proper rivet or eyelet length should be approximately equal to the material thickness plus the clinch allowance. Both rivets and eyelets are available commercially in 0.03-in. (0.8-mm) length increments. Two rules of thumb have been developed: (1) maximum length of clinch for full-tubular and bifurcated rivets should be figured at 100% of shank diameter; and (2) maximum length of clinch for semi-tubular rivets should be 50–70% of shank diameter to prevent buckling and ensure a tight set.

The hole diameter and method of producing the hole in the material also affect clinch diameter. The minimum clinching radius for tubular rivets or eyelets can be determined by multiplying the wall thickness by three. The clinching contour of riveting tools must be free of nicks and circular grooves. After hardening, the contour must be highly polished.

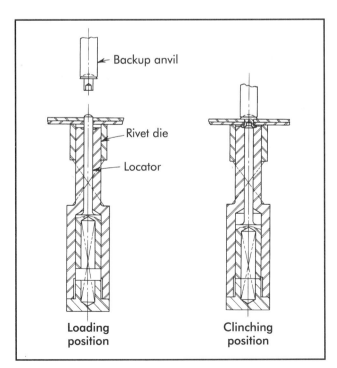

Figure 10-42. *Determination of rivet length for tubular rivets.*

## Equipment

Rivets can be deformed or set in many ways. Pressure may be applied continuously or with a series of hammer blows. Rivets can be manually or automatically inserted. Holes could be drilled prior to or during the riveting sequence. The tool used to apply deformation pressure may be a common hammer, pneumatic hammer (riveter or rivet gun), portable squeezing yoke, or stationary machine.

The final shape of the rivet becomes that of the tool (die) used to apply the deformation pressure. The shape may be flat (reflecting the contour of a hammer or a conventional bucking bar) or curved, as illustrated in *Figures 10-40* and *10-41*. The rivet die may be a simple bucking bar, interchangeable rivet set placed in the nozzle of an air hammer, or a complete forming die placed in a standard hydraulic or mechanical press.

Workholders or fixtures used to hold and locate workpieces being assembled by riveting include stationary, portable, and self-contained (riveting die fixture).

### Pneumatic Hammer

Conventional riveting is performed by placing the rivet in a predrilled hole, holding a pneumatic hammer against the head of the rivet, holding a bucking bar against the end of the rivet, then cycling or activating the hammer. Initiating pressure is exerted by the hammer while formation pressure is exerted by the bucking bar. Pneumatic hammers (rivet guns) are commercially available in a wide range of types and sizes. The nozzle of the rivet gun receives and retains the rivet set or die. Rivet sets are commercially available in a wide variety of shapes to mate with the many types of rivets commonly used.

### Portable Yoke Type

Portable yoke-type riveting equipment is commonly used for squeezing large rivets (0.19–1.00 in. [4.8–25.4 mm] in diameter), and consists of a yoke with a cylinder (air or oil) to provide the squeezing action. Equipment size can be minimized by using hydraulic cylinders and high pressures (5,000 lbf/in.$^2$ [34.5 MPa]). Equipment weight is minimized by using special high-strength, heat-treated steel. The cylinder advances an anvil to the rivet with a primary pressure of approximately 1,000 lbf/in.$^2$ (6.9 MPa). When resistance is encountered, the pressure increases to 5,000 lbf/in.$^2$ (34.5 MPa) for the rivet upsetting portion of the action. *Figure 10-38* showed a hydraulic riveting yoke.

## Stationary Machines

Stationary machines often are used for riveting. Workpieces being assembled are located with reference to each other (mated) and placed between the upper and lower elements of the machine.

Small stationary machines have a spring-loaded pin locator in the rivet die. Predrilled holes through the mated workpieces engage the pin locator, as illustrated in *Figure 10-43*. The upper riveting die (backup anvil) pushes the rivet into the holes, depresses the spring pin, and upsets the driven head on the lower anvil. The rivets are fed from a hopper to a feed track. The lower end of the feed track locates the rivet directly above the pin locator. The end of the track is split to allow the upper die to pick off the rivet for insertion.

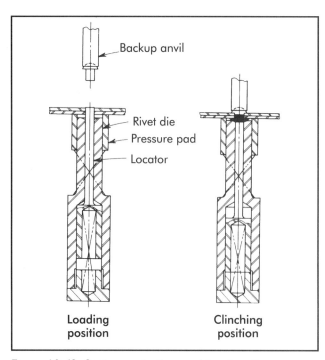

Figure 10-43. *Stationary riveting machine operation.*

Large stationary machines often combine both riveting and hole piercing. *Figure 10-39* illustrated the sequence of operations. The workpieces need not be predrilled, but must be located with reference to each other. The locating and holding fixture may be part of the machine. Extremely large stationary machines can hold and precisely locate large aircraft sections while thousands of the holes are pierced and rivets are driven. The entire production sequence is automatic, with the machine motions governed by tape control or CNC. Closed-circuit TV is used to monitor the operation.

## Spin Peening

Some materials to be assembled are brittle, while others require slender unsupported rivets or eyelets that will not withstand the single impact required to rivet or eyelet without distortion or cracking. Spin-peening machines may be used for delicate assembly operations. These machines deliver innumerable light blows while the hammer spins. The peening and spinning action of these machines is either mechanical or pneumatic.

## Stationary Holding Fixtures

Stationary holding fixtures are two or more parts located and pinned or clamped in position. They are usually freestanding fixtures where portable riveting equipment is used. Air-operated riveting hammers are used for smaller rivets, such as in aircraft or appliance work. Larger rivets used in structural work are deformed with hydraulic or air-operated riveting yokes (rivet squeezing action).

## Portable Fixtures

Portable riveting fixtures are used to locate two or more parts for transport to a stationary-type rivet unit or machine. The fixtures also hold the workpieces during the riveting sequence.

## Riveting Die Fixture

A riveting die fixture is completely self-contained for use in a punch press. This fixture consists of locating elements for two or more parts, and rivet buttons for driving one or several rivets in one stroke or hit of a press. The locating elements may be movable so parts can be positioned outside of the press and rivets can be placed in location. The assembled parts are placed in location in the riveting die, and all rivets are driven at once by the action of the press. *Figure 10-44* illustrates a riveting die fixture.

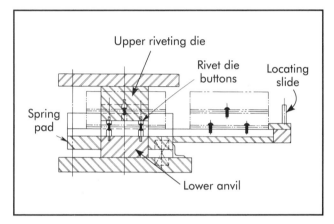

Figure 10-44. *Riveting die fixture.*

### Riveting Fixture Design

Beyond the primary requirements that the fixture precisely locate and hold workpieces, ease of loading and unloading is extremely important. Many parts can be riveted without clamping if the weight of the parts alone will keep the riveting surfaces in contact with one another. Light-gage panels require clamps because of the tendency to warp or twist. Clamps for low-volume production are usually hand-operated cam or toggle types. For high production, air-operated clamps of the same type are used. Locating pins and sheet holders (*Figure 10-45*) are preferred over clamps whenever possible. The area of application must be accessible to the riveting tool, so tilting fixtures are often used.

### STAPLING

*Stapling* is a joining operation using preformed U-shaped wire staples. Staples are made in a variety of shapes, wire sizes, leg lengths, and crown sizes. They are available in blunt-end, chisel-point, or divergent-point styles, and are cohered

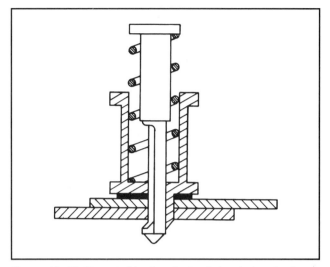

Figure 10-45. *Sheet holder for aligning holes during riveting.*

into strips and sticks. The sticks are loaded manually into staplers. *Figure 10-46* provides examples of staple nomenclature. Gun and hammer tackers are used to drive wire staples. Wire staples also can be driven and clinched by low-cost stapling machines.

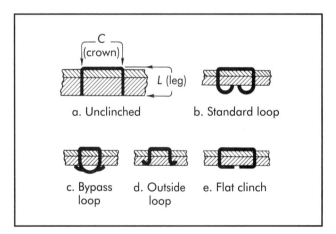

Figure 10-46. *Staple nomenclature.*

### WIRE STITCHING

*Wire stitching* is the process of joining two or more pieces of material with wire fed from a coil, cut to length, U-formed, driven through, and clinched by a specially designed machine called a stitcher. *Figure 10-47* illustrates the principle of wire-stitching machines. For low-cost, high-volume production, automatic wire-stitching machines are ideal for fastening components together. These machines are available to perform:

- carton or industrial-type stitching;
- carding, bagging, and labeling;
- book stitching, and
- metal stitching.

These machines vary in sizes from small bench to large floor models. Each machine draws wire from a coil, cuts it to proper length, forms it into a stitch, drives it through the material, and clinches it with every stroke of the machine.

## METAL STITCHING

*Metal stitching* is a method of fastening thin-gage metals and metals to nonmetals. Metal stitching fastens the work more economically since there is no need for other operations such as punching or drilling prior to fastening. Standard stitching machines can be tooled to form loop or flat clinches. Flat clinches are used when the stitched joint must carry heavy loads, such as in aircraft construction where joints are subject to severe stresses. The flat clinch is formed by an upward movement of the clinching die,

folding the legs flat against the bottom of the material. Loop clinches provide only point contacts with the material and, therefore, less strength than flat clinches. Loop clinching is formed by curling the legs of the wire in stationary solid dies. *Figure 10-48* illustrates flat and loop clinches.

## Other Applications

Metal stitches are used extensively by manufacturers of electronic components. Variable-resistor control and switch shafts are assembled into mounting bushings by wire retaining or snap

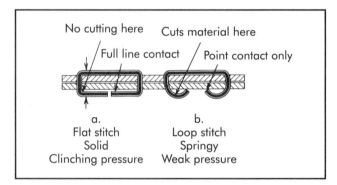

Figure 10-48. *Types of metal stitches.*

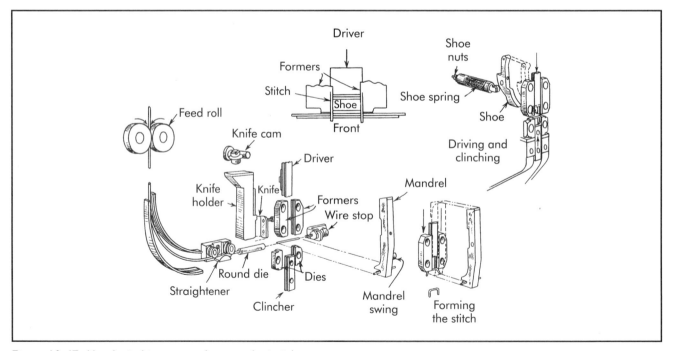

Figure 10-47. *Metal-stitching nomenclature and principles.*

rings. The rings are made of round or flat spring-tempered steel wire. The specially tooled machine draws the wire from a coil, cuts it to proper length, forms it to a U-shape, and drives it against a grooved die to form the round retaining ring into the circular groove of the shaft. *Figure 10-49* illustrates the parts and clinching dies of the retaining-ring assembly.

## STAKING

In staking, two or more parts are joined permanently by forcing the metal edge of one member to flow either inward or outward around the other parts. Staking is an economical method of fastening parts. The operation is completed with a single stroke of an arbor, kick, punch, air, or hydraulic press. *Figure 10-50* illustrates locking a ring to a shaft with center-punch staking. *Figure 10-51* illustrates joining together a bushing, bracket, and shell with a ring-staking punch. *Figure 10-52* illustrates the principles of inward staking by forcing the metal of a ring against the knurled portion of a shaft. To provide added rigidity and torque in assembly, spot-staking may be used. *Spot-staking* is essentially the same as ring staking except that three or more equally spaced chisel edges of the staking punch force metal into splined portions of the parts to be assembled. *Figure 10-53* illustrates a typical splined

Figure 10-49. *C-ring retaining-ring assembly.*

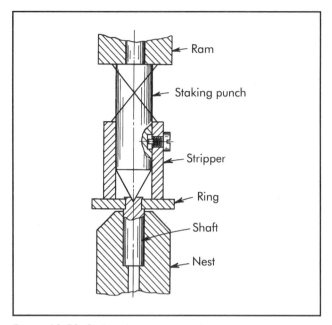

Figure 10-50. *Staking by center punch.*

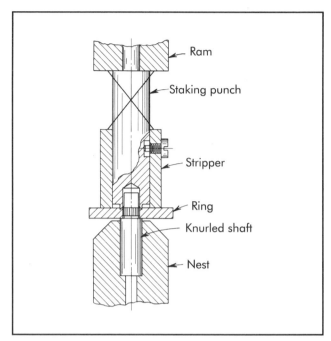

Figure 10-52. *Inward staking by ring punch.*

Figure 10-51. *Staking by ring punch.*

Figure 10-53. *Spot-staking punch and workpiece with splined hole.*

design and spot-staking punch to force metal into this spline. In some cases, it is desirable to use a combined spot- and ring-staking punch (*Figure 10-54*).

To facilitate efficient metal flow, the tips of punches must be free of nicks and circular grooves and, after hardening, these surfaces should be highly polished. Polishing the punch

tips in the direction of metal flow, instead of circular polishing, is desirable.

## TOOLING FOR ADHESIVE BONDING

The design of tooling for adhesive joining does not differ greatly from that used in the physical

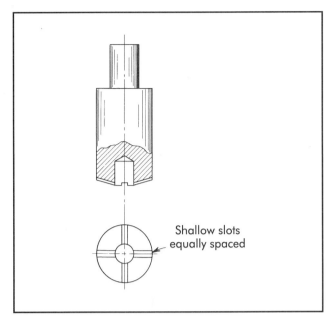

Figure 10-54. *Combined spot and ring-staking punch.*

Shallow slots equally spaced

joining process. In most cases, all standard jig and fixture components can be used.

Tooling can range in size from fixtures to hold small desktops to wing sections of a Lockheed C-5A transport plane.

The type of adhesive used is the main element to consider in the design of fixtures. Nine different types of adhesives are used today, all having different characteristics.

- Plural components (Class 1A) can cure at room temperature. Heating will increase the cure so fixtures must withstand heat to start chemical reaction.
- Heat-activated (Class IB) and film adhesives (Class V) require heat to start chemical reaction.
- Moisture-cure (Class IC) adhesives cure under atmospheric moisture. Due to this moisture, tooling must be made of nonrusting material.
- Evaporative adhesives (Class II) also require heat for curing. Again, fixtures must be able to withstand temperatures without distortion.
- Hot-melt adhesives (Class III) are applied in a molten state, which permits high-speed production. For higher production rates,

these fixtures must be easily loaded and unloaded with the use of hydraulic or pneumatic clamping.

- Delayed-tack adhesives (Class IV) are heat-activated to produce a tackiness that is retained upon cooling for periods of up to several days. Fixtures must control the piece-part configuration for periods of days without losing its intended holding pattern.
- Pressure-sensitive adhesives (Class VI), just as the name implies, apply masking tape, surgical tape, and labels through the use of pressure pads or pressure rollers to form a bond.

An understanding of adhesives is important because the use of plastics is ever-increasing. Since many composites are compromised when subjected to drilling or localized compression, adhesives provide a reliable solution.

## Curing Methods

Conventional methods of adhesive curing include convection and infrared-oven curing. These techniques can be troublesome due to temperature variation and abuse baking.

Cure response has been improved by induction, dielectric, and radiation (ultraviolet, microwave, electron beam, and infrared) curing.

- *Induction curing* uses a heating coil to energize the workpiece, rapidly increasing its temperature. The heat is conducted into the adhesive, initiating or accelerating the cure.
- *Dielectric curing* involves use of a varying electric field. Electric energy is converted into heat for the curing process.
- *Radiation curing* includes microwaves (waves shorter than visible light) and infrared radiation (wavelengths longer than visible light) to create an adhesive cure. Ultraviolet radiation and electron beams also can be used, although electron beam curing is primarily applied to laminating and coating operations.

High processing speeds and productivity gains are possible with these methods, but capital costs to implement the processes are high.

## Selection

The principal advantage of adhesive bonding is improved efficiency of production, and it is possible to select the adhesive that best meets individual job requirements for toughness and resistance to environmental factors.

The properties of the surfaces to be joined must be considered as well as the properties of the adhesive, since the substrate must uniformly coat the entire surface to create the strongest possible bond. The safety concerns of toxicity and dermatological hazards also should be addressed when selecting the appropriate adhesive.

## Joint Design

There are eight possible types of loading on a bond line:

1. shear,
2. peel,
3. tension/compression,
4. cleavage,
5. creep,
6. vibrational fatigue,
7. mechanical shock, and
8. thermal shock.

Adhesives generally are weak in peel and cleavage and stronger when subjected to shear and tension/compression. These characteristics should be considered when designing the workpiece.

## BIBLIOGRAPHY

Doyle, D. *Criteria for Proper Adhesive Selection: From Application to Viscosity.* SME Technical Paper AD90-450. Dearborn, MI: Society of Manufacturing Engineers.

Drain, K. and Schroeder, K. *New Developments in Structural Adhesives.* SME Technical Paper AD90-127. Dearborn, MI: Society of Manufacturing Engineers.

*Tool and Manufacturing Engineers Handbook*, Fourth Edition. 1983-1997. Dearborn, MI: Society of Manufacturing Engineers.

*Welding and Brazing.* Eighth Edition. 1971. Materials Park, OH: American Society for Metals.

## REVIEW QUESTIONS

1. The backup bar is critical in the design of welding fixtures. From *Figure 10-6*, determine the proper backup bars for the weld joints in *Figure A*. (Sketch side view of backup bar.)

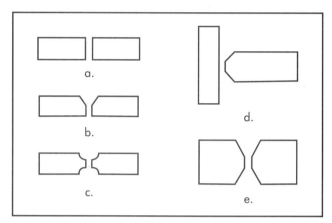

Figure A.

2. Determine the depth of heat penetration on plain carbon steel using *Equation 10-5* for induction heating. Answer a through e.
   a) 5 sec
   b) 10 sec
   c) 20 sec
   d) 45 sec
   e) 60 sec

3. Calculate the spot-welding electrode-tip diameter for the following material thicknesses.
   a) 0.062 in. (1.57 mm)
   b) 0.1875 in. (4.763 mm)
   c) 0.3125 in. (7.938 mm)
   d) 0.750 in. (19.05 mm)

4. Design a welding fixture in detail for *Figure B* to include a complete set of drawings and bill of material. The drawings must be complete to the extent that a toolmaker could make the fixture from the drawings. Use all standard jig and fixture components where applicable.

5. Name three methods of curing.

Figure B.

# Modular and Automated Tool Handling

## DESIGN

Since computer numerical controlled (CNC) machining centers can perform many operations, such as milling, drilling, tapping, turning, and boring, fixtures must be designed for all these operations. This requires cutter accessibility to all parts of the workpiece that need machining operations, as shown in *Figure 11-1*. Additionally, the closer a clamp is placed to a machined feature, the more the machining operation is restricted. Clamping can affect the tool diameter, cycle time, finish, and accuracy.

Probably the most important objective is to keep the fixture and clamping to a low profile to prevent interference with the ideal programmed pattern for the cutting tool. The programmer must raise the tool above the clamp, move it over the clamp, then drop it back down before resuming a cutting operation. Keeping the clamping as low as possible reduces the amount of travel to jump over the clamp, which results in savings to the cycle time. Also, it permits the tool to be chucked as short as possible.

## Accuracy

Most CNC machining centers are capable of holding extremely close tolerances, especially where all machining can be done with the same setup. For example, a machine may position to within 0.0005 in. (0.013 mm) of the programmed coordinates and return to the same position (repeatability) over and over within half that amount, or ±0.00025 in. (±0.0065 mm). Fixtures for CNC have an equal role in transferring this accuracy to the workpiece. While closer tolerances increase the cost of a fixture, they also can deliver a more accurate, consistent product.

Particular attention should be paid to flatness and parallelism when designing fixtures for CNC. *Figure 11-2* illustrates the effect on an 18.00-in. (45.7-cm) workpiece when held between vice jaws misaligned a total of 0.003 in. (0.08 mm). If a slot were milled down the center, it would be out of parallel by 0.009 in. (0.23 mm). This could be enough to scrap the workpiece.

## Rigidity

Accuracy, surface finish, and productivity are affected by rigidity. Cutting tools produce severe shock, pressure, and vibration on the workpiece, which must be alleviated by good fixturing.

At least one or two solid surfaces should be designed into the fixture, as shown in *Figure 11-3*,

Figure 11-1. *Cutter accessibility.*

Figure 11-2. *Angular misalignment.*

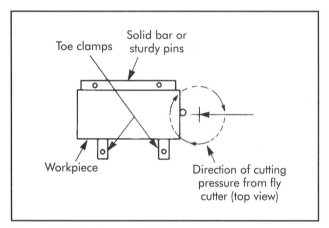

Figure 11-3. *Rigidity in tool design.*

to take the shock of the cutting tool. Clamps should be strong enough to hold the workpiece securely without distorting it.

## Speed and Ease of Workpiece Changing

The tool designer must consider loading and unloading the workpiece. Usually, the lot size and total quantity of parts to be machined will dictate the amount of money that can be spent on the fixture. Using standard fixture components can lower the build cost substantially.

In many cases, a simple clamp will work fine. This type of clamp can be manually guided when

close positioning is required. Many of the more sophisticated mechanical, pneumatic, and hydraulic clamping systems also can be used. These clamps move out of the way when released and permit easy removal of the workpiece. The fluid-operated systems provide the additional advantage of being activated jointly by a single lever, button, or programmed sequence.

Dedicated fixtures for each workpiece are not feasible. The process and time involved to provide a dedicated fixture from inception to being put on the job may be factors. As for storing large quantities of dedicated fixtures for possible reuse in the future, there is no practical solution. Either they are scrapped out and new ones are built as needed, or they are stored in space that could be better used for producing parts. Modular tooling systems were designed to solve both of these problems at the same time. These systems are called by other names, but regardless of the name, they are all kits of tooling components that can be used together in various combinations to locate and clamp workpieces for machining, assembly, and inspection operations.

The components of a typical starter kit are illustrated in *Figure 11-4*. A kit consists of mounting plates, angle plates, locators, clamps, and mounting accessories. Adapters also are available to permit the use of many standard and power-workholding devices. The method of assembling these components varies between systems.

Figure 11-4. *Starter kit of modular fixturing components. (Courtesy Fritz Werner Machine Tool Corp.)*

Fixtures made from modular tooling kits also can be used on standard machines and numerical-control (NC) machines as well as CNC machining centers. They can be positioned on U.S. or metric machine tables.

Modular tooling systems are valuable when confronted with a short lead-time or small production quantities that do not warrant the design and construction of a special jig or fixture.

## Construction

The first step in assembling a jig or fixture is to select a base large enough to handle the workpiece. Next, the main structure is constructed with riser blocks and reinforced with stop-thrust elements. Finally, the more specialized elements are added to properly locate and clamp the workpiece for machining.

An erector-set fixture is shown in *Figure 11-5* without a workpiece in place, and in *Figure 11-6* with a workpiece in place.

**Sample Parts**. An ideal method of constructing a jig or fixture with a modular tooling kit is to build it around a sample part. Simply position the sample part on the base and add locators, supports, and clamps as needed. This method reduces the construction time to a fraction of

Figure 11-6. *Erector-set fixture with a workpiece in place.* (Courtesy Flexible Fixturing Systems, Inc./Erwin Halder, Ltd.)

what it would take to design and build a dedicated jig or fixture to do the same job. The use of a sample part also can expedite frequent assembly and disassembly of jigs and fixtures.

**Template**. When no sample part is available, a jig or fixture can be assembled around a template of the part. Templates also are useful to check for interference that could occur when loading or unloading a part.

### Machining on Modular Fixtures

Jigs and fixtures assembled from modular tooling seldom need machining. Occasionally, limited machining may be required to produce a special component. Excessive machining, however, should always be avoided, since it will eliminate the economic advantages of the system. If, for some reason, the jig or fixture is impossible to construct without considerable machining, modular tooling should not be used.

### The Tool Assembler

Even though modular tooling is easily assembled, the need for tooling knowledge and experience is not diminished. Jigs and fixtures constructed with modular tooling kits must be

Figure 11-5. *Erector-set fixture.*

strong enough to withstand the machining forces imposed on them, built to use adjustable components, and often must accommodate an in-process part. The tool assembler must have the imagination and experience to foresee potential problems and plan accordingly. The selection of a tool assembler requires careful consideration.

### Pallet/Fixture Changers

The prevalence of just-in-time (JIT) production methods have made smaller lot sizes necessary and, consequently, more frequent changeover is required. The result is machines that may sit idle 30–40% of the time.

Idle time has lead to the use of manual pallet/fixture changers in vertical machining centers. One pallet slides away from the spindle to allow offline loading, unloading, cleaning, and setup work, while the machine continues making parts on another pallet (*Figure 11-7*). This process is effective on small or large lot sizes and allows changeover in less than one minute. The operator then is free to spend more time inspecting work while parts are being produced. New machining centers are now outfitted with pallet/fixture changers as standard equipment.

## ADVANTAGES OF MODULAR TOOLING

### Reduced Lead Time

Reduced lead time is the major advantage of a modular tooling system. Jigs and fixtures usually can be assembled in a few hours' time with readily available components, virtually eliminating lead time. Tooling often can be assembled in less time than it takes to prepare a tool-path program for the part.

### Adaptability

Tooling changes to accommodate new products, or revisions to existing products, are fast and easy with modular tooling. For companies experiencing frequent product changes, new products will not be held up waiting for changeover, since changes to existing tooling can be made immediately without interrupting production. Sometimes, tool trials show the need for revisions that

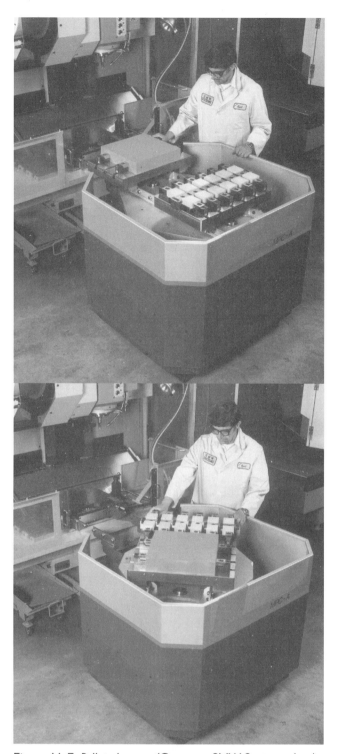

Figure 11-7. *Pallet changer.* (Courtesy SMW Systems, Inc.)

are difficult and time-consuming to make on a dedicated fixture. These changes can be made easily and quickly with modular tooling.

Even when a dedicated fixture is planned, modular tooling can be used to establish the basic design and tool clearances.

## Reusability

Although modular tooling kits may seem expensive, they usually pay for themselves in one to two years. At the completion of a production run, the modular jig or fixture can be completely dismantled and the components returned to the kit for reuse, whereas conventional tooling is usually stripped of any reusable parts and then scrapped for a fraction of its original cost. An example of this is shown in *Figure 11-8*. The fixture on the left will be useless at the end of the production run, whereas the modular fixture will be completely torn down and the components returned to the kit for reuse. Storage of the dedicated fixture for possible revision and reuse at a later date may be considered, but storage costs, when added to revision costs, usually make this an expensive proposition.

Figure 11-8. *Two methods of constructing a milling fixture.* (Courtesy Flexible Fixturing Systems, Inc./Erwin Halder, Ltd.)

## Backup Ability

Modular tooling can be swiftly assembled to temporarily replace a dedicated jig or fixture while it is being repaired, reworked, or revised. Just having backup tooling available for emergencies may make modular tooling well worth the investment.

## MODULAR TOOLING SYSTEM DESIGN

Typically, all of the major components of a modular tooling system feature a grid pattern of accurate locating holds or precision-spaced T-slots used to accurately position components to locate and clamp workpieces for machining. The major components include subplates, riser/tooling blocks, four-, six-, and two-sided tooling blocks, angle plates, and tooling cubes.

## Subplates

Subplates, or tooling plates (*Figure 11-9*), are being used at an increasing rate by progressive job shop and production manufacturers to increase productivity while lowering tooling costs. Subplates are adaptable to any machine table or pallet system and greatly increase the number of locating and clamping points available on the table surface.

Subplates are machined for an exact fit to the customer's machining center or machine, and subplate sets are available for machines using shuttle tables. The use of matched sets greatly reduces the need to indicate individual tables, thereby increasing the production capacity of multi-pallet machines.

Figure 11-9. *Subplates.* (Courtesy Mid-State Machine Products)

## Riser/Tooling Blocks

Most machining centers have unusable dead space between the centerline of the spindle and top of the machine table or pallet. This dead space varies between machines, but in each case it places limitations on the machine. To work around these limitations, operators will sometimes move the workpiece to the edge of the table to get more vertical quill movement. This arrangement, however, prevents machining the back of the workpiece without resetting the job, thereby eliminating the cost-saving potential provided by the indexable pallet.

Another method used by operators is to elevate the workpiece on blocks high enough to eliminate dead space. This method is costly in terms of setup time. Riser/tooling blocks (*Figure 11-10*) offer a workable alternative to the time-consuming double setups and eliminate the need for unstable parallels. The riser/tooling block is mounted squarely in the center of the machining-center pallet. The dead space is eliminated by the additional height of the block, thereby allowing the operator to make full use of the machine's in-

dexing capacity to machine up to five sides of a workpiece in one setup. The riser/tooling block's heavy-duty construction, along with qualified dowel holes or T-slots, ensures that setups will be solid and repeatable.

## Four-sided Tooling Blocks

Four-sided tooling blocks (*Figure 11-11*) are designed for use on horizontal machining centers and provide four identical surfaces for attaching workpieces or other components. When mounted on a rotary table or fourth axis, they can be indexed 90° to present four work setups to the cutting tool in rapid succession. In some cases, even the top surface is used to locate the workpieces for machining. Six-sided tooling blocks (*Figure 11-12*) are used in the same manner as the four-sided blocks, but with the addition of two identical mounting surfaces.

## Two-sided Tooling Blocks

Two-sided tooling blocks (*Figure 11-13*) are set up on the machine table in the same manner as four-way blocks, but provide more *Z*-axis travel. When mounted on a rotary table or fourth axis, they can be indexed 180° to present two work setups to the cutting tool in rapid succession. Two-sided tooling blocks are the logical solution to mounting workpieces too large to clear the

Figure 11-10. *Riser/tooling blocks.* (Courtesy Mid-State Machine Products)

Figure 11-11. *Four-sided tooling blocks.* (Courtesy Mid-State Machine Products)

Figure 11-12. *Six-sided tooling blocks.* (Courtesy Mid-State Machine Products)

Figure 11-13. *Two-sided tooling blocks.* (Courtesy Mid-State Machine Products)

spindle or coolant-chip shield if mounted on a column subplate. The open-frame design provides an access opening so the spindle can reach the back of the workpiece after indexing. Operations then can be performed on the front and back of a workpiece using the same setup.

## Angle Plates

Angle plates (*Figure 11-14*) are ideal for machining operations where a four-sided tooling block is neither required nor economically feasible. They are also suited for applications that require the fixture mounting near the front edge of the pallet, extra thick workpieces that take up most of the pallet, and workpieces that require their centerline to also be on the centerline of the pallet.

## Tooling Cubes

Tooling cubes (*Figure 11-15*) are designed for use on many of the newer machining centers equipped with pallet changers having limited weight capacities. The inside of the tooling cube is hollow and necked down at both ends to reduce weight while still providing sufficient rigidity to resist intense machining forces. Tooling cubes can be located quickly against edge stops with dowel pins or parallels. They usually are bolted to the

Figure II-14. *Angle plates.* (Courtesy Mid-State Machine Products)

Figure II-15. *Tooling cubes.* (Courtesy Mid-State Machine Products)

pallet through the center with threaded alloy bolts using either standard straps or a made-to-fit cover.

## Self-adjusting Fixturing Elements

The major components of a modular tooling system with self-adjusting fixturing elements include grid plates, pallets, consoles, tombstones, double angle plates, and angle plates in various sizes.

Mounting holes are multipurpose with 0.50-in. (12.7-mm) diameter bushings at the top and 1/2–13 threaded inserts at the bottom, or with 0.63-in. (16.0-mm) diameter bushings at the top and 5/8–11 threaded inserts at the bottom. They are located on a grid pattern with either 2.0000 in. ±0.0016 in. (50.800 mm ±0.041 mm) or 2.0000 in. ±0.0004 in. (50.800 mm ±0.010 mm) spacing between centers. Components can be attached using dowel pins to locate and bolts to clamp them with, or by using two shoulder bolts that both locate and clamp.

With conventional clamping, castings and forgings—which are often badly warped—usually are clamped at three points using strap clamps. The warped under-surface locates unevenly at these points, as shown in *Figure 11-16a*. When clamping pressure is applied, severe strains are produced (*b*). After machining, the surface will spring out flat when the clamps are released (*c*), resulting in scrap or rework.

With the S.A.F.E. system, the hardened steel elements "float" in their sockets to adjust to work-

Figure II-16. *Modular tooling system with self-adjusting fixturing elements.* (Courtesy Enerpac Group, Applied Power, Inc.)

piece irregularities and surface profile, as is shown in (d). Upper arms of the clamps adjust in the same manner during clamping (e), without causing distortion of the workpiece. When the workpiece is removed, the machine surface does not change.

## QUICK-CHANGE TOOLING

Machine tool productivity is greatly enhanced by sharp reductions in setup time, workholding, and tooling. Vast changes have been made to meet the demands of flexible manufacturing cells, automated parts-handling systems, and computer-integrated manufacturing. Quick-change tooling is one way to increase productivity and decrease setup time by presetting jobs and using offline setup methods, such as the pallet/fixture changers discussed earlier in this chapter.

### Components

Quick-change tooling consists of two parts: the clamping unit and the cutting head. The clamping unit mounts to the machine tool and acts as a receptacle for the interchangeable cutting unit (*Figure 11-17*).

To change tooling, the machine operator simply releases the locking system, changes the cutting unit, and locks the new tool into position. The operator then makes the offset adjustment according to the previously recorded data and continues machining the part. The total machine downtime is only about 30 seconds.

Figure 11-17. *Quick-change clamping unit and cutting head.*

### Strategies for Machining

Understanding the impact of quick-change tooling on productivity requires a review of the three basic functional areas of the manufacturing process: inventory planning and control; preproduction planning and setup reduction; and in-process machining.

#### Inventory Planning and Control

The primary objective of inventory planning and control is to maintain enough finished goods inventory to meet customer demand while keeping levels low enough to minimize costs. By producing smaller lots more frequently, inventory carrying costs can be reduced, as can shelf life and problems such as rust, contamination, and deterioration.

#### Preproduction Planning and Setup Reduction

The type of planning previously discussed necessitates setup reduction. In the preproduction planning phase of the manufacturing process, all elements are identified, organized, and scheduled in advance of the production run. The objective is to eliminate as many setups as possible and improve machine and operator efficiency during setup and in-process machining.

A setup reduction program can reap as much as a 75% reduction, as well as drastically reduce tool-change time during in-process machining. Such a program generally consists of a number of factors.

- Quick-change tooling: change of an entire pregaged cutting unit, as opposed to changing an individual insert or tool.
- Tool kitting: the identification, organization, and assembly of all tooling necessary to complete a production run or shift of operation.
- Pregaged tooling: once all tools are fitted with new cutting edges and assembled in the tool taxi, the "F" and "C" dimensions are measured and recorded in advance of production.
- Preproduction tool maintenance: all tool maintenance is performed in advance of the production run to avoid catastrophic tool failure and in-process tool maintenance.

- Advanced cutting-tool materials: materials such as ceramics and polycrystalline diamonds permit longer and faster production runs.
- Tool management software: this provides computerized tracking of tools and assists in the kitting process.

Preproduction planning is the least difficult and least expensive phase of the manufacturing process. However, it can provide a significant payback through reduced machine downtime.

### In-process Machining

Most of the factors that inhibit productivity can be eliminated through effective preproduction planning. While setup reduction programs most often address downtime resulting from tool change, tool maintenance, and part setup, the downtime incurred to adjust tool position for part accuracy is much less understood. This process is referred to as tool offsetting and, by traditional methods, requires an 8-minute, 17-step sequence of operations to adjust tool position and produce dimensionally accurate parts. Quick-change tooling reduces this process to a four-step procedure that takes about 30 seconds to perform.

### Presetting

To achieve maximum machine productivity, presetting should encompass three functions. The first area is workholding, in which chucks, chuck jaws, fixtures, pallets, etc., must be changed each time a new lot of parts is to be run.

The second area is toolholding, where part changeover necessitates the exchange of an entire toolholder but is composed of changes for tool wear and breakage during normal operation. This includes simple procedures, such as the indexing of an insert, in which tolerances of milled pockets and tools may not lend themselves to precision tool setting (*Figure 11-18*).

Typically, after indexing an insert, the tool is backed off a predetermined amount, a trial cut performed, the dimension checked, an offset entered into the machine (either manually or automatically), and a final cut taken. Since cutting pressures can affect metal removal, the subse-

Figure 11-18. *Toolholder presetting.* (Courtesy, ITW Woodworth)

quent part cycled, taking a cut at full depth also will require inspection to verify that the compensation factor was correct. In the event that the offset was incorrect, at best the process has to be repeated. At worst, the workpiece must be scrapped. Taking into account that many machines are equipped with multiple tools and spindles, and that tools frequently require dual offsets (for example, radial and depth), it is no wonder that production rates scarcely reach 40% of machine uptime.

The third area is part registry, which can be considered a natural extension of the capabilities to preset workholding devices. The concept of *part registry* can be defined as the accurate positioning of workpieces within a workholding device to a specified location, precisely simulating machine mounting conditions, yet prior to setting within the machine tool (*Figure 11-19*).

Part registry has grown in importance as the complexity and quality requirements (for example, spatial dimensions and true positioning) of machined parts has increased. This has had an impact on the setup time of high-value, intricate, and fragile workpieces, where the loading, fixturing, and handling of a part are recognized as critical elements of producing a component to specification. Consideration should be given to applications involving heavy, unwieldy parts in which machine loading is time-consuming and adds operator fatigue to the equation of machine productivity. Part registry also lends impetus to the trend toward chucking a part once for multiple

Figure 11-19. *Part registry.* (Courtesy ITW Woodworth)

machine processing and transfer between machines for improved quality and productivity.

## AUTOMATIC IDENTIFICATION

The term automatic identification (Auto ID) is applied to barcoding and other forms of keyless data entry. The goal of Auto ID is the reliable identification of physical objects.

Accuracy, speed, and reliability are reasons for implementing an Auto ID system. Automatic identification systems are fast. Not only is data input quick and dependable, but the information is provided in a virtual instant to almost any database or software package. This allows the user to make immediate and informed decisions. Without real-time information provided by Auto ID, quality control programs can only be partially realized.

Barcoding is easy to learn and use, and relatively inexpensive to implement. By reading barcodes, data can be acquired 100% faster and

10,000–15,000 times more accurately than by any other means of manual input.

## Barcodes

The concept behind barcode technology is identification. Any barcode symbolizes a distinctive mark of identification for products, comparable to that of a fingerprint. Barcoding can be applied to almost any aspect of the manufacturing enterprise. Quality control and tracking product life from production to customer can be streamlined and made more efficient with the use of barcoding.

The symbol itself, the information contained in the symbol, or the language used may be referred to as the barcode. The symbol is the actual physical barcode that appears on an object being scanned. The code refers to the information encoded in the symbol. The symbology is the language spoken by the symbol. Each barcode symbology has its own unique characteristics. For example, some symbologies can only encode numbers (numeric); others include letters and numbers (alpha-numeric). In addition to what a barcode symbology can encode, the symbology also can vary in density and character length of the printed code.

## Symbology for Manufacturing

The term "symbology" denotes each particular barcode scheme. To encode data, each symbology uses a unique set of rules that determines the type of information (numeric, alpha-numeric, or ASCII characters) that can be placed in the symbology. Each one-dimensional barcode is a design of wide and narrow bars and spaces called elements that represent the information. Over 60 different types of symbologies have been developed.

## Effective Tool Management, Reporting, and Tracking

Effective tool management is a critical factor for any industrial manufacturer. The accurate reporting of who regrinds tools, what departments or operators are using various tools, and tracking tools for inventory purposes can be made faster and more effective with barcoding.

Each tool type, identification number, description, and location is entered into a database. Identification numbers (barcodes) are assigned to each user/department of every tool or setup of items released from tool inventory. A barcode system has the capability to track items released to departments, work centers, job orders, or even operators. A numbering system, or the modification of one, would place the tools into logical groupings. For example, a two-digit prefix may indicate the tool category, such as 02 for straight-shank drill, 04 for reamers, and so on. The use of suffixes could be used for reporting purposes.

A barcoded tool-reporting system can effectively manage the regrinding, usage, and reporting of hundreds of tools and setups. Barcodes can be attached directly to large items. Small tools, such as drill bits, can have a barcode attached to the container/bin where they are stored. A pre-printed barcode menu sheet identifies what has been taken or disbursed. Operators, departments, or work centers may have an associated barcode to record normal issuance and returns. This enables supervisors/foremen to balance tool inventory and regrinding accordingly.

Additionally, tracking information found on traditional grinding-crib time cards can be scanned off of a preprinted barcode menu sheet. This speeds the collection of information by employees who regrind tools, allowing them to spend more time on the actual regrinding operation.

Information obtained can be reported in a timely manner with a tool-usage report. The report helps determine what to buy and when, and what to scrap. The report also can establish a reorder point for each tool and generate a report of all tools at or beyond their reorder levels.

Timely information is an essential component to improving daily tool management. In the few cases when a tool is not in the crib, all current locations and quantities of the tool type can be quickly ascertained. The system can provide invaluable assistance in reducing hidden inventory that exists in every shop and ensure the proper inventory for daily and weekly workloads.

## REFERENCES

Boyes, William E. 1989. *Handbook of Jig and Fixture Design*, 2nd ed. Dearborn, MI: Society of Manufacturing Engineers.

Brown, Charles R. 1991. "Strategies for Innovative Machining." SME Tech Paper TE91-145. Dearborn, MI: Society of Manufacturing Engineers.

DeLonghi, Charles S. 1991. "Quick Change Tooling for Today…and Tomorrow." SME Tech Paper TE91-144. Dearborn, MI: Society of Manufacturing Engineers.

Diehl, Werner K. 1988. "The Productive Advantages of Preset Workholding, Toolholding and Part Registration to Reduce Setup Time." SME Tech Paper TE88-125. Dearborn, MI: Society of Manufacturing Engineers.

Miller, Richard K. 1987. *Automated Guided Vehicles and Automated Manufacturing*. Dearborn, MI: Society of Manufacturing Engineers.

SMW Systems, Inc. *Here's How to Double the Output of Your Machining Center*. Video. Santa Fe Springs, CA: SMW Systems, Inc.

## REVIEW QUESTIONS

1. In what manufacturing situations are modular tooling systems most valuable?
2. How do pallet/fixture changers increase productivity?
3. List four major advantages of modular tooling.
4. Name the major components of a typical modular tooling kit.
5. Name the components of quick-change tooling.
6. Name the three basic functional areas of the manufacturing process.
7. To what does the term "bar code symbology" refer?
8. List at least four items that should be part of a tool management database.

# ◄12►

# The Computer in Tool Design

Over the last 25 years, many companies have completed the transition from manual drawings to computer-aided design (CAD). As a result, students and engineers are now familiar with these tools as a normal way of doing business. The emphasis today in manufacturing is how information moves from one location to the next.

The most important development in computer and CAD systems in the last few years has been the introduction of 3D solid-modeling software. Solid modeling allows a design engineer to create a precise mathematical and graphical representation of the final design.

All downstream applications, including tool design, can leverage the solid model to carry forward the original design intent. The use of 3D CAD also can significantly reduce the time required to generate 2D drawings. In many systems, the drawings generated from the solid model are linked with one another so changes to the model are automatically reflected in the drawing.

Although a 3D solid model generated by design engineering may be mathematically complete, the geometric dimensioning and tolerancing (GD&T) information necessary for manufacture is not included. Some CAD systems have addressed this by allowing a GD&T annotation to be incorporated into the 3D solid model. As CAD software evolves, this technique will be commonplace. This will virtually eliminate the need for 2D drawings and those from the critical design path for a tool development project in the future. Drawings may remain in shops, but more as a reference tool than the official design definition. For those not familiar with 3D solid modeling, refer to the description later in this chapter.

While CAD is the most common computer application being used in tool design, rapid prototyping, simulation, tolerance analysis, design document management, and 3D solid modeling are also important.

## COLLABORATIVE ENGINEERING

For any tool design to be a success, the designer has to be involved early with the product design engineering staff. The tool designer, however, cannot actually commit tooling for production until the product to be built is finalized. While the ideal environment would be for the product and tool design personnel to use a common database, tool designers are not typically aligned with design engineers. In modern multinational corporations, people can be located anywhere. Therefore, while a common database is desirable, it is not practical.

Tool designers and product design engineers can share information in ways that were not possible just a few years ago with the advent of the Internet and collaboration software. For example, people in multiple locations can review a product design simultaneously.

Various types of collaboration software products enable simultaneous viewing of CAD models or a representation of them. Collaboration software provides simultaneous sharing of interactive CAD sessions. Most of these systems allow any of the users in the session to take control of it from their location. Communication methods that provide sharing and access capabilities include the World Wide Web, allowing outside companies access through a company's firewall, or a combination. Given current bandwidth capacities, a typical collaboration session can be a combination of telephone audio and sharing screens over the Internet. Participants also can initiate data transfers during the session. Required security levels normally are dictated by the policies of the respective company.

## Engineering Design Transfer

The evolution of 3D solid modeling has given the tool designer use of the same tools as the product design engineer. In addition to interactive viewing and collaboration, actual data transfer is facilitated by the Internet. Traditional issues caused by dissimilar CAD systems still exist because tool and product designers often do not use the same CAD system. CAD data must be translated using some neutral format or a direct translator. The most common neutral formats are initial graphics exchange specification (IGES), standard for the exchange of product (STEP), and drawing exchange format (DXF) (*Figure 12-1*). DXF is a standard from the developers of AutoCAD. Due to the longevity and dominance of AutoCAD in the CAD world, DXF has become one of most widely used data exchange formats. DXF is an American standard code for information interchange (ASCII) text file containing both 2D and 3D geometric and topological information.

In 1980, the ANSI Y14.26 standards committees voted to adopt the IGES. Like DXF, IGES was designed to write to an ASCII text file. In 1984, the International Organization for Standardization (ISO) began trying to combine the European and U. S. standards into a single standard, STEP model data. This standard attempts to integrate European standards and IGES with standards that contain larger subsets of information. These data exchange formats have evolved. STEP now is primarily used for solids, and IGES and DXF for wire-frame and 2D. Because implementation of these formats vary among CAD systems, all translators should be

thoroughly tested with benchmark files before being used for production.

## RAPID PROTOTYPING AND MANUFACTURING

Rapid prototyping and manufacturing (RP&M) encompasses a group of technologies capable of directly producing physical objects from computer-generated data. This data, whether from CAD files, computer-aided tomography (CAT) scans, or 3D digitizers, serves to guide a mechanical system that builds a solid object in thin, horizontal cross-sections with each layer adhering to the previous one.

RP&M technology enables designers, engineers, and tool developers to join together liquid, powdered, extruded, and sheet materials to form prototypes, scaled models, functional models, casting patterns, shell molds, actual production parts, and short-run tooling (*Figure 12-2*).

Rapid-prototype parts and assemblies typically used in design engineering during the initial design stages also can communicate product designs to tool designers. Rapid-prototype models have been used to check tooling designs for fit and function.

RP&M technology was initially developed to reduce the lead time required to produce prototypes used to evaluate studies by engineers and designers. Now, this technology is used to reduce part costs and increase the speed of production and durability via new construction materials. The processes have been applied in the biomedical and microelectronics fields.

Stereolithography, or 3D printing, was the name given to the initial process. As other technologies developed, other terms followed, such as desktop manufacturing, toolless model-making, and automatic fabrication. Today, the generally accepted name is RP&M, though the term "solid free-form fabrication" also is used.

### Innovations

Increasingly, manufacturing industries need to reduce product development time to be competitive. RP&M is a fast approach to the design, prototype, and manufacture of new products. This

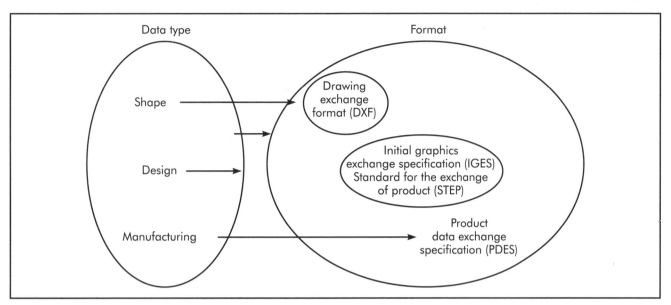

Figure 12-1. *The types of information contained in neutral data exchange formats. Data exchange standards are designed to support differing amounts of information relating to design and manufacturing.* (Courtesy Richard D. Irwin, Inc.)

Figure 12-2. (a) *A stereolithography process, and* (b) *the finished product.*

need to reduce product development time, together with the growth of computers in design offices, has motivated inventors to create new ways of producing physical objects from computer model data.

Rapid prototyping had its beginning in the late 1970s. The first system combined two emerging technologies—CAD and photo-curable polymers—into a machine that fabricated intricate geometries by building them one layer at a time. The first commercial RP systems became available in the late 1980s. Early applications were limited to basic prototyping, but with the fabrication of functional parts, RP&M has developed into a technology capable of improving visualization, shortening product development time, optimizing product quality, and discovering design flaws.

While some industries are just starting to use RP&M for the first time, others are acquiring their 10th system. A number of organizations have three or four entirely different systems from various vendors, since each fundamental technology produces pieces with different end properties at different costs. RP&M processes are dramatically altering the way companies design, model, prototype, tool up, and manufacture new products.

## Methodology

To create a 3D computer model of a part to be built, digital data from any of various modeling packages are converted into a standard stereolithography (STL) file format and transferred to the RP&M computer.

CAD software creates models in various forms: two-dimensional, two-and-one-half dimensional, three-dimensional wire-frame, three-dimensional surface, and three-dimensional solid. Yet, RP&M processes require CAD data as either solid models or 3D-surface wire-frame (*Figure 12-3*). Agreements are being developed that facilitate accepting other data formats in addition to STL

files. The STL file approximates the shape of a solid model using a mesh of small triangles called facets. The smaller the facet size, the more accurate the surface approximation, at the expense of increased file size and processing speed. The vertices are ordered to indicate which side of the triangle contains the mass. If one were to view the contents of an STL file, there would be a list of $X$, $Y$, and $Z$ coordinates that describe a surface mesh made of triangular facets (*Figure 12-4*).

Figure 12-4. *A typical computer screen shot of an STL file.* (Courtesy PrimeTech)

RP&M system software cuts the STL file data into thin horizontal cross-sections, generally 0.005–0.020 in. (0.13–0.50 mm) thick. The system-control software then uses the stack of digital cross-sections to produce each fused or joined layer of material, one on top of the next. A number of RP&M systems are commercially available in the United States, Europe, and Japan, with many others are under development.

- Stereolithography (*Figure 12-5*) uses an ultraviolet (UV) light source to cure liquid resin selectively, thus forming the desired model.
- Selective laser sintering (SLS) works with an infrared laser to sinter plastic powders and waxes.
- Fused deposition modeling (FDM) (*Figure 12-6*) employs an extrusion head to melt and deliver plastic or wax wire to build a model.

Figure 12-3. *Example of a 3D solid model.* (Courtesy Jergens, Inc.)

Figure 12-5. *Sample of a cured liquid-resin part.* (Courtesy DSM Somos)

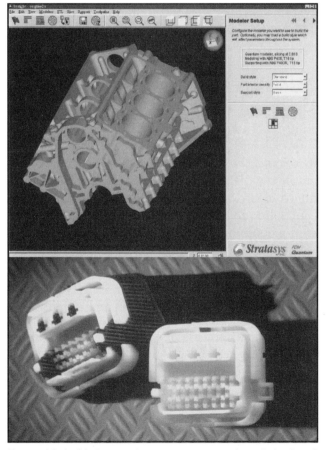

Figure 12-6. (a) *A typical computer screen shot of the fused deposition modeling process, and* (b) *a finished product.* (Courtesy Stratasys, Inc.)

• Direct-shell production casting (DSPC) uses an ink-jet nozzle to apply a liquid binder to ceramic powder. This process is unique in that it creates a shell mold from which a part can be cast, whereas the other processes produce the part directly.

Each has unique features and applications, making it unlikely that any one system will totally dominate the market. These systems can help manufacturers reduce time-to-market cycles and get their products out more quickly, an important factor in an increasingly competitive world market.

## 3D Printing

A newer generation of RP&M technology is generally referred to as 3D printing or desktop RP (*Figure 12-7*). These office-friendly machines can sit next to CAD workstations, office copiers, and fax machines. Some are used by the design engineer for concept modeling and early design review and approval. Rather than submit an RP&M job and wait for days, 3D printers bring the technology within reach, literally, of design groups running CAD solid-modeling software. While the term 3DP (three-dimensional printing) was trademarked for use with the technology developed at MIT, it is now commonly applied and accepted across multiple technologies. Companies

Figure 12-7. *A common 3D printing system.* (Courtesy Stratasys, Inc.)

with commercial licenses from MIT use a wide variety of proprietary base powders and liquid binders to form products from starch, metals, plaster, ceramics, and even medicines and real or synthetic bones.

## Applications

RP&M applications include the production of patterns for shell investment-cast functional metal parts, development of soft and hard tooling, human replicas, art, jewelry, perfume containers, all kinds of automotive and aerospace parts, and photoelastic part studies. The variety of parts and assemblies is extensive.

At Chrysler Corporation, RP&M was used to verify the form, fit, and function of parts. The Chrysler design team developed a new automatic transmission shift handle in the late 1980s. To verify the form of a part, a solid model was created in Chrysler's CAD system and the part built using stereolithography. The 3D model revealed several problems with the original design. The handle was too large for a person with small hands and uncomfortable for gear shifting. Following this discovery, three additional models were made at 8%, 10%, and 12% reductions in size with very little time delay. It was determined that the RP&M process saved $40,000 and 18 weeks of additional design time over the conventional proto-typing method. The traditional method would have required a functional prototype shift lever to be manufactured before the sizing problem would have been discovered (Jacobs 1992).

In another application of RP&M, Chrysler used it to prove the fit of a starter. The starter was designed for a European export version of the Jeep Cherokee with a turbo diesel engine. It was redesigned in France, but existed only as a CAD drawing. The CAD file was downloaded to an RP&M machine at Chrysler and a starter housing was built. Not only did the starter fit, but the battery cable and wire harness were able to be developed as well.

## Service Bureaus

Several service bureaus were founded and began providing service to manufacturers in the late 1980s. RP&M became available in Europe and Japan via this expanding base of service bureaus.

An increased need for RP&M gave rise to the establishment of service bureaus that would allow companies to benefit from the technology without actually purchasing and operating the equipment. Many companies, after utilizing service bureaus for a time, decide to purchase their own RP&M machines. Service bureaus allow potential users the opportunity to experiment before buying the technology, or simply to use the latest technology without investing in equipment that might become dated shortly thereafter.

However, if parts are built in-house, they may be turned around more quickly than at a service bureau. Additionally, control is maintained and secrets are kept inside the company. If a system is purchased, a budget for the people needed to operate the system and finish parts must be carefully researched. Consider the cost of training, system maintenance agreements, materials and other consumables, and any facility changes that the system might require.

## Cost Justification

Most RP&M users employ CAD software with solid-modeling capabilities. This limitation often requires the change to different software and incurring the associated costs of hardware and employee training. Solid-modeling software produces closed volumes and ensures a good data file. Solid modeling is the place to start if RP&M is in a company's future plans.

The decision to purchase RP&M systems depends on several factors, such as the number of parts to be built over a period of time. For example, if two parts a month are built at an average cost of $1,000/part, $24,000/year is spent on prototype parts. This may not be enough volume to justify the purchase of an RP&M system. However, if five parts a month are built at the same average cost per part, $60,000/year is spent. In this instance, it may make sense to consider the purchase of a system (Wohlers 1996). RP&M systems range in price from about $100,000–500,000, while 3D printers are offered at $40,000–70,000 in 2003 dollars.

## SIMULATION

A tool designer involved in building a complex machine or mold often will use simulation and analysis to study mechanisms. Full-motion characteristics of the mechanical design can be viewed and analyzed before valuable manufacturing resources are committed.

Simulation software has saved significant amounts of time and materials in the casting process as well. Software analyzes the flow of metal being poured into a mold and determines hot and cold spots and voids. The mold designer working with the design engineer can adjust the mold or casting process before the first part is ever made.

## TOLERANCE ANALYSIS

Tolerance analysis software is used to evaluate, optimize, and validate the capability of proposed designs. The primary purpose of tolerance analysis is to make sure that when all tolerances are summed, or stacked, in a worst-case situation, the final design is within the original product specifications.

## DESIGN DOCUMENT MANAGEMENT

Tool designs, as well as the physical tools themselves, must be controlled and stored so they can be easily retrieved. Changes in product design often can require a change to tooling as well. The information contained on tool design drawings is critical to the maintenance of tools and machinery. It is not unusual to request that all of the tooling associated with a product or project be identified. Document management systems used to manage engineering design drawings can be tailored for tool design as well.

The future direction for corporations will be to establish corporate-wide databases from existing isolated pockets of product design, tool design, and other information. The challenge is to set up information so future employees have access to lessons learned from previous projects.

## 3D SOLID MODELING

Without the ability to represent the third dimension, 2D CAD must depend on conventional practices, such as multiviews, to represent a 3D object indirectly. Because of this, the usefulness of the design process is limited. Digital 3D modelers create models with a closer one-to-one correspondence to the real object. Solid-based modeling can be created in the computer without the time and effort necessary to build real-world equivalents (*Figure 12-8*). Although the ultimate goal is to manufacture or construct the real object, the goal in the design process is to experiment and test numerous solid models in the early stages of design, which can speed the process considerably.

Three-dimensional modeling plays an important role in many manufacturing techniques, including CAD, computer-aided manufacturing (CAM), computer-integrated manufacturing (CIM), concurrent engineering (CE), and design for manufacturability (DFM). All are aimed at shortening the design cycle, minimizing material and labor expenditures, raising product quality, and lowering the cost of the final product. Central to

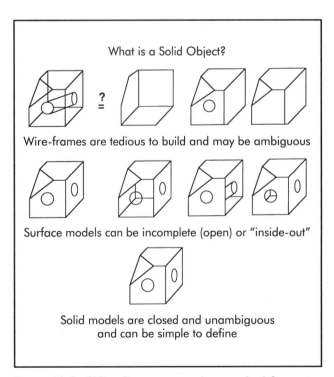

Figure 12-8. *CAD software systems have evolved from wire-frame geometries to surface models to the current generation of solid modelers. Solid models are closed (no holes in geometry) and unambiguous. (Courtesy Structural Dynamics Research Corporation)*

these goals is better communication within a company. By sharing a 3D database of information for a proposed product, more people can work simultaneously on various aspects of the design problem. The graphic nature of the database has convinced many that 3D modeling is a superior method of communicating many design intents.

## REFERENCES

Jacobs, P. F. 1992. *Rapid Prototyping and Manufacturing: Fundamentals of Stereolithography.* Dearborn, MI: Society of Manufacturing Engineers (SME).

Wohlers, T. 1996. *Rapid Guide to Rapid Prototyping.* Minneapolis, MN: Wohlers Associates, Inc./Stratasys, Inc.

## BIBLIOGRAPHY

Bedworth, D., et al. 1991. *Computer-Integrated Design and Manufacturing.* New York: McGraw-Hill, Inc.

Bertoline, G. R., et al. 1995. *Fundamentals of Engineering Design Graphics.* Burr Ridge, IL: Richard D. Irwin, Inc.

Conkol, G. K. (ed). 1994. *The Role of CAD/CAM in CIM: The Perspective.* Dearborn, MI: Computer and Automated Systems Association (CASA) of the Society of Manufacturing Engineers (SME).

Jacobs, P.F. 1996. *Stereolithography and other RP&M Technologies.* Dearborn, MI: Society of Manufacturing Engineers (SME).

Marks, P. (ed). 1994. *Process Reengineering and the New Manufacturing Enterprise Wheel: 15 Processes for Competitive Advantage.* Dearborn, MI: Society of Manufacturing Engineers (SME).

Mills, A. 1998. *Collaborative Engineering and the Internet.* Dearborn, MI: Society of Manufacturing Engineers (SME).

Orady, E. A. 1990. "Design and Development of an Expert System for Simultaneous Engineering." SME Technical Paper MS90-431. Dearborn, MI: Society of Manufacturing Engineers (SME).

——. 1989. "Design and Development of Manufacturing Evaluation System for Rotational Parts." pp. 51–56. Proceedings 15th Annual Conference of ASME. New York: American Society of Mechanical Engineers.

*Rapid Prototyping Systems: Fast Track to Product Realization.* 1994. Dearborn, MI: Society of Manufacturing Engineers (SME).

Schrader, G., and Elshennawy, A. 2000. *Manufacturing Processes and Materials.* Dearborn, MI: Society of Manufacturing Engineers (SME).

Shetty, D. 2002. *Design for Product Success.* Dearborn, MI: Society of Manufacturing Engineers (SME).

*The New Manufacturing Enterprise Wheel.* 1993. Dearborn, MI: Society of Manufacturing Engineers (SME).

*Tool Designer's Assistant.* 1993. St. Louis, MO: Carr Lane Manufacturing Co.

Zied, A. 1991. *CAD/CAM Theory and Practice.* New York: McGraw-Hill, Inc.

## REVIEW QUESTIONS

1. Identify the most significant trends concerning CAD in tool design.
2. What does the concept of "collaborative engineering" provide during the tool design process?
3. Illustrate the relationships between data types and associated formats.
4. What is the most widely used data exchange format?
5. What are the most common neutral formats used for data transfer?
6. Define the acronyms CAD, CAM, CIM, DFM, CE, RP&M, DXF, GD&T, CAT, and ASCII.
7. What constitutes the basic concepts of RP&M?
8. What does an STL file approximate?

# 13

# Geometric Dimensioning and Tolerancing

Geometric dimensioning and tolerancing (GD&T) methodology uses an international engineering language of dedicated symbols, rather than words, on blueprint drawings to detail a product in three-dimensional perspective. A precise mathematical lexicon describes the form, orientation, and location of a part's features within a zone of tolerance. Illustrations of basic geometric capabilities are shown in *Figure 13-1*.

## SYMBOLS AND DEFINITIONS

To properly gage geometric dimensioned and toleranced parts, it is first necessary to understand the terms and symbols mandated by the American Society of Mechanical Engineers (ASME) Y14.5M-1994 (R-1999) and the International Organization for Standardization (ISO), as shown in *Figure 13-2*.

*Maximum Material Condition* (MMC) is a modifier showing that the feature size contains the maximum amount of material, for example, the minimum diameter of a hole and maximum diameter of a shaft. MMC is designated by the symbol Ⓜ.

*Regardless of Feature Size* (RFS) is another modifier requiring that the tolerance of form, runout, or position be met regardless of where the feature lies within its size tolerance. In the ISO and ASME systems, there is no symbol for RFS, and it is assumed unless MMC or LMC is specifically designated. The RFS principle is understood to indicate that a geometric tolerance or datum reference applies at any increment of size of the feature within its size tolerance. No matter what the produced feature sizes are, RFS permits no additional positional, form, or orientation tolerance.

At *Least Material Condition* (LMC)—a modifier opposite to MMC—the feature size contains the minimum amount of material, for example, the maximum diameter of a hole and minimum diameter of a shaft. It is used to permit additional tolerance when the accuracy of location can be relaxed as the feature size approaches MMC. LMC is designated by the symbol Ⓛ.

*Projected tolerance zone* is used to control the 90° angle of a hole into which a pin, stud, screw, etc., will be assembled. It does this by extending a tolerance zone beyond the surface of the part to the functional length of the assembled pin, screw, stud, etc. A projected tolerance zone is indicated by the symbol Ⓟ.

*Basic dimension* is a theoretical value used to describe the exact size, shape, or location of a feature. A basic dimension is considered perfect. A tolerance is always required with a basic dimension to show the permissible variation from the stated value. A basic dimension is symbolized by boxing 1.000.

A *datum feature symbol* identifies the feature of a part from which functional relationships are established. It is symbolized by boxing the datum letter as shown in *Figure 13-3*.

Each feature requiring identification as the datum uses a different letter(s). To eliminate confusion, the letters I, O, and Q are not used as reference letters. Where the alphabet is used up, double letters, such as AA, BB, ZZ, etc., are used.

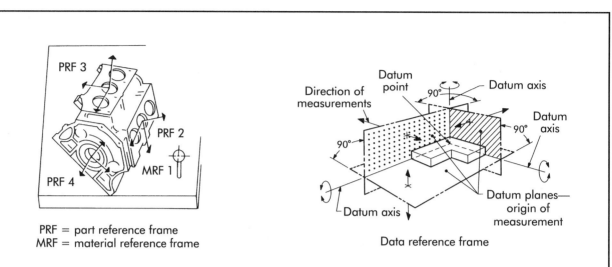

PRF = part reference frame
MRF = material reference frame

Data reference frame

| Basic Geometric Capabilities | | | | | | |
|---|---|---|---|---|---|---|
| Input Features | Relationship | | Feature Constructions | | | |
| | Shortest Distance | Acute Angle ∠ | Point | Normal Line | Parallel Line // | Normal Plane ⊥ |
| Point & Point | | N.A. | Midpoint | η = 2 | N.A. | Mid-plane |
| Point & Line | | N.A. | N.A. | N.A. | N.A. | N.A. |
| Point & Plane | | N.A. | N.A. | N.A. | N.A. | N.A. |
| Line & Line | | | Intersection | | N.A. | Mid-plane |
| Line & Plane | | | Intersection | N.A. | Projection | |
| Plane & Plane | | | N.A. | N.A. | Intersection | N.A. |

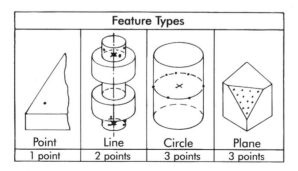

| Feature Types | | | |
|---|---|---|---|
| Point | Line | Circle | Plane |
| 1 point | 2 points | 3 points | 3 points |

Figure 13-1. *Basic geometric capabilities (ASME Y14.5M-1994).*

Figure 13-1. *continued.*

| Symbol for: | ASME Y14.5M | ISO |
|---|---|---|
| Straightness | — | — |
| Flatness | ⟗ | ⟗ |
| Circularity | ○ | ○ |
| Cylindricity | ⌭ | ⌭ |
| Profile of a line | ⌒ | ⌒ |
| Profile of a surface | ⌓ | ⌓ |
| All around | ⟲ | ⟲ |
| Angularity | ∠ | ∠ |
| Perpendicularity | ⊥ | ⊥ |
| Parallelism | // | // |
| Position | ⊕ | ⊕ |
| Concentricity (concentricity and coaxiality in ISO) | ◎ | ◎ |
| Symmetry | ≡ | ≡ |
| Circular runout | *↗ | ↗ |
| Total runout | *↗↗ | ↗↗ |
| At maximum material condition | Ⓜ | Ⓜ |
| At least material condition | Ⓛ | Ⓛ |
| Regardless of feature size | None | None |
| Projected tolerance zone | Ⓟ | Ⓟ |
| Tangent plane | Ⓣ | Ⓣ |
| Free state | Ⓕ | Ⓕ |
| Diameter | ∅ | ∅ |
| Basic dimension (theoretically exact dimension in ISO) | 50 | 50 |
| Reference dimension (auxiliary dimension in ISO) | (50) | (50) |
| Datum feature | *⟤Ⓐ | *⟤ or *⟤Ⓐ |
| Dimension origin | ⊕→ | ⊕→ |
| Feature control frame | ⊕ ∅0.5Ⓜ A B C | ⊕ ∅0.5Ⓜ A B C |
| Conical taper | ▷ | ▷ |
| Slope | ◹ | ◹ |
| Counterbore/spotface | ⊔ | ⊔ |
| Countersink | ∨ | ∨ |
| Depth/deep | ⩓ | ⩓ |
| Square | □ | □ |
| Dimension not to scale | <u>15</u> | <u>15</u> |
| Number of places | 8X | 8X |

*May be filled or not filled

Figure 13-2. *Geometric characteristic symbols (ASME Y14.5M-1994).*

| Symbol for: | ASME Y14.5M | ISO |
|---|---|---|
| Arc length | $\overgroup{105}$ | $\overgroup{105}$ |
| Radius | R | R |
| Spherical radius | SR | SR |
| Spherical diameter | S⌀ | S⌀ |
| Controlled radius | CR | None |
| Between | * ◄──► | None |
| Statistical tolerance | ⟨ST⟩ | None |
| Datum target | ⊘6/A1 or ⊙—⌀6/A1 | ⊘6/A1 or ⊙—⌀6/A1 |
| Target point | ✕ | ✕ |

*May be filled or not filled

Figure 13-2. *continued.*

Figure 13-3. *Datum feature symbol (ASME Y14.5M-1994).*

The *feature control frame* is a boxed expression containing the geometric characteristics symbol and the form, runout, location, orientation, or profile tolerances, plus any datum references and modifiers for the feature or datum. Proper symbols and applications are shown in *Figure 13-4.*

Datum references follow tolerances as shown in *Figures 13-5* and *13-6.* For convenience or conservation of drawing space, the feature control frame and datum feature identification symbol may be combined, as shown in *Figure 13-7.* A common datum, such as the axis or center plane of a part, can be established by the use of two datum letters separated by a dash (see *Figure 13-8*).

Feature control frames are placed on drawings in accordance with standard ASME Y14.5M-1994 (R-1999) drawing practices as noted in *Figure 13-9.*

The tolerance zone, shown in the feature control frame, represents the total allowable change from the desired form, attitude, runout, or location of the feature. Where the specified tolerance is to indicate the diameter of a cylindrical zone, the diameter symbol ⌀ is placed ahead of the tolerance in the feature control frame, as shown in *Figure 13-10.* Otherwise, the tolerance zone represents the total distance between two parallel lines, planes, or geometric boundaries.

Datum targets may be used wherever datum locations and sizes are required on castings, forgings, weldments, etc., and to aid in maintaining repeatability between manufacturing and inspection.

Older datum targets are designated by a circle divided into four quadrants, with the datum letter of the datum plane in the upper left and the number of the target in the lower right of the circle. The newer datum target is shown with a circle divided in half. The lower half denotes the specific datum and the upper half, when used, shows the datum size. These symbols and their proper applications are shown in *Figure 13-11.*

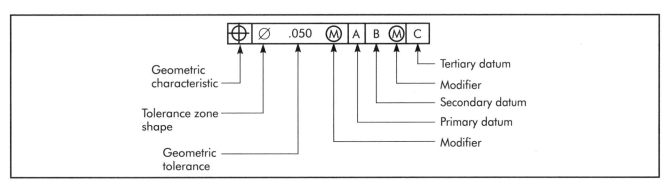

Figure 13-4. *Feature control frame.*

354

Figure 13-5. *Single datum reference following tolerance.*

Figure 13-6. *Multiple datum references following tolerance.*

Figure 13-7. *Combined feature control frame and datum feature symbol.*

Figure 13-8. *Common datum for axis or center plane.*

Figure 13-9. *Feature control symbols.*

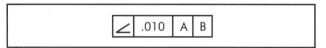

Figure 13-10. *Tolerance zone.*

*Virtual condition* is a constant (worst case) boundary generated by the collective effects of a size feature's specified MMC or LMC and the geometric tolerance for that material condition. Virtual condition is calculated by the following: holes (MMC – geometric tolerance = virtual condition) or (LMC + geometric tolerance = virtual condition); shafts (MMC + geometric tolerance = virtual condition) or (LMC – geometric tolerance = virtual condition).

Figure 13-11. *Datum targets.*

*Full Indicator Movement* (FIM) is the total indicator movement reading observed when properly applied to a part feature; it is the same as full indicator reading (FIR) and total indicator reading (TIR).

## Flatness

*Flatness* is a condition in which all elements of a surface are in the same plane. The flatness tolerance specifies a tolerance zone bordered by two parallel planes, within which the entire surface must lie. No datum is needed or proper with a flatness tolerance. When checking flatness, all elements of the concerned surface also must be within the specified size limits, if applicable, of the part to be acceptable. An example of flatness tolerance and its meaning are shown in *Figure 13-12*.

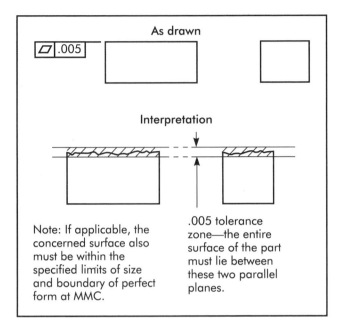

Figure 13-12. *Flatness tolerance and its meaning.*

## Straightness

*Straightness* is a condition where a portion of a surface or an axis is a straight line. The straightness tolerance specifies a cylindrical tolerance zone within which the entire axis or element must lie. No datum is needed or proper with a straightness tolerance. When checking the straightness of a surface element, all elements of the surface also must be within the specified size tolerance and boundary of perfect form at MMC. However, when checking the straightness of an axis or a center plane either RFS or MMC, a virtual condition results, and the boundary of perfect form may be exceeded up to the stated tolerance. However, each cross-sectional element still must be within the specified part size. Examples of straightness tolerances and their meanings are shown in *Figure 13-13*. A functional gage to check the part is shown in *Figure 13-14*.

## Circularity (Roundness)

With respect to a cylinder or cone, *circularity* is having all points of the surface intersected by any plane at right angles to an axis, equal distances from the axis.

With respect to the sphere, circularity is having all points of the surface intersected by any

Figure 13-13. *Straightness tolerance and its meaning.*

Figure 13-13. *continued.*

Figure 13-14. *Functional gage to check straightness of an axis MMC.*

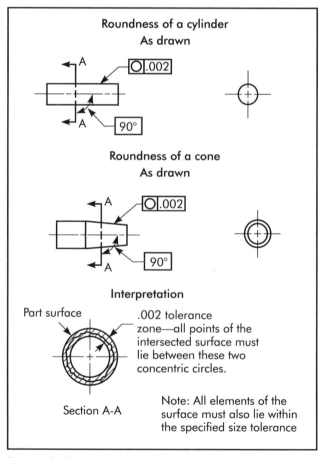

Figure 13-15. *Roundness tolerance and its meaning.*

plane passing through a common center, equal distances from that center.

The circularity tolerance specifies a tolerance zone bounded by two concentric circles, within which each circular element of the surface must lie. No datum is to be specified with a circularity tolerance. When checking circularity, all elements of the surface also must be within the specified size tolerance to be acceptable. Examples of circularity tolerances and their meaning are shown in *Figure 13-15.*

## Cylindricity

*Cylindricity* is having all points on the surface of a cylinder equal distances from a common axis. Cylindricity tolerance specifies a tolerance zone bounded by two concentric cylinders, within which the surface must lie. Datum is neither needed nor proper with a cylindricity tolerance. When checking cylindricity, all elements of the concerned surface also must be within the specified size tolerance and the boundary of perfect form at MMC. An example of a cylindricity tolerance and its meaning is shown in *Figure 13-16*.

## Profile of a Line or Surface

*Profile tolerancing* is used to specify an allowable deviation from a desired profile where other geometric controls cannot be used. The profile tolerance specifies a uniform boundary along the desired profile within which the elements of the surface or line must lie. Datums may or may not be necessary to establish a proper relationship of the profile to mounting surfaces for assembly purposes, etc. All profiles are defined using basic dimensions, and as such are not covered by Rule 1. Rule 1 regards the location of the profile. However, if, for whatever reason, conventional size dimensions and tolerances are associated with profile tolerancing, Rule 1 would apply and the

profile would be required to be within the specified size tolerance. Examples of profile tolerances and their meaning are shown in *Figure 13-17*.

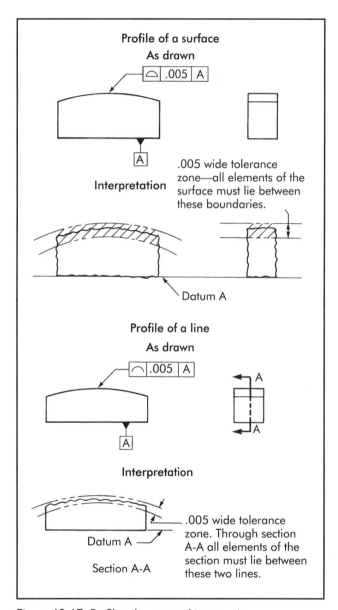

Figure 13-17. *Profile tolerance and its meaning.*

## Perpendicularity (Squareness)

*Perpendicularity* is the condition of a surface, median plane, or axis exactly 90° from a datum plane or datum axis. A perpendicularity tolerance always requires a datum, and is specified by one of the following:

Figure 13-16. *Cylindricity tolerance and its meaning.*

- a tolerance zone bordered by two parallel planes perpendicular to a datum plane or datum axis, within which the surface or median plane of the considered feature must lie;
- a tolerance zone bordered by two parallel planes perpendicular to a datum axis, within which the axis of the considered feature must lie;
- a cylindrical tolerance zone perpendicular to a datum plane, within which the axis of the considered feature must lie, or
- a tolerance zone defined by two parallel lines perpendicular to a datum plane or datum axis, within which all elements of the surface must lie.

A perpendicularity tolerance applied to a surface also controls the flatness of the surface to the extent of the stated tolerance and requires the surface to be within the stated limits of size. Examples of perpendicularity tolerances and their meanings are shown in *Figure 13-18*.

## Angularity

*Angularity* is a condition of a surface or axis at a specified angle (other than 90°) from a datum plane or axis. Angularity tolerance specifies a tolerance zone defined by two parallel planes at a specified basic angle from one or more datum plane or datum axis, within which the surface, axis, or center plane of the feature must lie. A datum is always required, and the desired angle is always shown as a basic angle. When checking angularity, all elements of the concerned feature also must be within the specified size limits of the part to be accepted. On surfaces, the angularity includes a control of flatness to the extent of the angularity tolerance. Examples of angularity tolerances and their meaning are shown in *Figure 13-19*.

## Parallelism

*Parallelism* is a condition of a surface, line, or axis equidistant from a datum plane or axis at all points. A parallelism tolerance always requires a datum, and can be specified by one of the following:

Figure 13-18. *Perpendicularity tolerance and its meaning.*

Figure 13-18. *continued.*

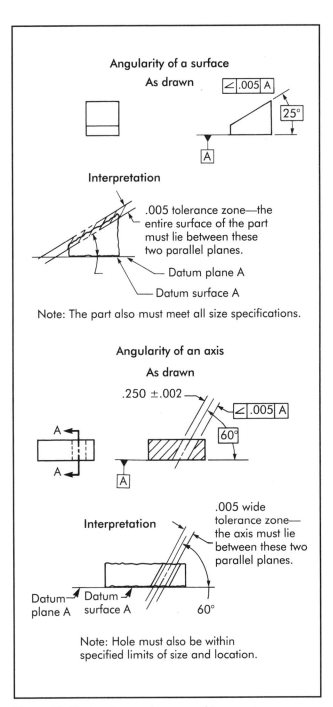

Figure 13-19. *Angularity tolerance and its meaning.*

- a tolerance zone bounded by two parallel planes or lines parallel to a datum plane or datum axis, within which the elements of the surface or the axis of the considered feature must lie, or
- a cylindrical tolerance zone whose axis is parallel to a datum axis, within which the axis of the considered feature must lie.

The parallelism tolerance applied to a surface also controls the flatness of a surface to the extent of the stated tolerance and requires the surface to be within the stated limits of size. Examples of parallelism tolerances and their meaning are shown in *Figure 13-20.*

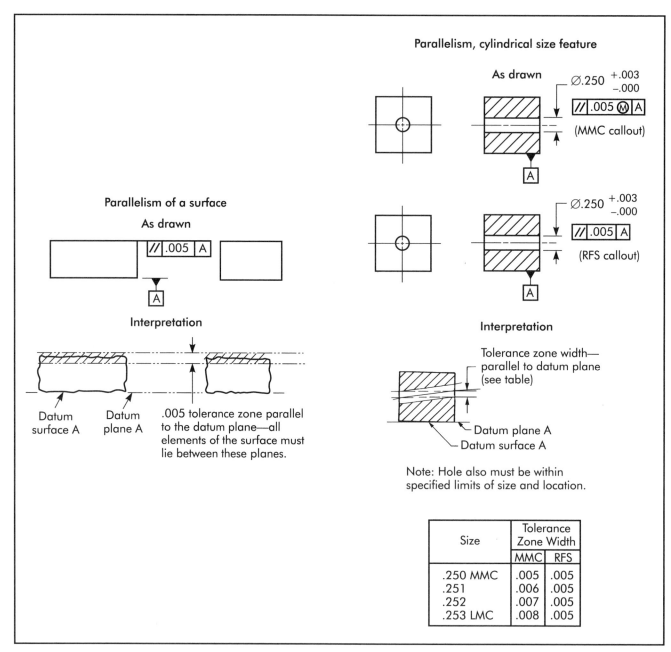

Figure 13-20. *Parallelism tolerance and its meaning.*

## Runout

*Runout* is the deviation from the desired form of a part surface of revolution when the part is rotated 360° around a datum axis. The runout tolerance specifies the maximum FIM allowed during the 360° rotation. Runout tolerance is always applied on an RFS basis and always requires a datum. It is used to maintain surface-to-axis control on a part (see *Figure 13-21*).

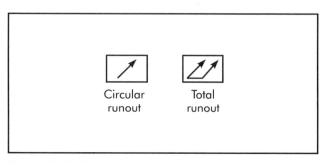

Figure 13-21. *Runout symbols.*

Circular runout controls only circular elements of a surface individually and independently from one another, but total runout provides composite control of all surface elements at the same time. When checking runout, all elements of the concerned surface must be within the specified size limits of the part to be acceptable. Examples of runout tolerances and their meaning are shown in *Figure 13-22*.

## Position

*Position* is a tolerance of location used to specify how far a feature of size may vary from the theoretically exact location on the part drawing.

A position tolerance defines a zone within which the axis or center plane of a feature of size can vary from this theoretically exact position. Position tolerances are applied only to size features, and datums are always required. A position tolerance is mainly used to maintain surface-to-surface control or axis-to-axis control on an MMC basis. True position tolerance theory and bonus tolerance are shown in *Figure 13-23*.

## Concentricity

*Concentricity* is another tolerance of location used to maintain axis-to-axis control on an RFS basis. It is a condition in which two or more features, in any combination, have a common axis. A concentricity tolerance specifies the condition in which the median points of all diametrically opposed elements of a figure of revolution (or

Figure 13-22. *Runout tolerance and its meaning.*

Figure 13-23. *Position tolerance theory—bonus tolerance.*

Figure 13-24. *Concentricity—a tolerance of location and its interpretation. It uses the center points of two elements set at opposing parallel 180° elements. The points formed at different cross sections must lie within the specified tolerance zone.*

## Symmetry

*Symmetry* is another tolerance of location. It is a condition in which the median points of all opposed or correspondingly located elements of two or more feature surfaces are congruent with the axis or center plane of a datum feature. When checking symmetry, all elements of the concerned feature also must be within the specified limits of the part to be acceptable. An example of a symmetry tolerance and its meaning is shown in *Figure 13-25.*

## THREE-PLANE CONCEPT

Datums are considered theoretically perfect points, axes, and planes, which are assumed to be exact for purposes of computation or reference. From that location, other features of a part may be determined. Datums are established by, or relative to, actual part features. When more than one datum feature is required on a part, it is advisable to establish a three-plane datum sys-

correspondingly located elements of two or more radially disposed features) are congruent with the axis (or center point) of a datum feature. When checking concentricity, all elements of the concerned surfaces also must be within the specified size limits of the part to be acceptable. An example of a concentricity tolerance and its meaning is shown in *Figure 13-24.*

Figure 13-25. *Symmetry—within the limits of size and RFS, all median points of opposed elements of the slot must lie between two parallel planes 0.020 apart, the two planes being equally disposed about datum plane B.*

tem consisting of three planes at right angles to each other. Where a primary datum plane is established by three points (minimum) somewhere on the most influential datum surface, a secondary datum plane is established by two points (minimum) on the second most influential datum surface, and a third datum plane is established by one point (minimum) on the least influential datum surface.

Where the size of a part, such as a cylinder, or width is used as a datum on a RFS basis, a datum axis or center plane must be established by making contact with feature surface extremities by using precision expanding chucks, mandrels, locators, etc.

## GENERAL RULES

General rules established by ASME Y14.5M-1994 (R-1999) provide users of geometrics with a better understanding of the system and provide for proper applications. A couple, in general, are important to understanding the principles in this standard, and should be reviewed carefully.

## Rule 1

When only a size tolerance is specified, Rule 1 states that the dimensional limits of an individual feature of size control form as well as size. Rule 1 does not control the relationships of features, and does not always have to apply. It may be avoided or disclaimed. Rule 1 states:

A. No element of the actual feature shall extend beyond a boundary of perfect form at MMC. This boundary is the true envelope implied by the drawing.

B. The actual local size of the feature at any cross section shall be within the specified tolerance of size.

C. The form control provisions of A do not apply to:

- commercial stock (bars, sheets, tubing, etc.) covered by industrial standards, and/or
- parts subject to free-state variation in the unrestrained condition (nonrigid).

*Figure 13-26* shows the allowable changes of form permitted by this rule.

## Rule 2

Rule 2 specifies that for all features of size, RFS applies with respect to the individual tolerance, datum reference, or both, where no modifier is specified. MMC or LMC must be specified on the drawing where it is required (refer to *Figures 13-27* and *13-28*).

### Rule 2a (Alternative Practice)

Rule 2a states that for a tolerance of position, RFS may be specified on the drawing with respect to the individual tolerance, datum reference, or both, as applicable.

## Screw Thread Rule

Each tolerance of attitude (orientation), form, runout, location, and/or datum reference specified for a screw thread applies to the pitch diameter. For gears and splines, a qualifying notation must be added (for example, major, minor, or pitch diameter). Where an exception to this practice is necessary, a notation must be added beneath the

Figure 13-26. *Rule 1.*

Figure 13-27. *Rule 2.*

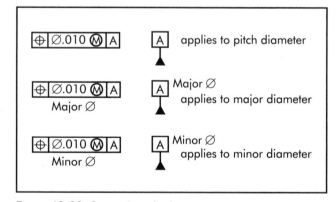

Figure 13-29. *Screw thread rule.*

Figure 13-28. *Rule 2 and 2a.*

feature control frame or datum feature symbol (*Figure 13-29*).

Datum features of size, which are controlled by a separate tolerance or position, are noted within the same feature control frame. They apply at their virtual condition to the effect of dif-

ferences in size between the applicable virtual condition of a datum feature and its MMC limit of size. When a virtual condition equal to MMC is the design requirement, a zero geometric tolerance at MMC is specified.

A virtual condition exists for a datum feature of size where its axis or center plane is controlled by a geometric tolerance. In such cases, the datum feature applies at its virtual condition even though it is referenced in a feature control frame at MMC or LMC. When a virtual condition equal to the MMC or the LMC is the design requirement, a zero tolerance at MMC or LMC is specified. An example is illustrated in *Figure 13-30*.

## REFERENCE

ASME Y14.5M-1994. *Dimensioning and Tolerancing*. Reaffirmed 1999. New York: American Society of Mechanical Engineers.

## BIBLIOGRAPHY

ANSI B4.4M-1981 (R1994). *Inspection of Workpieces*. Washington, D.C.: American National Standards Institute.

ANSI/IEEE 268-1992. *Metric Practice*. Washington, D.C.: American National Standards Institute.

ASME B1.2-2001. *Gages and Gaging for Unified Inch Screw Threads*. New York: American Society of Mechanical Engineers.

ASME B4.2-1999. *Preferred Metric Limits and Fits*. New York: American Society of Mechanical Engineers.

ASME B5.10-1994. *Machine Tapers—Self-holding and Steep Taper Series*. Reaffirmed 1999. New York: American Society of Mechanical Engineers.

ASME B46.1-1995. *Surface Texture (Surface Roughness, Waviness, and Lay)*. Reaffirmed 1999. New York: American Society of Mechanical Engineers.

ASME B89.3.1-1997. *Measurement of Out-of-roundness*. Reaffirmed 1999. New York: American Society of Mechanical Engineers.

ASME B94.6-1995. *Knurling*. Reaffirmed 1999. New York: American Society of Mechanical Engineers.

ASME B94.11M-1993. *Twist Drills*. Reaffirmed 1999. New York: American Society of Mechanical Engineers.

ASME Y14.1-1995. *Decimal Inch Drawing Sheet Size and Format*. Reaffirmed 1999. New York: American Society of Mechanical Engineers.

ASME Y14.2M-1998. *Line Conventions and Lettering*. Reaffirmed 1999. New York: American Society of Mechanical Engineers.

Figure 13-30. *Positional tolerance of zero at MMC.*

ASME Y14.3M-1999. *Multiview and Sectional View Drawings*. New York: American Society of Mechanical Engineers.

ASME Y14.5.1M-1994. *Mathematical Definition of Dimensioning and Tolerancing Principles*. Reaffirmed 1999. New York: American Society of Mechanical Engineers.

ASME Y14.6-1998. *Screw Thread Representation*. Reaffirmed 1999. New York: American Society of Mechanical Engineers.

ASME Y14.6aM-1998. *Screw Thread Representation (Metric Supplement)*. Reaffirmed 1999. New York: American Society of Mechanical Engineers.

ASME Y14.7.1-1998. *Gear Drawing Standards—Part 1: For Spur, Helical, Double Helical, and Rack*. Reaffirmed 1999. New York: American Society of Mechanical Engineers.

ASME Y14.7.2-1999. *Gear and Spline Drawing Standards—Part 2: Bevel and Hypoid Gears*. New York: American Society of Mechanical Engineers.

ASME Y14.8M-1996. *Castings and Forgings*. Reaffirmed 1999. New York: American Society of Mechanical Engineers.

ASME Y14.36M-1996. *Surface Texture Symbols*. Reaffirmed 1999. New York: American Society of Mechanical Engineers.

ASME Y14.38-1999. *Abbreviations and Acronyms Revision and Redesignation of ASME Y1.1-1989*. New York: American Society of Mechanical Engineers.

## REVIEW QUESTIONS

1. List 14 geometric characteristics and their USA symbols.
2. Match the terms with the appropriate leaders in *Figure A*.
3. What is meant by a virtual condition?
4. What is the virtual condition of the part in *Figure B* at MMC?
5. Name the three tolerances of location.
6. What is the virtual condition of the pin shown in *Figure C* at MMC?
7. Design a simple, functional gage to check the part in *Figure C*.

Figure A.

Figure B.

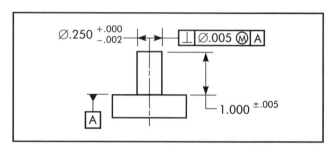

Figure C.

8. What is bonus tolerance?
9. Under what conditions are zero position tolerance methods used?
10. Design a simple, functional gage to check the part in *Figure D*.

Figure D.

# Bibliography

ASME Y14.5M-1994. *Dimensioning & Tolerancing*. New York: American Society of Mechanical Engineers, available from Society of Manufacturing Engineers (SME) (1-800-733-4763, www.sme.org).

Benedict, J., Veilleux, R., and Wick, C., eds. 1984. *TMEH Volume 2: Forming*. Dearborn, MI: Society of Manufacturing Engineers (SME) (1-800-733-4763, www.sme.org).

Benson, S. 1997. *Press Brake Technology*. Dearborn, MI: Society of Manufacturing Engineers (SME) (1-800-733-4763, www.sme.org).

Boyes, W. 1989. *Handbook of Jig and Fixture Design*. Second Edition. Dearborn, MI: Society of Manufacturing Engineers (SME) (1-800-733-4763, www.sme.org).

———. 1985. *Low-cost Jigs, Fixtures, and Gages for Limited Production*. Dearborn, MI: Society of Manufacturing Engineers (SME) (1-800-733-4763, www.sme.org).

Drozda, T. and Wick, C., eds. 1983. *TMEH Volume 1: Machining*. Dearborn, MI: Society of Manufacturing Engineers (SME) (1-800-733-4763, www.sme.org).

Forming Technologies Association of the Society of Manufacturing Engineers (SME). 1994. *Progressive Dies: Principles and Practices of Design and Construction*. Dearborn, MI: Society of Manufacturing Engineers (SME) (1-800-733-4763, www.sme.org).

Jacobs, P. 1995. *Stereolithography and Other RP&M Technologies: From Rapid Prototyping to Rapid Tooling*. Dearborn, MI: Society of Manufacturing Engineers (SME) (1-800-733-4763, www.sme.org).

———. 1992. *Rapid Prototyping & Manufacturing: Fundamentals of Stereolithography*. Dearborn, MI: Society of Manufacturing Engineers (SME) (1-800-733-4763, www.sme.org).

Kocherovsky, E. 1999. *HSK Handbook*. West Bloomfield, MI: Intelligent Concept. Available from Society of Manufacturing Engineers (SME) (1-800-733-4763, www.sme.org).

Krulikowski, A. 1999. *Fundamentals of GD&T Video Training Series*. 14 hours. Westland, MI: Effective Training, Inc., available from Society of Manufacturing Engineers (SME) (1-800-733-4763, www.sme.org).

———. 1996. *GD&T Self-Study Workbook*, 2nd Edition. Westland, MI: Effective Training, Inc., available from Society of Manufacturing Engineers (SME) (1-800-733-4763, www.sme.org).

Lange, K. 1985. *The Handbook of Metal Forming*. Dearborn, MI: Society of Manufacturing Engineers (SME) (1-800-733-4763, www.sme.org).

Leed, R. 2002. *Tool and Die Making Trouble-shooter*. Dearborn, MI: Society of Manufacturing Engineers (SME) (1-800-733-4763, www.sme.org).

Miller, G. 2002. *Tube Forming Processes: A Comprehensive Guide*. Dearborn, MI: Society of Manufacturing Engineers (SME) (1-800-733-4763, www.sme.org).

Neumann, A. 1999. *GD&T Trainer* CD-ROM. Dearborn, MI: Society of Manufacturing Engineers (SME) (1-800-733-4763, www.sme.org).

——. 1999. *Geometric Dimensioning and Tolerancing* Color Graphics Package CD-ROM. Dearborn, MI: Society of Manufacturing Engineers (SME) (1-800-733-4763, www.sme.org).

——. 1995. *Geometric Dimensioning and Tolerancing* Video Training Program. 24-30 hours. Dearborn, MI: Society of Manufacturing Engineers (SME) (1-800-733-4763, www.sme.org).

——. 1995. *Geometric Dimensioning and Tolerancing Workbook*. Dearborn, MI: Society of Manufacturing Engineers (SME) (1-800-733-4763, www.sme.org).

Oberg, E., et al. 2000. *Machinery's Handbook*, 26th Edition. New York: Industrial Press, Inc. Available from Society of Manufacturing Engineers (SME) (1-800-733-4763, www.sme.org).

Schrader, G. and Elshennawy, A. 2000. *Manufacturing Processes and Materials*. Dearborn, MI: Society of Manufacturing Engineers (SME) (1-800-733-4763, www.sme.org).

Shaw, C. 1996. *Metal Cutting Principles*. Oxford, United Kingdom: Oxford University Press, Inc. Available from Society of Manufacturing Engineers (SME) (1-800-733-4763, www.sme.org).

Smith, D. 2001. *Die Maintenance Handbook*. Dearborn, MI: Society of Manufacturing Engineers (SME) (1-800-733-4763, www.sme.org).

——. 1994. *Fundamentals of Pressworking*. Dearborn, MI: Society of Manufacturing Engineers (SME) (1-800-733-4763, www.sme.org).

——. 1990. *Die Design Handbook*, Third Edition. Dearborn, MI: Society of Manufacturing Engineers (SME) (1-800-733-4763, www.sme.org).

Society of Manufacturing Engineers. 2002. *Sheet Metal Forming Knowledge Base* CD-ROM. Dearborn, MI: Society of Manufacturing Engineers (SME) (1-800-733-4763, www.sme.org).

——. 1999. *Forging* video. 21 minutes. *Fundamental Manufacturing Processes* series. Dearborn, MI: Society of Manufacturing Engineers (SME) (1-800-733-4763, www.sme.org).

——. 1999. *Heat Treating* video. 30 minutes. *Fundamental Manufacturing Processes* series. Dearborn, MI: Society of Manufacturing Engineers (SME) (1-800-733-4763, www.sme.org).

——. 1999. *Measurement and Gaging* video. 28 minutes. *Fundamental Manufacturing Processes* series. Dearborn, MI: Society of Manufacturing Engineers (SME) (1-800-733-4763, www.sme.org).

——. 1999. *Punch Presses* video. 20 minutes. *Fundamental Manufacturing Processes* series. Dearborn, MI: Society of Manufacturing Engineers (SME) (1-800-733-4763, www.sme.org).

——. 1999. *Welding* video. 29 minutes. *Fundamental Manufacturing Processes* series. Dearborn, MI: Society of Manufacturing Engineers (SME) (1-800-733-4763, www.sme.org).

——. 1998. *TMEH Knowledge Base* CD-ROM. Dearborn, MI: Society of Manufacturing Engineers (SME) (1-800-733-4763, www.sme.org).

——. 1997. *Cutting Tool Geometries* video. 27 minutes. *Fundamental Manufacturing Processes* series. Dearborn, MI: Society of Manufacturing Engineers (SME) (1-800-733-4763, www.sme.org).

——. 1997. *Cutting Tool Materials* video. 25 minutes. *Fundamental Manufacturing Processes* series. Dearborn, MI: Society of Manufacturing Engineers (SME) (1-800-733-4763, www.sme.org).

——. 1997. *Sheet Metal Shearing and Bending* video. 21 minutes. *Fundamental Manufacturing*

*Processes* series. Dearborn, MI: Society of Manufacturing Engineers (SME) (1-800-733-4763, www.sme.org).

——. 1997. *Sheet Metal Stamping Dies & Processes* video. 20 minutes. *Fundamental Manufacturing Processes* series. Dearborn, MI: Society of Manufacturing Engineers (SME) (1-800-733-4763, www.sme.org).

——. 1997. *Sheet Metal Stamping Presses* video. 23 minutes. *Fundamental Manufacturing Processes* series. Dearborn, MI: Society of Manufacturing Engineers (SME) (1-800-733-4763, www.sme.org).

——. 1996. *Introduction to Workholding* video. 22 minutes. *Fundamental Manufacturing Processes* series. Dearborn, MI: Society of Manufacturing Engineers (SME) (1-800-733-4763, www.sme.org).

——. 1995. *Basic Holemaking* video. 24 minutes. *Fundamental Manufacturing Processes* series. Dearborn, MI: Society of Manufacturing Engineers (SME) (1-800-733-4763, www.sme.org).

——. 1995. *Successful GD&T Implementation* video. 27 minutes. Dearborn, MI: Society of Manufacturing Engineers (SME) (1-800-733-4763, www.sme.org).

——. 1994. *Rapid Tooling, Rapid Parts* video. 40 minutes. *Manufacturing Insights* series. Dearborn, MI: Society of Manufacturing Engineers (SME) (1-800-733-4763, www.sme.org).

——. 1988. *Progressive Dies: Proven Design Principles and Applications* video program. Dearborn, MI: Society of Manufacturing Engineers (SME) (1-800-733-4763, www.sme.org).

Trent, E. 2000. *Metal Cutting*. Burlington, MA: Butterworth Heinemann. Available from Society of Manufacturing Engineers (SME) (1-800-733-4763, www.sme.org).

Veilleux, R. and Wick, C., eds. 1985. *TMEH Volume 3: Materials, Finishing & Coating*. Dearborn, MI: Society of Manufacturing Engineers (SME) (1-800-733-4763, www.sme.org).

Wick, C., et al., eds. 1984. *Tool and Manufacturing Engineers Handbook*, 4th edition. Volume 2, *Forming*. Dearborn, MI: Society of Manufacturing Engineers.

Winchell, W. 1996. *Inspection and Measurement in Manufacturing*. Dearborn, MI: Society of Manufacturing Engineers (SME) (1-800-733-4763, www.sme.org).

# Index